Advances in Experimental Mechanics VII

Edited by
R.A.W. Mines
J.M. Dulieu-Barton

Advances in Experimental Mechanics VII

Selected, peer reviewed papers from 7th International Conference on
Advances in Experimental Mechanics,
7th-9th September 2010,
School of Engineering,
University of Liverpool, UK

Edited by

R.A.W. Mines
University of Liverpool, UK

J.M. Dulieu-Barton
University of Southampton, UK

ttp **TRANS TECH PUBLICATIONS LTD**
Switzerland • UK • USA

Trans Tech Publications Ltd
Laubisrutistr. 24
CH-8712 Stafa-Zurich
Switzerland
http://www.ttp.net

Volumes 24-25 of
Applied Mechanics and Materials
ISSN 1660-9336

Full text available online at http://www.scientific.net

Distributed worldwide by

Trans Tech Publications Ltd
Laubisrutistr. 24
CH-8712 Stafa-Zuerich
Switzerland

Fax: +41 (44) 922 10 33
e-mail: sales@ttp.net

and in the Americas by

Trans Tech Publications Inc.
PO Box 699, May Street
Enfield, NH 03748
USA

Phone: +1 (603) 632-7377
Fax: +1 (603) 632-5611
e-mail: sales-usa@ttp.net

BRITISH SOCIETY FOR
STRAIN MEASUREMENT

Organising Committee

Dr Bob Mines, University of Liverpool (Conference Chair)
Professor Janice Barton, University of Southampton
Dr Venky Dubey, Bournemouth University
Dr Jerry Lord, NPL
Professor Fabrice Pierron, ENSAM, France
Dr Simon Quinn, University of Southampton
Mr John Edwards, BSSM
Ms Biana Gale, BSSM
Mrs Maria White, University of Liverpool

Technical Committee

Prof. Augusto Ajovalasit, University of Palermo, Italy
Prof. Leslie Banks-Sills, Tel Aviv University, Israel
Prof. James Barber, University of Michigan, USA
Prof. Janice Barton, University of Southampton, UK
Dr Steve Bate, Serco Assurance, UK
Prof. Adib Becker, University of Nottingham, UK
Dr Christian Berggreen, DTU, Denmark
Dr Richard Burguete, Airbus UK Ltd, UK
Dr Paul Cunningham, Loughborough University, UK
Prof. Isaac Daniel, Northwestern University, USA
Dr Geoff Dutton, Rutherford Appleton Laboratory, UK
Prof. Josef Eberhardsteiner, Vienna University of Technology, Austria
Dr Carol Featherston, Cardiff University, UK
Prof. Emmanuel Gdoutos, Democritus University of Thrace, Greece
Prof. Michael Gilchrist, University College, Dublin, Ireland
Prof. Michel Grediac, Blaise Pascal University, France
Dr Salih Gungor, The Open University, UK
Prof. Karen Holford, Cardiff University, UK
Dr David Hollis, LAVision UK Ltd, UK
Dr Zhihong Huang, University of Dundee, UK
Prof. Jon Huntley, Loughborough University, UK
Prof. Daniel Inman, Virginia Tech University, USA
Prof. Pierre Jacquot, EPFL, Switzerland
Dr Alastair Johnson, German Aerospace Centre, Germany
Dr Arthur Jones, University of Nottingham, UK
Dr Stefan Kaczmarczyk, University of Northampton, UK

Sponsors

Airbus

AWE

Instron

Imetrum

LaVision

Co-Sponsors

Association Française de Mécanique

Engineering Integrity Society

European Association for Experimental Mechanics

German Society for Experimental Stress Analysis

Institution of Mechanical Engineers

Institute of Physics

   -   Applied Mechanics Group

Japanese Society for Experimental Mechanics

Society for Experimental Mechanics

Italian Association for Stress Analysis

PREFACE

The 2010 Annual British Society for Strain Measurement Conference is the seventh in the series on Advances in Experimental Mechanics. This is the fifth time the papers have been collected as a volume of Applied Mechanics and Materials. The 61 papers in this volume represent the diverse nature of Experimental Mechanics. The fourteen sessions address different disciplines, e.g. aerospace, mechanical and civil engineering, and state of the art technologies, e.g. biomechanics, novel sensors, and composite and cellular materials. The papers come from both academia and industry with half of the contributions coming from outside the UK, reflecting the international flavour of this event. We thank the authors and the referees for the hard work and dedication in the preparation and review of the papers.

We are particularly grateful to the University of Liverpool for agreeing to host the event and the support of their staff in the local arrangements. We would also like to thank the organising committee and in particular John Edwards and Biana Gale, of the BSSM, for their efforts in helping us compile this volume.

Bob Mines, University of Liverpool
Janice Dulieu-Barton, University of Southampton
September 2010

Table of Contents

Committee   v
Sponsors   vii
Preface   ix

Keynote Lecture 1

Mechanical and Transport Properties of Concrete at High Temperatures
A. Galek, H. Moser, T. Ring, M. Zeiml, J. Eberhardsteiner and R. Lackner 1

Keynote Lecture 2

Experimental Characterisation of Parameters Controlling the Compressive Failure of Pultruded Unidirectional Carbon Fibre Composites
O.T. Thomsen and K.K. Kratmann 15

BSSM Measurements Lecture

Aspects of Uncertainty Analysis for Large Nonlinear Computational Models
K. Worden, W.E. Becker, M. Battipede and C. Surace 25

Session 1: Structural Health Monitoring

Detection of Cracking in Gear Teeth Using Acoustic Emission
R. Pullin, A. Clarke, M.J. Eaton, K.M. Holford, S.L. Evans and J.P. McCory.. 45

Active Sensor Arrays for Damage Detection
P.H. Malinowski, T. Wandowski and W.M. Ostachowicz 51

Towards a Hybrid Infrared Approach for Damage Assessment
R.K. Fruehmann, J.M. Dulieu-Barton and S. Quinn 57

Session 2: Vibration

Designing a Hollow Langevin Transducer for Ultrasonic Coring
P. Harkness, A. Cardoni, J. Russell and M. Lucas.. 65

Experimental Modal Analysis of an Automotive Powertrain
C. Delprete, A. Galeazzi and F. Pregno 71

Investigation into the Damping and Stiffness Characteristics of an Elevator Car System
I. Herrera, H. Su and S. Kaczmarczyk 77

Analysis of Sensors for Vibration and Nip Forces Monitoring of Rubber Coated Rollers
M.C. Voicu, R. Schmidt, B. Lammen, H.H. Hillbrand and I. Maniu... 83

Session 3A: Composite and Cellular Materials 1

Evaluation of Edge Cracks in Cross-Ply Laminates Using Image Correlation and Thermoelastic Stress Analysis
G.P. Battams, J.M. Dulieu-Barton and S.W. Boyd 91

Mechanical Behavior of Syntactic Foams for Deep Sea Thermally Insulated Pipeline
D. Choqueuse, P. Davies, D. Perreux, L. Sohier and J.Y. Cognard.. 97

Analysis of the Strain and Stress Fields of Cardboard Box during Compression by 3D Digital Image Correlation
J. Viguié, P.J.J. Dumont, P. Vacher, L. Orgéas, I. Desloges and E. Mauret 103

Correlation between Full-Field Measurements and Numerical Simulation Results for Multiple Delamination Composite Specimens in Bending
C. Devivier, D. Thompson, F. Pierron and M.R. Wisnom.. 109

Towards a Planar Cruciform Specimen for Biaxial Characterisation of Polymer Matrix Composites
M.R.L. Gower and R.M. Shaw 115

Session 3B: Novel Sensor Technology

Measurement of Mechanical Strain Using Chromatic Monitoring of Photoelasticity
C. Garza, A.G. Deakin, G.R. Jones, J.W. Spencer and K.K.B. Hon.. 123

Development of a Methodology to Assess Mechanical Impulse Effects Resulting from Lightning Attachment to Lightweight Aircraft Structures
C.A. Featherston, M.J. Eaton, S.L. Evans, K.M. Holford, R. Pullin and M. Cole.......................... 129

A Laser Speckle Method for Measuring Displacement Field. Application to Resistance Heating Tensile Test on Steel
C. Pradille, M. Bellet and Y. Chastel.. 135

Characterisation of Full-Field Deformation Behaviour Using Digital Imaging Techniques
Y.H. Tai, M. Zanganeh, D. Asquith and J.R. Yates... 141

Digital Holographic Interferometry by Using Long Wave Infrared Radiation (CO_2 Laser)
I. Alexeenko, J.F. Vandenrijt, M. Georges, G. Pedrini, T. Cédric, W. Osten and B. Vollheim....... 147

Session 4: Civil Engineering

Size of the Fracture Process Zone in High-Strength Concrete at a Wide Range of Loading Rates
R.C. Yu, X.X. Zhang, G. Ruiz, M. Tarifa and M. Cámara 155

Practical *In Situ* Applications of DIC for Large Structures
N.J. McCormick and J.D. Lord 161

From the Inside, Out Use of Optical Measuring Techniques for Wind Turbine Development
A. Stanley and M. Klein... 167

Details of Temperature Compensation for Strain Measurements on NPL Bridge - Demonstrator for SHM
E.N. Barton and B. Zhang... 173

Effect of Loading Rate on the Fracture Behaviour of High-Strength Concrete
G. Ruiz, X.X. Zhang, R.C. Yu, E. Poveda, R. Porras and J. del Viso...................................... 179

Session 5A: Non Linear Behaviour

Numerical Prediction of the Response of Metal-to-Metal Adhesive Joints with Ductile Adhesives
N.G. Tsouvalis and K.N. Anyfantis 189

Experimental Characterization of the Viscoplastic Material Behaviour of Thermosets and Thermoplastics
M. Kästner, S. Blobel, M. Obst, K. Thielsch and V. Ulbricht 195

Strain Evolution Measurement at the Microscale of a Dual Phase Steel Using Digital Image Correlation
H. Ghadbeigi, C. Pinna, S. Celotto and J.R. Yates 201

Biaxial Ratcheting Response of SS 316 Steel
R. Suresh Kumar, C. Lakshmana Rao and P. Chellapandi 207

On the Mutual Interactions of Monotonic and Cyclic Loading and their Effect on the Strength of Aluminium Alloys
Z.L. Kowalewski and T. Szymczak 213

Session 5B: Fracture and Damage

Validation of Acoustic Emission (AE) Crack Detection in Aerospace Grade Steel Using Digital Image Correlation
R. Pullin, M.J. Eaton, J.J. Hensman, K.M. Holford, K. Worden and S.L. Evans....................... 221

Towards the Derivation of Stress Intensity Factors by Parametric Modelling of Full-Field Thermoelastic Data
R.I. Hebb, J.M. Dulieu-Barton, K. Worden and P. Tatum 227

A SEM-Based Study of Structural Impact Damage
M.T.H. Sultan, A. Hodzic, W.J. Staszewski and K. Worden 233

Application of Real-Time Photoelastic Analysis to Single Fibre Fragmentation Tests
S. Blobel, K. Thielsch, M. Kästner and V. Ulbricht 239

Characterization of Fiber Bridging in Mode II Fracture Growth of Laminated Composite Materials
K.N. Anyfantis and N.G. Tsouvalis 245

Session 6A: Residual Stresses

Evaluation of the Impact of Residual Stresses in Crack Initiation with the Application of the Crack Compliance Method Part I, Numerical Analysis
G. Urriolagoitia-Sosa, B. Romero-Ángeles, L.H. Hernández-Gómez, G. Urriolagoitia-Calderón,
J.A. Beltrán-Fernández and C. Torres-Torres 253

Evaluation of the Impact of Residual Stresses in Crack Initiation with the Application of the Crack Compliance Method Part II, Experimental Analysis
G. Urriolagoitia-Sosa, B. Romero-Ángeles, L.H. Hernández-Gómez, G. Urriolagoitia-Calderón,
J.A. Beltrán-Fernández and C. Torres-Torres 261

Mapping Residual Stress Profiles at the Micron Scale Using FIB Micro-Hole Drilling
B. Winiarski and P.J. Withers 267

Session 6B: Biomechanics

Acetabular Component Deformation under Rim Loading Using Digital Image Correlation and Finite Element Methods
H. Everitt, S.L. Evans, C.A. Holt, R. Bigsby and I. Khan 275

Skin Thermal Effect by FE Simulation and Experiment of Laser Ultrasonics
C.H. Li, S.A. Li, Z.H. Huang and W.B. Xu 281

Biomechanics and Numerical Evaluation of Cervical Porcine Models Considering Compressive Loads Using 2-D Classic Computer Tomography CT, 3-D Scanner and 3-D Computed Tomography
J.A. Beltrán-Fernández, L.H. Hernández-Gómez, G. Urriolagoitia-Calderón, A. González-Rebatú
and G. Urriolagoitia-Sosa.. 287

Session 7: Manufacturing Processes

Spot Weld Strength Determination Using the Wedge Test: *In Situ* Observations and Coupled Simulations
R. Lacroix, J. Monatte, A. Lens, G. Kermouche, J.M. Bergheau and H. Klöcker 299

Improving Fatigue Performance of Alumino-Thermic Rail Welds
M. Jezzini-Aouad, P. Flahaut, S. Hariri, D. Zakrzewski and L. Winiar...................................... 305

The Effect of Ultrasonic Excitation in Metal Forming Tests
S. Abdul Aziz and M. Lucas 311

Modeling of Coating Stress of Plasma-Sprayed Thermal Barrier Coatings
M. Arai.. 317

Session 8A: Impact and Dynamics

Tensile Properties of Die-Cast Magnesium Alloy AZ91D at High Strain Rates in the Range between 300 s^{-1} and 1500 s^{-1}
I.R. Ahmad and D.W. Shu 325

The Dynamic Buckling of Stiffened Panels –A Study Using High Speed Digital Image Correlation
C.A. Featherston, J. Mortimer, M.J. Eaton, R.L. Burguete and R. Johns 331

Finite Element Model Updating of a Thin Wall Enclosure under Impact Excitation
O.S. David-West, J. Wang and R. Cooper 337

What we Can Learn about Stick-Slip Dynamics
   F. Di Liberto, E. Balzano, M. Serpico and F. Peruggi 343

Determination of High Strain-Rate Compressive Stress-Strain Loops of Selected Polymers
   T. Yokoyama and K. Nakai.. 349

Session 8B: Identification

Identification of the Shape of Curvilinear Beams and Fibers
   M.L.M. François, B. Semin and H. Auradou 359

Construction of Shape Features for the Representation of Full-Field Displacement/Strain Data
   W.Z. Wang, J.E. Mottershead, A. Patki and E.A. Patterson 365

An Innovative Own-Weight Cantilever Method for Measuring Young's Modulus in Flexible
Thin Materials Based on Large Deflections
   A. Ohtsuki 371

Identification of the Mechanical Properties of Superconducting Windings Using the
Virtual Fields Method
   J.H. Kim, F. Nunio, F. Pierron and P. Vedrine 379

Measuring the Static Modulus of Nuclear Graphite from Four-Point Flexural
Strength Tests and DIC
   J.D. Lord, N.J. McCormick, J.M. Urquhart, G.M. Klimaytys and I.J. Lingham 385

Session 9: Composite and Cellular Materials 2

Experimental and Numerical Buckling Analysis of Delaminated Hybrid
Composite Beam Structures
   M.M.N. Esfahani, H. Ghasemnejad and P.E. Barrington... 393

Effects of Bonded Splice Joints on the Flexural Response of Pultruded Fibre
Reinforced Polymer Beams
   G.J. Turvey... 401

Assessment of Quasi-Static and Fatigue Loaded Notched GRP Laminates Using
Digital Image Correlation
   W.R. Broughton, M.R.L. Gower, M.J. Lodeiro, G.D. Pilkington and R.M. Shaw 407

Comparison of the Drop Weight Impact Performance of Sandwich Panels with Aluminium
Honeycomb and Titanium Alloy Micro Lattice Cores
   R. Hasan, R.A.W. Mines, E. Shen, S. Tsopanos, W. Cantwell, W. Brooks and C.J. Sutcliffe 413

Viscoelastic Characterization of Short Fibres Reinforced Thermoplastic in
Tension and Shearing
   A. Andriyana, L. Silva and N. Billon... 419

Session 10: Education

The Role of Experimental Stress Analysis at Graduation and Post Graduation Courses -
A Brazilian Case
   R.J.P.C. Miranda, P. Domingues, L.M. Zamboni and J.C. Salamani 427

Keyword Index... 433

Author Index 437

Keynote Lecture 1

Applied Mechanics and Materials Vols. 24-25 (2010) pp 1-11
© (2010) Trans Tech Publications, Switzerland
doi:10.4028/www.scientific.net/AMM.24-25.1

Mechanical and transport properties of concrete at high temperatures

Artur Galek, Harald Moser, Thomas Ring, Matthias Zeiml, and Josef Eberhardsteiner

Institute for Mechanics of Materials and Structures (IMWS)

Vienna University of Technology

Karlsplatz 13/202, 1040 Vienna, Austria

e-mail: {artur.galek,harald.moser,thomas.ring,matthias.zeiml,josef.eberhardsteiner}

@tuwien.ac.at

Roman Lackner

Material-Technology Innsbruck (MTI), University of Innsbruck

Technikerstraße 13, 6020 Innsbruck, Austria

e-mail: roman.lackner@uibk.ac.at

Keywords: Concrete, Cement paste, High temperature, Strains, Permeability

Abstract

When concrete structures are subjected to fire loading, temperature-dependent degradation of the material properties as well as spalling of near-surface concrete layers has a considerable effect on the load-carrying capacity and, hence, the safety of these structures. Spalling is caused by inter-acting thermo-hydro-chemo-mechanical processes with both mechanical and transport properties playing an important role. Within experimental research activities at the IMWS, these properties are subject of investigation, i.e., (i) the strain behavior of concrete under combined thermal and mechanical loading and (ii) the permeability increase of temperature-loaded concrete and cement paste.

1 Introduction

Interacting thermo-hydro-chemo-mechanical processes lead to changes/degradation of both the mechanical as well as the transport behavior of concrete with increasing temperature. Regarding the former, a decrease of both strength and stiffness of concrete can be observed (see, e.g., [9, 11]). The strain behavior of heated concrete exhibits a path dependence when mechanical and thermal load are applied simultaneously (see, e.g., [4, 13]). This is attributed to the differing and non-linear behavior of the material phases of concrete (i.e., cement paste and aggregate). Regarding the transport behavior of concrete, numerous fire experiments have demonstrated the role of transport processes such as the advection of water vapor regarding the spalling behavior of concrete at elevated temperatures (see, e.g., [2, 6]). Since dense concrete mixes with low permeability proved to be more prone to spalling, parameters related to the pore structure of concrete (such as the permeability) were identified as key parameters determining the durability of concrete structures in case of fire loading. The beneficial effect of adding polypropylene (PP) fibers (see, e.g., [5, 12]), considerably increasing the permeability of concrete at elevated temperatures (see, e.g., [3, 14]), supports this conclusion.

In this paper, the mechanical as well as the transport behavior of concrete subjected to high temperatures is investigated experimentally. Section 2 deals with the strain behavior of concrete subjected to combined thermal and mechanical loading, presenting new experimental results monitoring the longitudinal and radial strains of uniaxial loaded specimens. In Section 3, experimental results for the temperature-dependent permeability of concrete and cement paste are presented.

2   Mechanical behavior

As concrete is a heterogeneous material, the different thermal behavior of the two main constituents, cement paste and aggregates, as well as the interaction between the two phases affects the macroscopic mechanical behavior at elevated temperatures. During heating, dehydration of the cement paste results in shrinkage whereas the aggregates expand (the magnitude of thermal expansion depends on the type of aggregate). These strain incompatibilities together with thermo-chemical damage of the material phases result in degradation of the mechanical properties. The

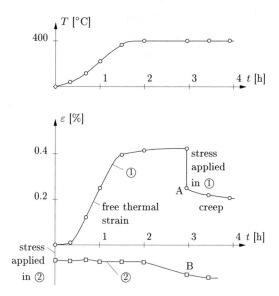

Figure 1: Illustration of path dependence of combined thermal and mechanical loading of concrete [13]: points A and B exhibit the same temperature and stress level (experimental results obtained with $T_{max} = 400°C$ and $\Sigma = 0.45\, f_{c,0}$; $f_{c,0}$: compressive strength of concrete at room temperature)

macroscopic strain behavior of concrete is a combination of the thermal strains of the two phases as well as their interaction. When concrete is heated without mechanical load[1], expansion of the aggregates induces tensile plastic deformations within the cement paste, the resulting macroscopic strains mainly follow the expansion of the generally stiffer aggregates (see, e.g., [4, 10]). When compressive loading is applied prior to thermal loading, the tensile plastic deformations within

---

[1]The respective strain state is referred to as free thermal strain (FTS) state.

the cement paste are reduced, leading to (i) less degradation of the macroscopic mechanical material properties (see, e.g., [11]) and (ii) a path dependence in the strain behavior, i.e., a different macroscopic strain state compared to the case of heating without mechanical load and subsequent compressive loading to the same extent (see Figure 1).

2.1 Materials and methods

In order to identify the described path dependence in the strain behavior for given concrete mix designs (see Table 1), concrete cylinders (100 mm in diameter and 200 mm long) were subjected to combined thermal (radiant) and mechanical (compressive) loading. The experiments were conducted at different load levels $s = \Sigma/f_{c,0}$ [–] (i.e., the ratio between stress and compressive strength at room temperature). The experimental setup is shown in Figure 2. Specimen and steel pistons used for mechanical loading are surrounded by a radiant oven, enabling for simultaneous mechanical and thermal loading. Both longitudinal and radial displacements are monitored by displacement sensors connected to the specimen via temperature-resistant steel rods. In longitudinal direction, the relative displacement between two steel rings mounted to the specimen is recorded.

Table 1: Mix design of investigated concrete without and with polypropylene (PP) fibers

cement CEM I [kg/m^3]	290
additive (fly ash) [kg/m^3]	50
water* [kg/m^3]	185 / 190
polypropylene (PP) fibers* [kg/m^3]	0 / 1.5
siliceous aggregates [kg/m^3]	1859
fraction 0–4 mm [mass-%]	36
fraction 4–8 mm [mass-%]	17
fraction 8–16 mm [mass-%]	34
fraction 16–22 mm [mass-%]	13
aggregate mineralogy:	
quartz [mass-%]	68
feldspar [mass-%]	21
carbonate [mass-%]	11
water/cement-ratio* [–]	0.64 / 0.66
water/binder-ratio*,† [–]	0.56 / 0.58
initial density* [kg/m^3]	2384 / 2391
slump* [mm]	430 / 410
air content* [%]	1.0 / 2.5
28d strength* [MPa]	+ / 40.7

* Left value corresponds to plain concrete, right
  value corresponds to concrete with PP-fibers.
† The amount of additives is weighted by 0.8 [7].
+ Concrete without PP-fibers is about to
  be tested.

Prior to testing, the specimens were stored under water for 28 days and subsequently, at ambient conditions until testing (approximately 20 °C and 60 % relative humidity). During testing, mechanical loading was applied first (at load levels s of 0, 10, 20, 30, 40, 50 and 60 %, respectively and subsequently, the temperature in the oven was increased at a rate of 1 °C/min and both longitudinal and radial displacements were recorded, giving access to the longitudinal and radial strains, respectively. In general, three samples were tested at every load step.

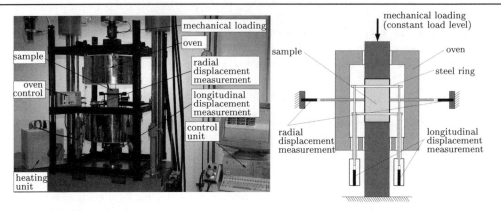

Figure 2: Experimental setup for strain experiments on concrete subjected to combined thermal and mechanical loading

2.2 Results

Figure 3 shows experimental results for concrete with PP-fibers, i.e., the longitudinal and radial strains for different load levels. The results for the longitudinal strains are in good agreement with experimental data available in the literature[2] (see Figures 3(a,c,e)). The curves for $s = 0$ correspond to free thermal strain (E_{th}), having the same magnitude in longitudinal and radial direction. A period of almost constant strain increase up to approximately 550 °C is followed by a steep increase in strain rate, corresponding to transformation of the siliceous aggregates (from α- to β-quartz which is associated with an increase in volume by 2%). Above approximately 600 °C, the strains remain almost constant.

When the specimen is mechanically loaded prior to thermal loading, the magnitude of strains and (with increasing load level) the characteristics of the evolution of strains changes. In longitudinal direction (see Figures 3(a,c,e)), the total strain is smaller for the mechanically-loaded case compared to the unloaded case. This can be explained by delayed elastic strains (in consequence of a temperature-dependent decrease of the stiffness of concrete) as well as plastic strains (in consequence of a decrease of the strength of concrete). The difference between the mechanically-unloaded and mechanically-loaded case increases with increasing load level, resulting even in contraction of the specimen for higher load levels at higher temperature. With increasing load level this transition from expansion to compaction occurs earlier, e.g., at lower temperatures. For $s = 0.6$, no significant longitudinal expansion of the sample can be observed since compressive loading and the resulting compaction overshadows the thermal expansion almost completely. Moreover, for load levels $s \geq 0.3$, the specimen failed before the maximum testing temperature of 800 °C was reached (and the experiment was stopped) since the compressive strength decreases with increasing temperature. At higher load levels ($s \geq 0.4$), however, the variation in the results increases significant. The strains in radial direction (see Figures 3(b,d,f)) exhibit a similar behavior as the longitudinal strains. With increasing load level, the difference to the mechanically-unloaded case ($s = 0$) increases. Moreover, the initiation of failure of the specimen can be seen, e.g., in Figure 3(f) by a fast increase of the radial strains. As already mentioned, the failure temperature decreases with increasing temperature.

---

[2]To the authors' knowledge, no experimental results are available in the literature for radial strains.

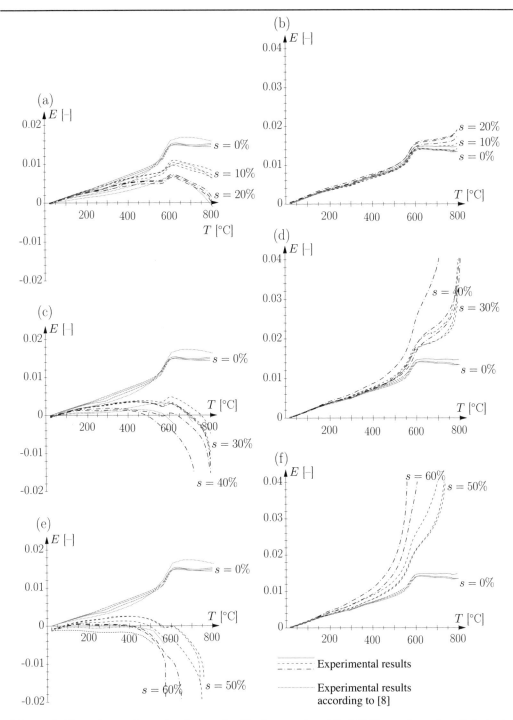

Figure 3: Experimental results for strain E of concrete with PP-fibers: (a,c,e) radial strain for load level $s = 0, 10, 20, 30, 40, 50$ and 60 %; (b,d,f) longitudinal strain for load levels $s = 0, 10, 20, 30, 40, 50$ and 60 % as function of temperature

Based on the directly measured strain E shown in Figure 3, Figure 4 shows the difference between E and the strain at $s = 0$ (i.e., the free thermal strain E_{th}). In contrast to Figure 3 the average value for every load step is presented in Figure 4. The influence of combined mechanical and thermal loading is directly visible and the obtained results are in good agreement with experimental results from literature [10]. The difference $E - E_{th}$ increases linearly with increasing load level, except of the results for $s = 0.4$ and $s = 0.5$ where a considerable variation of the experimental results was observed and, hence, the averaged results must be used with caution.

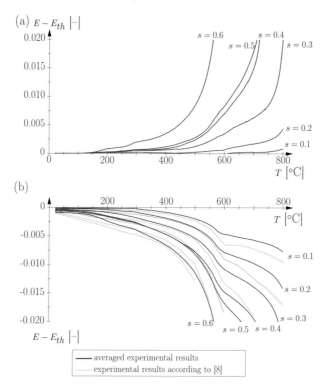

Figure 4: Experimental results for strain $E - E_{th}$ of concrete with PP-fibers: (a) radial and (b) longitudinal strains for load level $s = 0$, 10, 20, 30, 40, 50, and 60 % as function of temperature

3   Transport behavior

As already mentioned, the transport parameters of concrete influence the spalling behavior under fire loading and, hence, the structural safety of concrete structures. The permeability of concrete governs the ability of concrete to advect water vapor originating from vaporization of pore water in consequence of heating of concrete during fire loading. Hereby, the magnitude of pore-pressure built up is directly related to the permeability of concrete and its evolution during temperature loading. In this section, experimental investigations of the effect of thermal loading as well as the influence of addition of PP-fibers on the permeability of concrete and cement paste are presented.

3.1 Materials and methods

The permeability tests were performed on concrete and cement-paste samples with and without PP-fibers (see Tables 1 and 2 for the mix designs). The mix design of the cement paste was derived from the concrete mix design by exclusion of the aggregates (except for the aggregate fraction < 0.25 mm) and upscaling of all other quantities to obtain 1 m^3 of cement paste. Hence, a PP-fiber content of 4.0 kg/m^3 in the cement paste corresponds to 1.5 kg/m^3 PP-fibers in the concrete mix. For both fiber-reinforced mixtures (concrete and cement paste), monofilament polypropylene fibers (PP-fibers) with a length of 6 mm and a diameter of 18 μm were added during mixing.

Table 2: Mix design of investigated cement paste without and with PP-fibers

cement CEM I [kg/m^3]	782
additive (fly ash) [kg/m^3]	135
water [kg/m^3]	355
polypropylene (PP) fibers* [kg/m^3]	0 / 4.0
siliceous aggregates < 0.25 mm [kg/m^3]	471
water/cement-ratio [–]	0.45
water/binder-ratio† [–]	0.40
initial density [kg/m^3]	1743 / 1747
slump* [mm]	200 / 195

* Left value corresponds to plain cement paste, right value corresponds to cement paste with PP-fibers.
† The amount of additives is weighted by 0.8 [7].

The concrete specimens used for the permeability tests were produced from large concrete blocks which were cast and stored under water for 28 days. After storage, cylinders with a diameter of 150 mm were drilled from the concrete block and cut into discs with a height of 50 mm. The cement-paste specimens were cast in plastic pipes with a diameter of 50 mm and a height of 100 mm. After 28 days of water curing, the cylinders were cut into discs with a height of approximately 30 mm. Prior to the permeability tests, all concrete and cement-paste specimens were subjected to the same heating procedure: heating-up with a heating rate of 1 °C/min, storage at the target temperature for a minimum duration of 12 h, cooling with a maximum cooling rate of 1 °C/min. The rather slow temperature increase and decrease was chosen to avoid microcracking due to large temperature gradients within the specimens. The permeability tests were performed for eight different temperature levels and a minimum number of three samples was tested at each pre-heating temperature: 20, 80, 105, 140, 200, 300, 400 and 600 °C. After cooling to ambient temperature, the specimens were built into steel rings, and the lateral side was sealed in order to ensure one-dimensional air flow.

The experimental setup and the schematic illustration of the experimental device are shown in Figures 5 and 6. In order to cover the wide range of the permeability of concrete and cement-paste specimens subjected to temperature loading, the volume of the air reservoir can be varied. Within the experiment, the permeability was determined from the gas flux through the specimen which was determined either from the pressure decrease in the test-chamber volume, $p_t(t)$ (see [14] for details) or from the pressure decrease in the reservoir, $p_r(t)$ (with constant pressure p_t in the test chamber, i.e., at the top of the specimen (see [8]):

1. *Decreasing-pressure experiment (DPE):*

   The pressure in the system was increased to a target pressure between 8 and 9 bar. After stationary flow was established at that target pressure, the air supply was closed and the

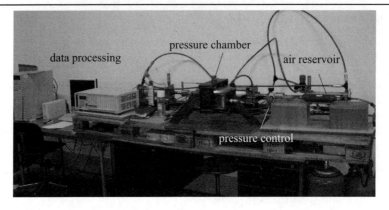

Figure 5: Experimental setup used for permeability tests

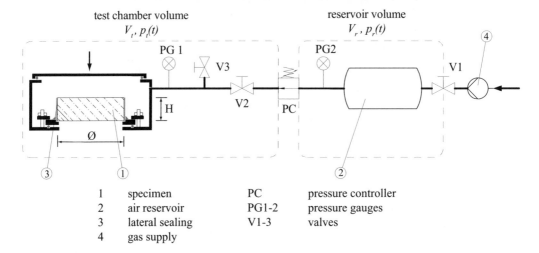

1	specimen	PC	pressure controller
2	air reservoir	PG1-2	pressure gauges
3	lateral sealing	V1-3	valves
4	gas supply		

Figure 6: Schematic illustration of experimental device used for permeability tests

decrease of pressure by air flow through the specimen was recorded as a function of time, $p_t(t)$. In order to assume quasi-stationary conditions, the pressure drop must occur sufficiently slow which can be actively influenced by the size of the additional reservoir.

2. *Constant-pressure experiment (CPE):*

   Hereby, a pressure of approximately 9 bar was maintained in the pressure reservoir whereas the desired target pressure in the test chamber (ranging from 5 to 8 bar) was set by adjusting the pressure controller. Upon reaching stationary flow conditions, the gas supply was closed and the pressure histories $p_t(t)$ and $p_r(t)$ were recorded (see Figure 6).

From monitoring the pressures in the system, the flux and, consequently, the permeability was determined as a function of pressure (more specifically as a function of $1/p$) and the intrinsic permeability (i.e., the permeability which is independent of the fluid passing through the specimen) was determined by linear regression (see Figure 7 and [14] for details).

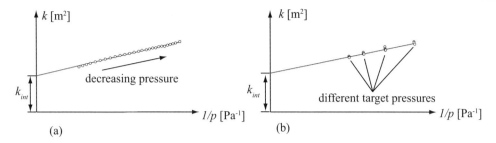

Figure 7: Illustration of identification of intrinsic permeability from regression analysis of (a) decreasing-pressure experiments (DPE) and (b) constant-pressure experiments (CPE)

3.2 Results

Figure 8 shows the residual permeability of concrete and cement paste without and with PP-fibers. At low pre-heating temperatures, no fiber effect is observed, neither for concrete nor for cement paste. In both cases, the respective experimental values are almost equal. However, the permeability of cement paste is smaller than the respective values of concrete. This can be attributed to the existence of interfacial transition zones (ITZ) between cement-paste matrix and aggregates, having higher permeability than the cement paste [1]. The fiber effect can be observed between pre-heating temperatures of 140 and 200 °C, for both concrete and cement paste. In case of cement paste, this effect is stronger (leading to a larger jump in k_{int}) since the PP-fibers are present in the cement paste only (with a higher amount per m^3 of cement paste than per m^3 of concrete). In case of concrete, also the (non-permeable) aggregates as well as ITZ contribute to the macroscopic permeability, therefore the fiber effect is smaller for the composite than for the cement paste alone. With further increase in pre-heating temperature, the fiber effect is reduced since temperature-induced damage of the concrete phases overshadows the effect of melting of the PP-fibers. The extent of this trend, however, is different for concrete and cement paste, with higher differences between cement paste without and with PP-fibers compared to the respective values for concrete. This leads to the conclusion that temperature-induced damage is higher in case of concrete where

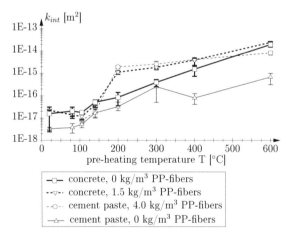

Figure 8: Intrinsic permeability of concrete and cement paste without and with PP-fibers as function of pre-heating temperature (symbols represent average value, lines mark minimum and maximum value)

strain incompatibilities between cement paste and aggregate contribute to damage of the composite. In case of cement paste, on the other hand, dehydration (leading to an increase in pore space as well as cracking) is the only source of damage, therefore the fiber effect persists to a larger extent.

4   Conclusions and ongoing work

Both mechanical and transport properties of concrete are strongly affected by temperature loading. In this paper, the strain behavior of concrete subjected to combined thermal and mechanical loading as well as the permeability of concrete were investigated experimentally.

The experimental strain results showed a dependence on the load level applied prior to thermal loading. The obtained experimental results are in good agreement with results from the literature [10]. The increase of mechanical loading in combination with thermal loading leads to a decrease of the longitudinal strains. A linear dependency between the strain difference $E(T, s) - E_{th}(T, 0)$ and the thermal strain $E_{th}(T, 0)$ was observed which is in agreement with observations presented in [13]. The additionally-monitored radial strains complete the strain data.

Experimental investigation of the permeability of concrete and cement paste without and with PP-fibers showed an increase of the permeability with increasing temperature. Moreover, the PP-fiber effect (i.e., a permeability jump at the melting temperature of the fibers) was found in both materials, the magnitude of this effect and its evolution with temperature, however, were different for concrete and cement paste.

The presented experimental investigations are interconnected with research activities concerning material modeling of the behavior of concrete at elevated temperatures. In case of the mechanical behavior, a suitable material model (accounting for the heterogeneous nature of concrete) is designed, capturing the experimentally-obtained dependence of the strain of heated concrete on the load level. In case of the transport behavior of concrete, a network model is currently designed, taking information on the size and arrangement of pores in concrete into account. In both cases, further experimental investigation will serve as basis for the modeling work in order to capture the main mechanisms in heated concrete.

5   Acknowledgements

This research was conducted with financial support by the Austrian Ministry for Transport, Innovation and Technology (bm.vit) within the KIRAS-project (Austrian security research program) 813794 "Sicherheit von Hohlraumbauten unter Feuerlast" ("Safety of underground structures under fire loading").

References

[1] D. P. Bentz. Fibers, percolation, and spalling of high-performance concrete. *ACI Materials Journal*, 97(3):351–359, 2000.

[2] T. Z. Harmathy. Effect of moisture on the fire endurance of building elements. Technical Report STP 385, American Society of Testing and Materials, 1965.

[3] P. Kalifa, G. Chéné, and C. Gallé. High temperature behaviour of HPC with polypropylene fibres: from spalling to microstructure. *Cement and Concrete Research*, 31:1487–1499, 2001.

[4] G. A. Khoury, B. N. Grainger, and P. J. E. Sullivan. Strain of concrete during first heating to 600°C. *Magazine of Concrete Research*, 37(133):195–215, 1985.

[5] W. Kusterle, W. Lindlbauer, G. Hampejs, A. Heel, P.-F. Donauer, M. Zeiml, W. Brunnsteiner, R. Dietze, W. Hermann, H. Viechtbauer, M. Schreiner, R. Vierthaler, H. Stadlober, H. Winter, J. Lemmerer, and E. Kammeringer. Brandbeständigkeit von Faser-, Stahl- und Spannbeton [Fire resistance of fiber-reinforced, reinforced, and prestressed concrete]. Technical Report 544, Bundesministerium für Verkehr, Innovation und Technologie, Vienna, Austria, 2004. In German.

[6] C. Meyer-Ottens. *Zur Frage der Abplatzungen an Betonbauteilen aus Normalbeton bei Brandbeanspruchung [Spalling of normal–strength concrete structures under fire loading]*. PhD thesis, Braunschweig University of Technology, Braunschweig, Germany, 1972. In German.

[7] Österreichisches Normungsinstitut. *Beton – Teil 1: Festlegung, Herstellung, Verwendung und Konformitätsnachweis [Concrete – Part 1: Specification, production, use and verification of conformity]*, 2004. In German.

[8] P. Paulini and F. Nasution. Air permeability of near surface concrete. In F. et al. Toutelemonde, editor, *Proceedings of the 5th International Conference on Concrete Under Severe Conditions: Environment and Loading (CONSEC'07)*, pages 241–248, Paris, 2007. Laboratoire central des ponts et chaussees (LCPC).

[9] L. T. Phan and N. J. Carino. Review of mechanical properties of HSC at elevated temperatures. *Journal of Materials in Civil Engineering*, 10(1):58–64, 1998.

[10] U. Schneider. *Ein Beitrag zur Frage des Kriechens und der Relaxation von Beton unter hohen Temperaturen [Contribution to creep and relaxation of concrete under high temperatures]*. Habilitation thesis, TU Braunschweig, Braunschweig, Germany, 1979. In German.

[11] U. Schneider. Concrete at high temperature – a general review. *Fire Safety Journal*, 13:55–68, 1988.

[12] U. Schneider and J. Horvath. Abplatzverhalten an Tunnelinnenschalenbeton [Spalling of concrete for tunnel linings]. *Beton- und Stahlbetonbau*, 97(4):185–190, 2002. In German.

[13] S. Thelandersson. Modeling of combined thermal and mechanical action in concrete. *Journal of Engineering Mechanics (ASCE)*, 113(6):893–906, 1987.

[14] M. Zeiml, R. Lackner, D. Leithner, and J. Eberhardsteiner. Identification of residual gas-transport properties of concrete subjected to high temperatures. *Cement and Concrete Research*, 38(5):699–716, 2008.

Keynote Lecture 2

Applied Mechanics and Materials Vols. 24-25 (2010) pp 15-22
© (2010) Trans Tech Publications, Switzerland
doi:10.4028/www.scientific.net/AMM.24-25.15

Experimental Characterisation of Parameters Controlling the Compressive Failure of Pultruded Unidirectional Carbon Fibre Composites

O.T. Thomsen[1, a], K.K. Kratmann[2,b]

[1]Department of Mechanical and Manufacturing Engineering, Aalborg University, Pontoppidanstræde 101, DK-9220 Aalborg East, Denmark

[2]Fiberline Composites A/S, Barmstedt Alle 5, DK-5000, Middelfart, Denmark

[a]ott@me.aau.dk, [b] kkr@fiberline.com

Keywords: UD CFRP, Pultrusion, Compressive failure, Kinkband formation models, Fourier transform misalignment analysis (FTMA), Digital Image Correlation (DIV), modified Iosipescu test

Abstract. The classical kink-band formation models predict that the compressive strength of UD carbon fibre reinforced composite materials (UD CFRP) is governed by fibre misalignment as well as of the mechanical shear properties. A new image analysis procedure for experimental determination of the fibre misalignment, the Fourier transform misalignment analysis (FTMA), has been developed. Moreover, a modified asymmetric Iosipescu test specimen geometry has been developed and validated for accurate measurement of the composite material shear properties without parasitic effects due to axial splitting. In the test procedure the shear strain distribution is measured using Digital Image Correlation (DIC) and the results calibrated based on FEA modelling results. Using the measured properties as input, the predictions of the classic compressive strength models have been compared with measured compressive strengths. Finally, an alternative approach to the classical kink band equilibrium has been proposed and demonstrated to provide more accurate predictions than the classical models.

Introduction

Generally the failure of composites is a complex mix of competing failure mechanisms. This is because of the in-homogeneity of the composite materials where failure can occur in any of the constituents, their interfaces and by interaction between them. Depending on constituents, interfaces and loading scenarios, different failure mechanisms may be lead to failure of the material. An overview of the most common compressive failure mechanisms in unidirectional (UD) fibre composite materials is presented in Fig. 11, Hahn and Williams [1] and Fleck [2].

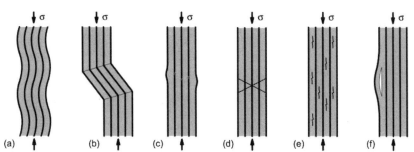

Fig. 1. Compressive failure modes of fibre composites. (a) Elastic micro buckling. (b) Fibre kinking. (c) Fibre crushing. (d) Shear band formation. (e) Matrix cracking. (f) Buckle delamination.

The competition between the failure mechanisms can be visualised by failure mode maps as the one shown in Fig. 2, Fleck [2].

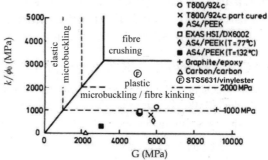

Fig. 2. Failure mode map by Fleck [2], with the shear modulus G along the first axis and shear strength k normalised with respect to the fibre misalignment f_o on the second axis. Fiberline's pultruded CFRP lamella have been added marked with ⓕ.

As can bee seen from Fig. 2, the primary parameters governing the compressive strength of UD composites are the fibre misalignment angle f_o, the shear strength k, and the shear modulus G of the composite material. It is seen that a very small values of f_o compared k, and a G value below 3GPa will trigger elastic microbuckling. By increasing the fibre misalignment the failure mode will switch to plastic micro buckling / fibre kinking. The shear modulus G can then be increased by changing or modifying the resin system, which might also improve the shear strength. If so the failure mode can switch to fibre crushing. Most engineering composites fail by plastic microbuckling / fibre kinking. This is also the case for the vinylester based UD CFRP lamella manufactured by Fiberline Composites A/S. This has been verified by simple 3-point bending tests, and a micrograph from the failure zone is shown in Fig. , from which fibre kink band formation is clearly seen.

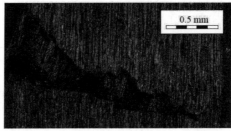

Fig. 3. Post mortem micrograph of UD CFRP tested in 3 point bending, showing kink band formations on the compressed side (left).

This work has focused on the on the elastic perfectly plastic kink band model by Budiansky [3], where the critical compressive stress s_c can be determined as:

$$\sigma_c = \frac{\tau_y}{\gamma_y + \phi_0} \tag{1}$$

where τ_y is the shear yield stress, γ_y is the shear yield strain, and f_o is the initial fibre misalignment. Assuming linear elastic behaviour in shear (i.e. $G=\tau_y/\gamma_y$) Eqn. (1) can be formulated in terms of the shear modulus G.

$$\sigma_c = \frac{G}{1 + \phi_0/\gamma_y} \tag{2}$$

Eqns. (1) and (2) are often written in the same equation and both referred to as the elastic perfectly plastic model.

The elastic perfectly plastic kink band model has also been the subject of several suggestions for improvements, including Budiansky and Fleck [4] who extended the model to incorporate strain hardening, kink band inclination, combined external compressive and shear loading, and finally finite fibre stiffness. However, but the simple perfectly plastic formulation [3] remains a central part of recent studies by Soutis et al. [5], and Lee and Soutis [6]. [7], showing excellent agreement between measured and predicted strengths. See also Kratmann [8] for an overview. Taking the classical kink band model by Budiansky [3] as the basis for this work, the principal parameters controlling the kink band formation in UD CFRP laminates, i.e. the initial fibre misalignment f_o and the composite material shear properties, defined by the shear modulus G and the shear yield strain γ_y. were determined experimentally. For evaluation of the models, the compressive strength was also measured.

The FTMA Method – A New Method for Fibre Misalignment Measurement

Of the parameters controlling fibre kinking, the fibre misalignment is probably the most difficult to quantify, as no standardised measuring techniques exist. Fibre waviness/misalignment is considered a manufacturing defect or imperfection, and its characteristics depend on the specific manufacturing process involved. Examples hereof are the differences between the repetitive waviness introduced by braiding or weaving, local waviness of a UD prepreg being layed out in a mould, and the random waviness in pultruded UD composites as sketched in Fig. .

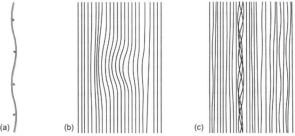

(a) (b) (c)

Fig. 4. Different types of fibre waviness/fibre misalignment generally involved with different types of manufacturing processes. (a) Fibre waviness from braiding and weaving. (b) Wavy regions in UD prepregs or UD tapes. (c) Random waviness in pultrusion [8].

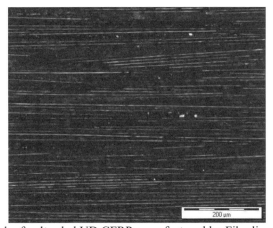

Fig. 5. Micrograph of pultruded UD CFRP manufactured by Fiberline Composites A/S.

The pultruded UD composites studied in this work are considered to be randomly misaligned as the micrograph in Fig. 5 shows, i.e. although a few fibres follow each other; fibres next to those have a significantly different orientation. The fibre waviness/misalignment is 3-dimensional by nature. Dedicated measuring techniques developed specifically for the quantification of fibre misalignment and waviness are limited, see [9], [10], [11], [12]. In this work a novel image analysis procedure named Fourier transform misalignment analysis (FTMA) for measuring fibre misalignment in unidirectional fibre composites has been developed [13]. The FTMA method, which is a 2D technique measuring the fibre directions in section planes prepared by poslishing, measures angles of individual fibre segments isolated by Fourier noise filtering. Working with relatively low magnification micrographs (10× optical magnification) thousands of fibres can be identified, their angle measured and a statistical fibre misalignment distribution generated for the analysed plane, see Fig. 6 (sample analysis results). Areas of 30mm × 38mm have been analysed using the FTMA method, identifying approximately 300,000 fibres in less than 30 minutes using on an Intel E8400, 2×3GHz, 4 GB RAM running Windows XP 64-bit. From these analyses it is assessed that the FTMA method, is probably the most precise and computationally efficient analysis tool for measuring fibre misalignment that are currently available.

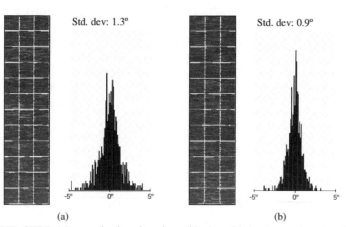

(a)   (b)

Fig. 6 FTMA of UD CFRP (a) open bath pultrusion, (b) closed injection pultrusion (Fiberline). The histograms represent the distribution of single fibre orientations, with the mean orientation equal to 0°. The white lines in the micrographs represent the domain borders.

Determination of the Composite Material Shear Properties

The Iosipescu shear test has been used to characterise the shear properties of pultruded unidirectional carbon fibre reinforced composites. The testing was based on the ASTM standardised Iosipescu shear test [14], which was modified with respect to the geometry, see Fig. 7. The proposed asymmetric test specimen geometry with horizontally aligned fibres in combination with GFRP tabs prevents axial splitting from occurring, and thereby provides more consistent measurements with less experimental scatter. The shear modulus and shear yield strain have been determined using the modified Iosipescu shear test combined with Digital Image Correlation (DIC), see Fig. 8 (sample DIC measurements of in-plane shear strains). The conventional and the proposed Iosipescu test specimen geometries have been tested and validated using both linear and non-linear finite element analyses. The obtained results have been compared and statistically evaluated. It has been shown that the modified test specimen geometry yields the same absolute results as the conventional Iosipescu geometry, but without change in the stress distribution during the test and with less experimental scatter. For full details see [15].

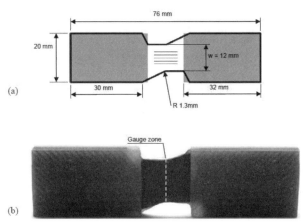

Fig. 7. Sketch with dimensions and photo of the modified asymmetric Iosipescu specimen geometry. The grey areas of the sketch indicate the tabs, and the horizontal lines in the gauge area the fibre orientation [15].

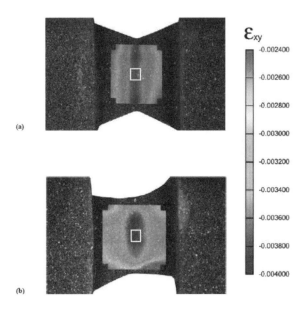

Fig. 8. (a) DIC measurement of ε_{xy} distribution for tabbed symmetric specimen. (b) DIC measurement of ε_{xy} distribution for proposed tabbed asymmetric specimen. The white boxes represent the area from which the shear strain is determined as the average value [15].

Determination of Compressive Strength

Two variants of pultruded unidirectional carbon fibre reinforced polymer (UD CFRP) have been investigated experimentally [16]. The only difference between the variants was the resin system, which for one variant was a vinylester system (Var1), and for the other an improved resin system (Var2). For both material systems the fibre misalignment, shear stress-strain relationships and compressive strength have been determined experimentally. For the measurement of the compressive strength three different test methods were used. A simple four point bending test (ISO 14125 [17]) which is used by Fiberline for quality control. To supplement this Mechanical Combined Loading (MCL) compression tests [18] have been carried out by Risø DTU, National

Laboratory for Sustainable Energy, Denmark. The MCL-test is a pure compression test, compliant with ISO 14126 [19], where the end-to-shear load ratio is kept fixed throughout the test by a mechanical mechanism. Finally, a four point bending test of a sandwich beam with skins made from the pultruded CFRP lamellae was developed specifically for this investigation [16], see Fig. 9.

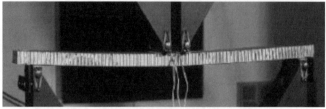

Fig. 9. Photo of the setup used for 4 point bending of UD CFRP sandwich specimens [16].

Fig. 10. Visualisation of the 2D variation of the domain mean angle (fibre misalignment angle) over the gauge area of a Var1 sandwich specimen.

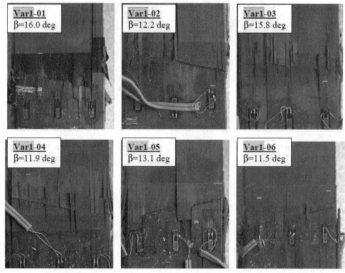

Fig. 11. Photographs of compressive kink band failure with clear inclination angle obtained by four point bending of a sandwich beam made from the Var1 material system [16].

It was found that the Var2 material system, which exhibits pronounced strain hardening in shear, is approximately 10% stronger than the Var1 system, which exhibits approximately elastic perfectly plastic shear behaviour. Using as input the fibre misalignment measurements based on the FTMA method (see Fig. 10 for sample visualisation of results), and the CFRP shear properties measured using the modified Iosipescu test setup, the compressive strengths were estimated. Examples of typical compressive kink band failure are shown in Fig. 11 for the Var1 material system. It was found that the classical compressive strength model by Budiansky [3] should not be used in its linearized form, and that the extended model by Budiansky and Fleck [4] also fails to predict the compressive strength, although designed to account for strain hardening. The classical compressive strength models [3], [4] are based on kink band equilibrium considerations, and an alternative approach to treat the experimental data was proposed in [16]. A good correlation with the experimentally recorded compressive strength values was found, and it was further found that transverse kink band stresses and the kink band inclination angle can not be neglected for composites exhibiting pronounced strain hardening in shear.

Conclusions

The work presented has treated the compressive failure of UD CFRP composite materials manufactured using pultrusion. Special emphasis has been focused on accurately measuring the properties governing the compressive failure, according to the classical strength models by Budiansky [3] and Budiansky and Fleck [4]. To achieve this, a new method for fast and accurate measurement of the fibre misalignment, the FTMA method, has been developed, validated and benchmarked against other existing methods. In addition, a modified Iosispescu asymmetric test specimen has been developed and validated for accurate measurement of UD composite material shear properties without parasitic effects due to axial splitting. In the test procedure the shear strain distribution is measured using Digital Image Correlation (DIC), and the results calibrated based on FEA modelling results. Finally, the predictions of the classic compressive strength models have been compared with measured compressive strengths, using the measured material properties as input. A good correlation with the experimentally obtained compressive strength values was found, and it was further found that transverse kink band stresses and the kink band inclination angle can not be neglected for composites exhibiting pronounced strain hardening in shear.

Acknowledgements

The main part of the work presented has been conducted as part of an Industrial Ph.D. programme carried out in collaboration between Fiberline Composites A/S, Denmark and the Department of Mechanical and Manufacturing Engineering, Aalborg University, Denmark, and co-sponsored by the Danish Agency for Science, Technology and Innovation.

References

[1] H.T. Hahn, J.G. Williams. Failure mechanisms in unidirectional composites. In: Composite materials: testing and design (seventh conference). ASTM STP 893. American Society for Testing and Materials, Philadelphia (1986), 115-139.

[2] N.A: Fleck NA. Compressive Failure of Fibre Composites. Advances in Applied Mechanics, Vol. 33 (1997), 43-119.

[3] B. Budiansky. Micromechanics. Computers & Structures, Vol. 16 (1983), 3-12.

[4] B. Budiansky, N.A. Fleck. Compressive failure of fibre composites. Journal of the Mechanics and Physics of Solids; Vol. 41, (1993), 183-211

[5] C. Soutis, F.C. Smith, F.L. Matthews. Predicting the compressive engineering performance of carbon fibre-reinforced plastic. Composites: Part A, Vol. 31, (2000), 531-536.

[6] J. Lee, C. Soutis. Thickness effect on the compressive strength of T800/942C carbon fibre-epoxy laminates. Composites: Part A, Vol. 36, (2005), 213-227.

[7] J. Lee, C. Soutis. A study on the compressive strength of thick carbon fibre-epoxy laminates. Composite Science and Technology, Vol. 67, (2007); 2015-2026.

[8] K.K. Kratmann, *Evaluation of Compressive Failure of Pultruded Unidirectional Carbon Fibre Composites*. Ph.D. Thesis, Special Report No. 67, Department of Mechanical Engineering, Aalborg University, ISBN 87-91464-21-8 (2010).

[9] A.R. Clarke, G. Archenhold, N.C. Davidson. A novel technique for determining 3D spatial distribution of glass fibres in polymer composites. Composites Science and Technology, Vol. 55, (1995), 75-91.

[10] A.R. Clarke, G. Archenhold,, N.C. Davidson, N.A. Fleck. Determining the power spectral density of the waviness of unidirectional glass fibres in polymer composites. Applied Composite Materials, Vol. 2 (1995), 233-243.

[11] S.W. Yurgartis. Measurement of small angle fiber misalignments in continuous fiber composites. Composites Science and Technology, Vol. 30 (1987),:279-293.

[12] C.J. Creighton, M.P.F. Sutcliffe, T.W. Clyne. A multiple field image analysis procedure for characterisation of fibre alignment in composites. Composites Part A: Applied Science and Manufacturing,Vol. 32, (2001), 221-229.

[13] K.K. Kratmann, M.P.F. Sutcliffe, L.T. Lilleheden, R. Pyrz and O.T. Thomsen O.T., A novel image analysis procedure for measuring fibre misalignment in unidirectional fibre composites. Composites Science and Technology, Vol. 69, (2009), 228-238.

[14] ASTM D 5379-93. Standard Test Method for Shear Properties of Composite Materials by the V-notched Beam Method. American Society for Testing and Materials, W. Conshohocken, PA (1993).

[15] K.K. Kratmann, K.K., O.T. Thomsen, O.T., L.T. Lilleheden, R. Pyrz. Measurement of the Shear Properties of Unidirectional Carbon Fibre Composites by Using a Modified Iosipescu Test Specimen. Submitted (2010).

[16] K.K. Kratmann, K.K., O.T. Thomsen, O.T., L.T. Lilleheden, R. Pyrz. Experimental Evaluation of the Classical Compressive Strength Models for Unidirectional Carbon Fibre Composites. Submitted (2010).

[17] ISO 14125. Fibre-reinforced plastic composites - Determination of flexural properties. International Organization for Standardization, Geneva, Switzerland (1998.).

[18] J.I. Bech, S. Goutianos, T.L. Andersen, R.K. Torekov and P. Brøndsted. A new static and fatigue compression test method for composites. Strain (Early on-line), doi: 10.1111/j.1475-1305.2008.00521.x

[19] ISO 14126. Fibre-reinforced Plastic Composites – Determination of Compressive Properties in the In-plane Direction. International Organization for Standardization, Geneva, Switzerland. (1999)

BSSM Measurements Lecture

Applied Mechanics and Materials Vols. 24-25 (2010) pp 25-41
© (2010) Trans Tech Publications, Switzerland
doi:10.4028/www.scientific.net/AMM.24-25.25

Aspects of Uncertainty Analysis for Large Nonlinear Computational Models

K. Worden[1,a], W.E. Becker[1,b], M. Battipede[2,c] and C. Surace[3,d]

[1] Dynamics Research Group, Department of Mechanical Engineering,
University of Sheffield, Mappin Street, Sheffield S1 3JD, UK

[2] Department of Aeronautical and Space Engineering, corso Duca degli Abruzzi 24,
Politecnico di Torino, 10129 Torino, Italy

[3] Department of Structural and Geotechnical Engineering, corso Duca degli Abruzzi 24,
Politecnico di Torino, 10129 Torino, Italy

[a]k.worden@sheffield.ac.uk, [b]w.becker@sheffield.ac.uk, [c]manuela.battipede@polito.it,
[d]cecilia.surace@polito.it

Keywords: Nonlinear models, uncertainty analysis, Bayesian sensitivity analysis.

Abstract. This paper concerns the analysis of how uncertainty propagates through large computational models like finite element models. If a model is expensive to run, a Monte Carlo approach based on sampling over the possible model inputs will not be feasible, because the large number of model runs will be prohibitively expensive. Fortunately, an alternative to Monte Carlo is available in the form of the established Bayesian algorithm discussed here; this algorithm can provide information about uncertainty with many less model runs than Monte Carlo requires. The algorithm also provides information regarding sensitivity to the inputs i.e. the extent to which input uncertainties are responsible for output uncertainty. After describing the basic principles of the Bayesian approach, it is illustrated via two case studies: the first concerns a finite element model of a human heart valve and the second, an airship model incorporating fluid structure interaction.

Introduction

The early stages of development of a new engineering structure from design to commissioning can be exceedingly costly. One of the major drivers in this cost is the need for physical prototyping and experimental testing of the prototype to make sure it meets design requirements. It is argued that these costs could be reduced considerably by more emphasis on computer modelling in the early stages of product development – a sort of virtual prototyping. However, one of the major problems anticipated in the attempt to replace testing with modelling is that computer models will not generally provide a complete representation of the in-the-field structural behaviour. Assuming that adequate computer resources are available; the main reason for lack of model fidelity will be *model uncertainty*. In this context there are two types of uncertainty: *aleatory* and *epistemic*. Aleatory uncertainty, sometimes called 'irreducible' uncertainty is associated with the truly random and unknowable features of the operating environment of the structure e.g. the temperature experienced by the structure. Epistemic uncertainty, sometimes called 'reducible' uncertainty is associated with incomplete knowledge of the structure; information which could in principle be learned by future experimentation. If predictions are to be made on the basis of a model of a structure or system, it is critical to take account of aleatory uncertainty in order to encompass the full range of possible behaviours under random conditions. It is also important to assess the contribution of uncertain model parameters to the overall response uncertainty; in this way future experimentation can be directed towards learning more about the uncertain parameters which contribute most to uncertainty in the predictions. These two problems can be accommodated respectively by *uncertainty propagation* and *sensitivity analysis*. The current paper will discuss how the required analysis can be carried out, particularly in relation to a principled Bayesian framework. The discussion is illustrated through the use of two case studies: the first concerning a human heart valve model and the second, an airship model incorporating fluid structure interaction.

Uncertainty Propagation in Structural Dynamics

This paper will focus on a particular aspect of uncertainty analysis which has special relevance in structural dynamics. In fact, one can argue that the analysis of uncertainty raises three main questions in dynamics. First, an appropriate *model* of uncertainty is a vital element in the design and modelling of high-value engineering structures and systems - this is the problem of *quantification*. There are numerous theoretical frameworks which allow a specification of uncertainty, like: probability theory, possibility theory, Dempster-Shafer etc., and an excellent discussion of the relations between these approaches can be found in [1]. The broader problem of quantification involves selecting the appropriate theoretical framework and then assigning a quantitative measure of uncertainty or risk. Given that there is more than one way of assigning a measure of risk, the problem arises of translating between them. This problem of normalisation is broadened into one of *fusion* by posing the question: given two assignments of uncertainty from different frameworks, how can one refine each estimate in the light of the extra information from the other? Equally important in the design process is a prescription for deciding how a measure of uncertainty on the inputs, or specification of a problem, will affect the outputs or results. This is the problem of *propagation*. Part of this problem is to find which parameters contribute most to output uncertainty; these parameters will require the most effort to estimate in the *a priori* specification of the problem.

The main concern of this paper is the problem of uncertainty propagation and in particular, the propagation through nonlinear systems. Nonlinear systems present a specific problem in uncertainty analysis because they are not guaranteed to be structurally stable; they may bifurcate when small changes are made to the inputs or parameters. In contrast, linear systems are structurally stable and small perturbations of the inputs or parameters will induce small perturbations of the outputs. In fact this property provides the main line of attack for uncertainty propagation through linear systems; to quote [2], 'If the coefficients of variation of the involved quantities are small, then the perturbation method can be directly applied to the majority of problems...' (It is important to point out here that [2] is not subject to this restriction and proposes approaches which can deal with the situation when the coefficients of variation are moderate to large.)

It may actually be the case that the solution to one of the problems suggests or fixes the solution to one of the others; as an example consider the specification of uncertainty in terms of an *interval*. This is a situation in which nothing is known of a variable x except an upper bound $\bar{x}$ and a lower bound $\underline{x}$; the variable can then be expressed by the interval $[\underline{x}, \bar{x}]$. The adoption of this model for uncertainty, in a sense, fixes the strategy for uncertainty propagation. The reason is that there is a type of arithmetic for such uncertain quantities which mirrors all the usual operations for *certain* or *crisp* numbers; for example, addition of two interval numbers is governed by,

$$[a, b] + [c, d] = [(a + c), (b + d)] \qquad (1)$$

Multiplication of two interval quantities is given by,

$$[a, b] * [c, d] = [\min(ac, ad, bc, bd), \max(ac, ad, bc, bd)] \qquad (2)$$

(The latter definition takes into account the effect of negative values). Multiplication of an interval number by a positive scalar c is given by,

$$c * [a, b] = [ca, cb] \qquad (3)$$

Finally, if an interval number is the input to a monotonically-increasing function F, then the result is given by,

$$F([a,b]) = [F(a), F(b)] \tag{4}$$

One can immediately see how uncertainty based on an interval prescription could be propagated through an engineering model, say a Finite Element (FE) model. A general crisp FE model is specified in terms of some initial quantities e.g. the mass and stiffness matrices for a dynamic calculation (the assembly of such matrices is in itself non-trivial, but that is another matter). A calculation can then proceed towards some desired output e.g. a set of natural frequencies. The algorithm for the calculation can ultimately be broken down into a sequence of the usual mathematical operations. Now, if the initial specification for the mass and stiffness matrices was in terms of intervals (matrices of intervals), in order to encode uncertainty as to their values, in principle, one could carry the calculation of interest forward by replacing each crisp operation by its interval analogue. The result of this procedure would (if successful) be an interval containing upper and lower bounds on the natural frequencies consistent with the uncertainties in the initial parameters or inputs to the model. In fact this type of approach, based on interval analysis, has been implemented with some success in terms of modal analysis and FE analysis [3]. One of the major problems with interval analysis is that it is inherently conservative; during long sequences of calculations, particularly if these are recursive, the uncertainty encoded in the width of the interval, can grow considerably. The final intervals from a long calculation are guaranteed to include the true range in the outputs engendered by the uncertainty in the inputs, but they are often much larger - so much larger in some cases, that they are useless. Another problem with interval analysis is that it represents the uncertainty in possibly the coarsest fashion imaginable. A much better theory of uncertainty is probability theory; this accounts for randomness in the most informative manner possible in direct contrast to interval analysis [4].

The specification of a (continuous) random variable X in probability theory is in terms of the probability density function (pdf) for the variable, $p(x)$. The usual definition of this quantity is that the probability that a random variable X will fall in the range $[x, x + dx]$ is given by,

$$P(X \in [x, x + dx]) = \int_{-\infty}^{\infty} p(x)dx \tag{5}$$

with the standard meanings for the various symbols. The pdf encodes all possible knowledge of the behaviour of X. So if probability is perfect, it begs the question - why not always use probability theory? The answer is, one does if one can, but there are practical issues which raise difficulties. The most immediate problem is that it is not always straightforward to obtain the pdf for a variable. If the pdf is not derivable from the underlying physics of the problem, it will need to be estimated or elicited from experts, and this can be a problem [4].

Suppose that the problem of interest is to compute an output y for a problem, given d input parameters $\{x_1, \cdots, x_d\}$. For example, one may wish to know the maximum value of stress over a component modelled by FE with the various inputs being the material properties etc. which require prior specification. Regardless of the complexity of the calculation, it can be regarded as a function of the inputs,

$$y = f(x_1, \cdots, x_d) \tag{6}$$

which results in the output of interest. So far, all of this is deterministic; however, suppose that the set of inputs is uncertain. If the inputs are random, the best specification one can have is the multivariate joint pdf (jpdf) $p(x_1, \cdots, x_d)$, whose definition is a natural generalisation of (5). If the jpdf of the inputs is known, it may still be an intractable problem to translate this into a pdf (specification of uncertainty) for the output y. To explain things in the simplest terms, suppose that one has a univariate problem $y = f(x)$ (maximum stress only depends on Young's modulus, say). If the pdf of the input $p_x(x)$ is known, the pdf of the output $p_y(y)$ follows from a transformation law,

$$p_y(y) = p_x(f^{-1}(y)) \left| \frac{df^{-1}(y)}{dy} \right| \tag{7}$$

and this is all very well as long as one can write down the function f explicitly. If f is only specified implicitly by a large (potentially nonlinear) computer model, this will not be possible. A further complication is that very few problems indeed have a single input. In the absence of an analytical solution, one approach to the propapation problem is that of *Monte Carlo* (MC) analysis [5]. The idea is simply stated: one generates many random samples of possible inputs consistent with their jdpf and then propagates the individual crisp input sets through the algorithm or computation; the result is a set of outputs which one can use in order to estimate their pdf. The problem is that, if the input set is multi-dimensional, it may require very many samples to be drawn in order to cover all possibilities. This is fine as long as the calculation is quick; however, the implication is that one is dealing with a large complex calculation which may well be expensive to run in terms of computer resources. In the latter case, MC analysis may be prohibitively expensive.

One way out of this dilemma is to use a *surrogate model* or *emulator* (also called *metamodels*, *response surfaces, fast-running models* etc. under various circumstances). Again, the idea is simple to state: suppose the true (expensive) model specification is as in (6), a surrogate model is simply an alternative model $y_s(x_1, \cdots, x_d)$ which behaves like the true system on a given set of inputs, but is structurally much simpler. The most basic examples of such a surrogate model might be a straight-line regression fit to a polynomial, or a polynomial fit to a transcendental function. The key idea here is exposed by the use of the term *regression*; the basis of regression analysis is that one can specify a parametric model for an input-output process and then *estimate* or *learn* the parameters of the model by minimising the model error on some known input-output data (usually called training data). This general regression problem is central to the well-explored discipline of *machine learning* [6]. The key observation is that the regression model can typically be learned using a small number of input-output sets, and this will require a smaller number of runs of the true model, certainly many less than a MC analysis would require. Once the surrogate model or emulator is established, it can then be used to generate the large number of input-output sets required by MC analysis in order to characterise the output uncertainty. A further critical advantage of the emulator is that, in contrast to the true model, it will often have a simple closed-form structure which could in principle be used for the analogue of equation (7), and this offers the possibility of avoiding MC completely. The idea of the emulator will be central to the next section's discussion of Bayesian sensitvity analysis.

Bayesian Sensitivity Analysis

The discussion of the last section was centred on the idea of uncertainty propagation; given a series of uncertain inputs to a computation, how does one characterise the uncertainty on the output? As discussed in the introduction there are other pressing problems in dynamics which are associated with uncertainty, one of them being *sensitivity analysis* (SA) [7]. In a sense, this is a refinement of the propagation problem where one wishes to determine (qualitatively) which input variables (with their associated uncertainties) are responsible for the output uncertainty. The SA technique which will be discussed and applied here is based on Bayesian probabilistic approach as detailed in [8]. (The approach is well-established within the probability and statistics community, the current paper is concerned with arguing the benefits of adopting it in the engineering context.) Each uncertain input parameter is represented as a probability distribution, and an emulator is fitted using multiple runs of the true model. In this case, the emulator is based on a powerful regression paradigm from machine learning called a *Gaussian process* [9]. From this emulator, statistical quantities relating to sensitivity and uncertainty can be inferred directly – for example, output uncertainty distributions and parameter sensitivities. A further advantage of using Gaussian process regression is that the uncertainty of the emulator fit is itself quantified.

Gaussian Process Regression. As discussed above, any computer model can be regarded as a function of its inputs: $f(x)$. (From this point, the multivariate input set will be specified as a boldface vector; this releases subscript indices to indicate instances of training data.) Although the function is (usually) deterministic and governed by known mathematical relationships, it is often complex to the point of mathematical intractability. From a practical point of view, $f(x)$ can then be regarded as an unknown function, given that one does not know the output for a given set of inputs until one has actually run the model - or an emulator if the true model is computationally expensive. As indicated above, the emulator discussed here will be the *Gaussian process* (GP).

GPs are an extension of the standard multivariate Gaussian probability distribution, they are essentially a probability density function *over functions*. Whereas most forms of regression return a crisp value $f(x)$ for any given x, a GP returns a specification for a Gaussian probability distribution. This means that prediction using the model automatically provides estimates of the confidence interval on the output. GPs also adhere to the Bayesian paradigm for probability, that is, a number of prior assumptions are made about the function being modelled, and then training data (samples from the model) are used to update and evaluate a posterior distribution over functions. A key assumption is that the model is a smooth function of its inputs – it is this that allows extra information concerning the response to be gained at reduced computational cost. Clearly if a function is smooth (in a sense, slowly-varying), information from a data point (input) will allow inferences on the behaviour of the function on neighbouring points.

The basic elements of GP regression will be discussed here, a much more extensive treatment can be found in [10]. For any set of n input point $\{x_1, \cdots, x_d\}$ (the subscripts now provide a training point label), each of dimension d, the prior beliefs about the corresponding outputs (if a vector of outputs is considered, see later) is specified by a multivariate normal distribution. This in itself is not informative enough, complete specification of the Gaussian distribution requires some prior assumptions about its mean and covariance. The mean is assumed to have the form of a regression fit through the training data,

$$E\{f(x)|\beta\} = h(x)^T\beta \tag{8}$$

where $h(x)$ is a specified regression function of x, and β is the corresponding vector of coefficients. The vertical line in Eq.8 indicates that this expression is a conditional expectation. For simplicity, $h(x)$ is chosen here to be $(1, x)$, representing a linear regression,

$$E\{f(x)|\beta\} = \beta_0 + \beta_1 x_1 + \cdots \beta_d x_d \tag{9}$$

(this can be extended to higher-order polynomial fits if required). Note that a straightforward regression approach would propose a relationship of the form,

$$y = f(x) = \beta_0 + \beta_1 x_1 + \cdots \beta_d x_d \tag{10}$$

The coefficients for a standard least-squares regression model of the form (10) would usually be obtained by minimising the sum of the squared prediction errors over the training data; the result would be a rather minimal linear model in the context of the $h(x)$ discussed here. Eq. (9) is rather more sophisticated than this for reasons which will become clear later. In addition to a prior specification for the mean for the GP model, one requires covariance information. The covariance between output points for the GP model is assumed to be,

$$\text{cov}\{f(x), f(x')|\sigma^2, B\} = \sigma^2 c(x, x') \tag{11}$$

where σ^2 is a scaling factor and B will be a diagonal matrix of length-scales, representing the roughness of the output with respect to the individual input parameters. These quantities are specified before any training on the data and are referred to as *hyperparameters*. The covariance function used here is chosen to be a squared-exponential function of the form,

$$c(x, x') = \exp\{-(x - x')^T B(x - x')\} \tag{12}$$

The posterior distribution for the GP is found by conditioning the prior distribution on the training data y, and integrating out (or marginalising over) the hyperparameters σ^2 and β. (Note that the output here is considered to be univariate and the vector is an ordered array of all of the output values in the training data; it is thus of dimensions $n \times 1$.) The process of marginalisation is central to the Bayesian approach; essentially one integrates the model over all possible values of the hyperparameters according to their probability. This means that the hyperparameters themselves require a prior specification for their probability density; (this will not be discussed further here, the curious reader can consult [12] for more detail). The integrals involved are usually all of a Gaussian form, and although the expressions are almost always very complicated, the results can be given in closed-form. The result is a Student's t-process, conditional on B and the training data,

$$[f(x)|B, y] \sim t_{n-q}\{m^*(x), \hat{\sigma}^2 c^*(x, x')\} \tag{13}$$

where,

$$m^*(x) = h(x)^T \hat{\beta} + t(x)^T A^{-1}(y - H\hat{\beta}) \tag{14}$$

$$c^*(x, x') = c(x, x') - t(x)^T A^{-1} t(x') + \\ (h(x)^T - t(x)^T A^{-1} H)(H^T A^{-1} H)(h(x')^T - t(x')^T A^{-1} H)^T \tag{15}$$

$$t(x)^T = (c(x, x_1), \cdots, c(x, x_n)) \tag{16}$$

$$H^T = (h(x_1), \cdots, h(x_n)) \tag{17}$$

$$A = \begin{pmatrix} 1 & c(x_1, x_2) & \cdots & c(x_1, x_n) \\ c(x_2, x_1) & 1 & & \vdots \\ \vdots & & \ddots & \\ c(x_n, x_1) & \cdots & & 1 \end{pmatrix} \tag{18}$$

$$\hat{\beta} = (H^T A^{-1} H)^{-1} H^T A^{-1} y \tag{19}$$

$$\hat{\sigma}^2 = \frac{1}{n-d-3} y^T \{A^{-1} - A^{-1} H(H^T A^{-1} H)^{-1} H^T A^{-1}\} y \tag{20}$$

$$y = (f(x_1), \cdots, f(x_n))^T \tag{21}$$

Note that the determination of the emulator (12) is basically an exercise in machine learning and as such its quality is critically dependent on the number and distribution of training data points in the input space, and the values of the hyperparameters. The diagonal matrix of roughness parameters B cannot generally be integrated out analytically and must be evaluated using *maximum a posteriori* (MAP) estimation – this calculation typically represents the most computationally intensive part of the process. It is useful to dwell a little on what has happened through the last sequence of equations. Arguably the most important part of the GP model is the mean of the model as this is what one takes as the prediction on a given x. The prior specification of the model is given by Eq.(8) and carries no information yet about the training data. After incorporation of the training data (and integration over the hyperparameters), the posterior form of the mean is given by Eq.(14) and this can be written in a form which exposes its meaning a little more clearly; one writes,

$$m^*(x) = h(x)^T \hat{\beta} + \Sigma(x)\hat{e} \tag{22}$$

where $\Sigma = t(x)^T A^{-1}$ and $\hat{\epsilon} = y - H\hat{\beta}$. The latter term is essentially the vector of residual errors over the training set for a linear regression model. If the input-output process were truly linear and there was no measurement (or other) uncertainty, this term would be zero and Eq.(22) would collapse onto a perfect linear regression model. If the process is not linear and/or there is measurement uncertainty, the second term in Eq.(22) 'switches on' and modifies the linear regression core. The second term then acts as a sort of (non-polynomial) interpolant; it carries information about the training data through both terms and it also carries information about the assumed smoothness of the model through the matrix Σ which is built from data dependent on the covariance function. This combination of regression and interpolation in the posterior mean, means that it has much greater predictive power than a linear regression alone. In fact, the GP is even more sophisticated than this; as one also has a posterior variance for the model, one can establish natural confidence bounds on predictions.

The dependence of the emulator on training data means that some model runs are always required. As discussed above, the advantage of the Bayesian sensitivity approach is that, typically far fewer runs are needed to train the emulator than would be needed for a full MC analysis. The main reason for this is that the emulator carries information about the smoothness of the data and therefore knowing the function at one point in the input space means one has information about the likely values at neighbouring points; MC analysis does not allow for inference of this nature, the points at which the function is evaluated are considered isolated from each other. To deal with the sampling of training data in as principled a manner as possible, ideas of experimental design are applied and a *maximin Latin hypercube* design (maximin LHD) is used here [11]. A Latin hypercube design divides input space into regions of equal probability and distributes points in a *Latin-square* across "probability-space". A maximin LHD improves on this by additionally maximising the minimum distance between input points, thus optimising the space-filling properties of the design.

Inference for Sensitivity Analysis. Having established an emulator, the next parts of the process are associated with establishing how uncertainty propagates through the emulator and finding the sensitivities of interest [8,9,12]. Several quantities can be inferred from the posterior distribution-over-functions described above, that are relevant to sensitivity analysis. Fundamental quantities such as the mean and variance of the output distribution can be evaluated and this provides an estimate of the output uncertainty. In addition to total output variance, one can determine *main effects*, *interactions* and other sensitivity measures for the input parameters, based on their contribution to, or responsibility for, output variance. If a small amount of uncertainty (low variance) on an input parameter generates a high degree of uncertainty (high variance) on the output, then the output is clearly sensitive to that parameter in terms of uncertainty propagation.

The function $f(x)$ can be decomposed as follows, into *main effects* and *interactions*,

$$y = f(x) = E(Y) + \sum_{i=1}^{d} z_i(x_i) + \sum_{i<j} z_{ij}(x_i, x_j) + \cdots + z_{12\ldots d}(x) \tag{23}$$

where the expectation operator E is defined here with respect to the probability distribution of the inputs and,

$$z_i(x_i) = E(Y|x_i) - E(Y) \tag{24}$$

$$z_{ij}(x_i, x_j) = E(Y|x_i, x_j) - z_i(x_i) - z_j(x_j) - E(Y) \tag{25}$$

with higher-order terms being defining in an analogous fashion. Here $z_i(x_i)$ represents the *main effect* of the input variable x_i; $z_{ij}(x_i, x_j)$ is the *first order interaction* of the input variables x_i and x_j and further terms represent higher-order interactions; Y is the random variable corresponding to the function output y and therefore $E(Y)$ is the expected value of the output y considering all possible combinations of inputs. The main effect of an input can be thought of as the effect (on the output) of varying that parameter over its input range, averaged over all the other inputs. Interactions describe the effect of varying two or more parameters simultaneously, additional to the main effects of both variables. Plotting main effects serves as a visual indication of the influence of particular inputs and interactions, showing (albeit qualitatively) the variance of the output with respect to individual input parameters and the nonlinearities associated with those responses. This will become clear when the case study material is discussed later.

One can infer posterior mean values for main effects and interactions by simply substituting the posterior mean into the definitions detailed above. The necessary tool is the conditional expectation defined by,

$$E(Y|x_p) = \int_{\chi_{-p}} f(x)p(x_{-p}|x_p)dx_{-p} \tag{26}$$

where the subscripts p and $-p$ indicate subsets of the inputs indexed by a set of labels p and the complement of p respectively and $p(x_{-p}|x_p)$ represents the multivariate probability density function of the input parameters indexed by $-p$ conditional on those indexed by p. The set χ_{-p} is that spanned by the input variables indexed by $-p$. The posterior mean of the main effect can be derived by substituting (26) into (14) in place of $f(x)$, the result is,

$$E^*\{E(Y|x_p)\} = R_p(x_p)\hat{\beta} + T_p(x_p)e \tag{27}$$

where E^* is the expectation with respect to the posterior distribution now (i.e. that governing the random variable Y, and,

$$R_p(x_p) = \int_{\chi_{-p}} h(x)^T p(x_{-p}|x_p)dx_{-p} \tag{28}$$

$$T_p(x_p) = \int_{\chi_{-p}} t(x)^T p(x_{-p}|x_p)dx_{-p} \tag{29}$$

$$e = A^{-1}(y - H\hat{\beta}) \tag{30}$$

Although this results in a series of matrix integrals, a Gaussian or uniform distribution over the input parameters allows these to be evaluated analytically. Expressions for interactions can be similarly derived using their respective definitions.

An alternative approach to summarising sensitivity information is in terms of *variance and sensitivity indices*; such methods are widely used [12]. This approach involves quantifying the proportion of output variance for which individual input parameters are responsible. In particular, sensitivity can be measured by conditional variance,

$$V_i = \text{var}\{E(Y|x_i)\} \tag{31}$$

This is the expected value of the contribution of the input variable x_i to the output variance. Note that this is also the variance of the main effect of x_i, hence it is known as the *main effect index* (MEI). This can be extended to measure conditional variance of interactions of inputs, i.e.,

$$V_{ij} = \text{var}\{z_{ij}(x_i, x_j)\} \tag{32}$$

and so on for higher-order interactions. Although this approach provides detailed insight into the effects of combinations of inputs on output uncertainty, it can be time-consuming to examine all possible interaction permutations for models with many input dimensions. An alternative sensitivity measure [12], describes the output variance that would remain if one were to learn the true values of all inputs except x_i,

$$V_{ti} = \text{var}(Y) - \text{var}\{E(Y|\mathbf{x}_{-i})\} \tag{33}$$

This last measure, called the *total sensitivity index* (TSI), measures the variance caused by an input x_i and any interaction of any order including x_i. It allows a more holistic view of the uncertainty attributed to each input, but does not give any details as to how it is distributed between main effects and interactions. Between the MEIs and TSIs, a detailed view of the sensitivities of inputs and their interactions can be gained.

To calculate posterior means and variances of the above quantities from the emulator, the following equation is obtained [12],

$$E^*(V_p) = E^*(E\{E(Y|\mathbf{x}_p)^2\}) - E^*\{E(Y)^2\}$$

$$= \hat{\sigma}^2[U_p - tr(A^{-1}P_p) + tr\{W(Q_p - S_pA^{-1}H - H^TA^{-1}S_p^T + H^TA^{-1}P_pA^{-1}H)\}]$$

$$+tr(e^TP_pe) + 2tr(\hat{\beta}S_pe) + tr(\hat{\beta}^TQ_p\hat{\beta}) - \text{var}^*\{E(Y)\} - \{E^*(E(Y))\}^2 \tag{34}$$

where,

$$U_p = \int_{\mathcal{X}_p}\int_{\mathcal{X}_{-p}}\int_{\mathcal{X}_{-p}} c(\mathbf{x},\mathbf{x}^*)p(\mathbf{x}_{-p}|\mathbf{x}_p)p(\mathbf{x}'_{-p}|\mathbf{x}_p)p(\mathbf{x}_p)d\mathbf{x}_{-p}d\mathbf{x}_{-p}d\mathbf{x} \tag{35}$$

and the terms P_p, Q_p and S_p have a similar integral structure. The symbol $\mathbf{x}^*$ denotes a vector comprised of $\mathbf{x}_p$ and $\mathbf{x}'_{-p}$. The presentation of these quantities is not just gratuitous, the motivation is to give some idea of the complexity of the sort of calculation that can nonetheless be carried out analytically in some cases of importance; this is one of the remarkable strengths of the Bayesian approach. The full details of the calculation of V_i and V_{Ti} can be found in [12]. All the quantities of interest presented here can be computed using the software package GEM-SA [11], and this was used in the case studies which now follow.

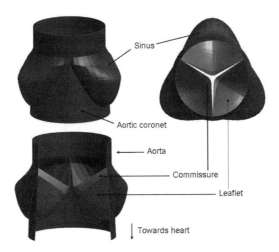

Figure 1. Layout of the aortic valve.

Case Study 1: A Heart Valve Model

Background. The aortic valve is situated in between the left ventricle and the aorta in the heart. Its function is to ensure one-way flow of oxygenated blood from the heart to the rest of the body. The valve consists of three leaflets and three sinuses, arranged as shown in Fig.1. When ventricular blood pressure is greater than aortic blood pressure (in systole), the leaflets open; conversely when the opposite is true (diastole) the leaflets are forced against each other, restricting any blood flow. In a full life span, the valve opens and closes around 3.7 billion times [13]. However, the aortic valve is prone to a variety of diseases and it is not uncommon that it requires replacement. Implanting a replacement that performs consistently under such demanding circumstances is a very challenging affair.

Modelling. A reasonable starting point in prosthesis design is to acquire a comprehensive understanding of the natural component to be replaced; one approach to this is via physical modelling. However, the uncertainty involved in modelling a heart valve is substantial – geometry, material properties and loading can only be specified to within fairly wide ranges. Furthermore, the complicated nature of the material properties is an additional source of epistemic uncertainty - indeed, some work has been done to investigate this [14]. Clearly a comprehensive uncertainty analysis of a typical heart valve FE model would provide useful insight into the robustness of the simulation. For the purposes of this study, a 3-dimensional dynamic FE model of the heart valve was created parametrically in ANSYS and solved in the LS-Dyna package. The approach was to use Belytschko-Lin-Tsai shell elements to model the entire structure, encompassing the sinus, leaflets and a section of aorta at either end. An illustration of the mesh is shown in Fig.2. The simulation was performed over the initial opening movement of the valve: a total time of 0.1s, starting from the point where the pressure on the leaflet becomes positive in the z-direction. The model consisted of around 3200 elements, depending on the set of input parameters used.

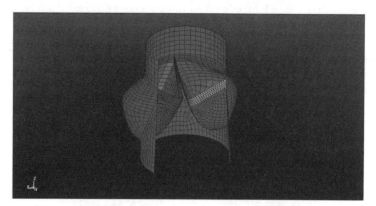

Figure 2. Two-thirds of aortic valve mesh (typical parameter set).

The geometry is described with a relatively small number of parameters [13], each of which is quoted as a range. Some finer geometrical points in the model are left to the modeller's discretion, as describing exact dimensions of a naturally varying biological system is meaningless. The thickness of the tissue in the leaflet also varies from commissures to leaflet tips, however it was considered sufficient in this investigation to have only two regions of different thickness – one for the commissures and the other for the rest of the leaflet.

The material of the heart valve is anisotropic and nonlinear, and is different between the leaflet and the sinus. The leaflet is comprised of a series of collagen fibres running largely in the circumferential direction, joined by loose connective tissue. At low strains these fibres are not under tension and resistive force is due purely to the connective tissue, however for greater extensions the fibres provide a stronger elastic resistance – this is responsible for the hyperelastic nature of the material. The aortic tissue is also hyperelastic, but due to the largest strains occurring in the leaflet, it is usually considered sufficient to model it as linear. Although both leaflet and aortic tissue is hyperelastic and anisotropic, for the purposes of this investigation the material was assumed to be elastic, albeit having separate elastic moduli between leaflet, commissure and sinus. One justification for this is that existing uncertainty data on distributions of elastic moduli used for modelling aortic valves is scarce enough – hyperelastic parameters even more so. Furthermore, hyperelastic material models introduced too many input dimensions for the scope of the study.

A number of recent aortic valve simulations have made use of fluid-structure interaction techniques to more realistically model blood/leaflet interaction [15,16], however in the interests of reducing computational expense, for this investigation the loading was applied as a set of simple pressure curves, assumed to be uniform across the leaflet [16]. The respective pressures on the leaflet, and the aortic and ventricular sides of the valve were obtained from the pressure curves in the cardiac cycle, which are well established quantities. Although this is not a completely accurate representation of the leaflet loading (since this is an interactive relationship that depends on the position of the leaflets), it was considered adequate for the purposes of this uncertainty analysis.

Table 1. Chosen Input parameters for sensitivity analysis.

Parameter	Name	Mean	Standard Deviation	Data source
Elastic modulus of sinus (MPa)	Es	2.45	0.8	Lit. Review
Thickness of sinus shells (mm)	Ts	1.3	0.325	Lit. Review
Elastic modulus of leaflet (MPa)	El	2.05	0.65	Lit. Review
Thickness of leaflet shells (mm)	Tl	0.35	0.0085	Lit. Review
Thickness of commissure shells (mm)	Tc	1	0.25	Lit. review/Thubrikar
Radius of base (mm)	Rb	12.7	0.675	Thubrikar
Angle of coaptation (degrees)	Φ	31	3	Thubrikar
Leaflet separation in relaxed state (ratio)	$Fltrad$	6	2	Subjective

Sensitivity Analysis. A full holistic sensitivity analysis of a FE model is impractical, even with the Bayesian approach outlined above, since the number of uncertain parameters (geometrical, material, loading and numerical) can be extended almost without limit. It was therefore considered enough to pick eight input parameters to examine, that were found to vary substantially from one simulation to the next in a comprehensive literature review on the topic. Ranges could then be collected for each parameter based, in a loose sense, on the "belief" expressed in the existing literature, in the absence of formal statistical data. Other parameters were collected from studies quoted in [13]. The input parameters are presented in Table 1. Each parameter was assigned a normal distribution for mathematical tractability.

Only a brief illustration of the results of the analysis can be given here, the reader is referred to [17] for much more detail. The methods used in this investigation require that the output of the model be a scalar quantity; this meant that a certain subjectivity crept into the analysis, a prior selection of a small number of quantities of interest was made and a separate sensitivity analysis was carried out for each. The output parameters considered are shown in Table 2, representing: maximum stress, speed of valve opening (which is one of the methods used to validate aortic valve models [18]), maximum displacement of leaflets and leaflet buckling (which appears to important in the mechanisms of valve failure [19]). In each case, the input ranges were used to create a maximin Latin hypercube design of 250 training points. This input matrix was then run through LS-Dyna to evaluate the model 250 times and provide corresponding output points. Separate sensitivity analyses were then run for each output of interest using GEM-SA, which generates main effect plots as well as MEIs and TSIs.

Table 2. Output parameters chosen for analysis.

Output Parameter	Description
Sigmaxmax	Maximum stress in any part of the model, at any time step
Thalfop	Time when leaflet tips pass between coaptation points (point at which leaflet is likely to buckle most)
Topen	Time for leaflet tips to reach aortic radius
Dispmax	Maximum displacement of leaflet tips (vector sum)
Wig	Percentage buckling at Thalfop (defined by angles successive elements make with each other in the x-y plane)

Results. The posterior expected value of the mean and variance for each output parameter is shown in Table 3. This gives an overall idea of the uncertainty in each output, given the uncertainty assigned to each input. To illustrate the comparative uncertainty between outputs, the standard deviation has also been presented as a percentage of the mean. It is evident that some of the uncertainties here are substantial – *Thalfop*, which represents the speed of opening of the valve, varies over 100% inside the 95% confidence limit. Similarly, the maximum stress value varies by around one order of magnitude.

Table 3. Output uncertainties.

Parameter	Posterior mean (μ)	Posterior standard deviation (σ)	σ/μ (%)
Dispmax	0.0125	0.00160562	12.8
Thalfop	0.00194	0.00122988	63.4
Sigmaxmax	635878	354868.99	55.8
Wig	0.161268	0.01573881	9.8

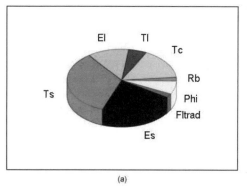

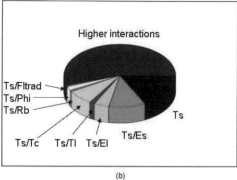

(a)   (b)

Figure 3. *Thalfop* sensitivity indices: (a) Total effect indices; (b) Interactions involving *Ts*.

A more detailed analysis shows that the uncertainties can be the result of high-order interactions. For example, less than one third of the uncertainty in the opening time (*Thalfop*) is explained by first-order interactions. An examination of the total effect indices gives more insight, as illustrated in Fig.3a. Notably, the thickness and elastic modulus of the sinus and their interactions with other variables (Fig.3b) are responsible for a large proportion of the output variance – this supports the supposition that the flexibility of the sinus is partly responsible for the ease of opening of the valve [19]. Another first-order effect worth mentioning is that interactions between sinus thickness and commissure thickness play an important role (7.5%), possibly because the thicker commissure needs a flexible sinus to allow it to reverse its curvature.

In order to illustrate the information in the main effects, Fig.4 shows a main effect plot for the leaflet stiffness and reveals that there is a significant nonlinearity in the response of *wig* to *El*. It is evident that below an elastic modulus of around 500 kPa the level of bucking increases dramatically.

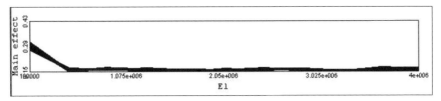

Figure 4. Main effect plot of *El* for output *Wig.*

Discussion. Only a small fraction of the available results have been given here; even so, one can see the value of the analysis. Table 1 shows that for the four parameters investigated there is very substantial variation as a result of input uncertainty, with some standard deviations being over 50% of the mean. It should be remembered also that this is only due to the uncertainty in 8 input parameters, whereas in reality many more inputs are subject to variance, suggesting that these estimates are if anything, slightly conservative. This highlights a general problem in modelling biomechanical systems – often data is scarce, and aside from that, aleatoric uncertainties are unavoidable. At best, quantitative results from this model could only be quoted to quite vague ranges, given the assumed input uncertainty. The sensitivity analyses also show that there are likely to be complex interactions at work in a typical heart valve model.

Case Study II: Fluid Structure Interaction - An Unmanned Airship Model

For a range of aerial applications, lighter-than-air aircraft – *airships* – are enjoying a resurgence in popularity, in particular in the unmanned version. In order to combat some of the problems encountered by conventional airship designs and expand the range of operable weather conditions, a new design has been proposed in [20], called the Elettra Twin Flyers (ETF) (Fig.5). The design innovations are twofold – firstly there are two gas envelopes rather than one, positioned side by side. This allows a smaller profile for the equivalent lift. The two balloons are connected by a rigid structure that also acts as a platform for affixing a payload (e.g. surveillance equipment) and operational equipment. Secondly, the ship is moved in all six degrees of freedom by a series of directional propellers; the advantage being that the ship can be manoeuvred effectively without the requirement of forward motion. The innovations are intended to give the ship greatly increased manoeuvrability at low velocities, and allow the ETF to hover with the prow orientated in any direction, even in adverse weather conditions. This case study is concerned with the structural analysis of a model corresponding to the prototype design of the ETF.

The model is complex and only the briefest summary of its salient features will be given here, the reader is referred to [21] for much more detail (that reference also surveys the relevant background literature on the modelling of airships). In terms of overall scale, the model investigated here is 36m in length, this being the length required for sufficient buoyancy to support the total weight of the structure and payload (approximately 3 tonnes).

Material. In a sense, the material properties of the structure are not of great interest here as the main uncertainty discussed later will relate to the direction of aerodynamic loading; however, they are mentioned here as the materials used are quite novel and are themselves subject to uncertainty, and this is important for static stress analysis [21]. The frame material is a sandwich layup, consisting of layers of carbon/epoxy T300 15k/976 composite, either side of an Ultracor® honeycomb core [22]. This has the benefit of low density and high stiffness and resulted in a quasi-isotropic material at the macroscopic level, to simplify modelling. However, it was found in [23], that the material properties of the composite varied significantly with changes in temperature. Since the airship is potentially operating in conditions of +/-50°C (dependent on altitude, location and weather), the uncertainties in material properties were expected to have a substantial effect on the response of the structure to loading.

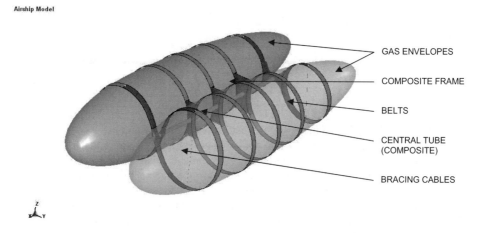

Figure 5. Illustration and arrangement of materials in the ETF

Modelling. The modelling of the airship was performed to reveal the stress and deformation of the airship under aerodynamic loading conditions. Control of the structure is currently based on the assumption that the structure is rigid; this is an important point because even minor deformations and rotations of the arms on which the driving propellers are located can have effects on the control system that are hard to predict. One main objective of the FE analysis was thus to quantify deformation in certain parts of the structure under a variety of loading conditions. In terms of structural integrity, both linear and nonlinear static models were constructed in the programme of study in order to study structural stresses, and the results are presented in [21]; however, the results shown here are from a full Fluid Structure Interaction (FSI) model considered the effect of gusts of wind incident at different angles to the airship. As in the case of the heart-valve model, the FE software LS-Dyna was used; the solver being capable of handling highly nonlinear and transient simulations, and the coupling of Lagrangian and Arbitrary-Lagrangian-Eulerian (ALE) meshes required for FSI.

Aerodynamic Loading. The forms and magnitudes of aerodynamic loading can vary greatly, from the response of the structure in steady-state flow, to ramped or stepped loading (i.e. from gusts of wind in still air). The gust is an important aspect in the study of the interaction of any aircraft with the real atmosphere and was considered to be crucial in this study. The form of gust model used is the classic constant-gradient gust referred to for example in the European Aviation Safety Agency regulations [24]. The velocity profile of this linear gust is known to closely follow that of a natural gust. Since the pressure distribution on a non-trivial structure due to a particular gust of wind is far from obvious (especially at certain angles), the FSI model was created with an Arbitrary-Eulerian-Lagrangian (ALE) mesh overlapped with the Lagrangian mesh used for the "dry" model. The possibility of structural deformation that would change the boundary conditions of a standard CFD model necessitates this approach. The ALE mesh combines the benefits of the Lagrangian and Eulerian mesh approaches, respectively that the mesh is allowed to move in space (thus limiting the size of the mesh necessary); and that the material can flow through the mesh, allowing the very large deformations that are characteristic in fluid flow. An extensive discussion on the ALE method is found in [25]. The two meshes were coupled together using a penalty-based approach [26], Fig. 6 illustrates the arrangement of the two meshes. The ALE mesh is cylindrical because it allows investigation of different incident gust angles; the airship can simply be rotated in the mesh without the need for extra pre-processing.

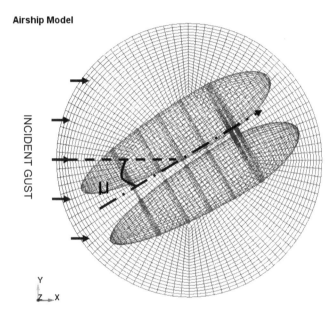

Figure 6. Illustration of cylindrical ALE mesh overlapping Lagrangian airship mesh, viewed from above; μ = angle between gust direction and ship axis.

The air is modelled in terms of compressible, viscous flow without considering turbulence; this represents a first approximation of the airflow around the structure for use in uncertainty analysis. Although there are many conceivable uncertainties associated with the definition of the gust hitting the airship, in this analysis only the wind angle was investigated. After adding the fluid mesh, the run time of the model became very substantial, so for this stage of the investigation the other uncertainties were treated as known to save on analysis time, though they may be investigated in future analyses.

Results and Discussion. To show the effects of the gust direction on an output of the model, Fig. 7 (left) shows the main effects on final displacement between opposite propellers in the x-direction of the model, (this is an important quantity, being a measure of the deformation of the structure and thus potential degradation of the control system). Extracting a meaningful quantity from this model is not trivial because the impact of the gust causes structural oscillations, and all displacement and stress values are a function of time. The variation in displacement is reasonably linear and quite small. This is attributed to the rigidity of the structure caused by tension in the balloons. The displacement is increasingly negative as the gust is angled more towards the front of the airship, likely because the balloon is pushed towards the rear of the ship, causing the structure to contract slightly.

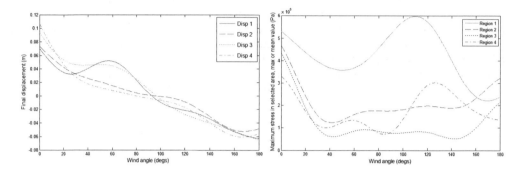

Figure 7. Posterior means of: displacements 1-4 (left), stresses in regions 1-4 (right).

Other outputs of the model are less well-behaved. A significant difficulty with this model is that a small change in gust angle can drastically change the output of the model. Fig.7 (right) shows the variation of stress with gust angle for four chosen regions in the model. Each of these regions represents an area of observed stress concentration in the frame. The response of stress is shown to be highly nonlinear in relation to the gust angle. This can sometimes cause problems with an emulator approach since a highly-structured response requires a higher density of training data (however, a check of the emulator accuracy using cross-validation showed that it was acceptable here). Finally, in a model like the one presented above, the number of uncertainties can be extended almost without limit. Loading uncertainties such as wind pressure distribution, velocity, air density and viscosity are could all vary (in fact, co-vary) due to environmental conditions or natural variation. Considering all these uncertainties in such a model is currently impractical, but it is worth remembering that considering only a small set of the model uncertainties will result in an underestimate of the true output uncertainty.

Summary

It does not seem appropriate to present a long summary here. The objective of the paper was simply to give an overview of some issues relating to uncertainty propagation through large (computationally-expensive) structural models. In fact, much of the discussion centred on the more refined problem of sensitivity analysis. The main thesis of the paper is that an established Bayesian approach to the sensitivity analysis, based on a Gaussian process emulator or surrogate model, can yield insight into the problem without the restrictive requirement of performing many model runs. (More correctly, the approach needs many less model runs than Monte Carlo.) In order to illustrate the sort of information which is readily available from the Bayesian analysis, two case studies were presented. Between the two case studies, one sees many of the aspects of engineering models which present difficulties for the modeller.

Acknowledgements

The authors would like to sincerely thank their collaborators on various aspects of the research presented here: Dr Jen Rowson, Dr Graeme Manson and Dr Jeremy Oakley of the University of Sheffield; Dr Alaster Yoxall of Sheffield Hallam University and Professor Piero Gili of the Politecnico di Torino. The authors would like to thank Dr Mark Kennedy for the use of the GEM-SA software and Ove Arup for the use of LS-Dyna. The activity regarding the airship model presented in this article is part of the project entitled "Innovative Solutions for Control Systems,

Electric Plant, Materials and Technologies for a Non-Conventional Remotely-Piloted Aircraft" funded by the Piedmont Region of Italy. The heart-valve model work is part of the research from the Sheffield Dynamics Group EPSRC Platform Grant and the funding from EPSRC is gratefully acknowledged.

References

[1] G.J. Klir, and R.M. Smith: *Annals of Mathematics of Artificial Intelligence* Vol.32 (2001) pp.5-33.

[2] I. Elishakoff, and Y. Ren: *Finite Element Methods for Structures with Large Stochastic Variations* (Oxford Texts in Applied and Engineering Mathematics, 2003).

[3] B. Lallemand, A. Cherki, T. Tison and P. Level; Journal of Sound and Vibration Vol.220 pp.353-364.

[4] A. O'Hagan and J.E. Oakley; *Reliability Engineering and System Safety* Vol.85 (2004) pp.239-248.

[5] Y.A. Shreider; *Method of Statistical Testing: Monte Carlo Method* (Elsevier, Amsterdam; London, (1964).

[6] C.M. Bishop; Pattern Recognition and Machine Learning (Springer-Verlag, 2007).

[7] A. Saltelli, K. Chan and E.M. Scott; *Sensitivity Analysis* (Wiley, New York, 2000).

[8] J.E. Oakley, and A. O'Hagan; *Journal of the Royal Statistical Society, Series B* Vol.66 (2004) pp.751-769.

[9] M.C. Kennedy, C.W. Anderson, S. Conti and A. O'Hagan; *Reliability Engineering and System Safety* Vol.91 (2006) pp.1301-1309.

[10] C.E. Rasmussen and C.K.I.Williams; Gaussian Processes for Machine Learning (MIT Press, 2006).

[11] *Gem-SA Homepage.* http://ctcd.group.shef.ac.uk/gem.html (accessed: 27/03/10).

[12] T. Homma and A.K. Saltelli; *Reliability Engineering and System Safety* Vol.52 (1996) pp.1-17.

[13] M. Thubrikar; *The Aortic Valve* (CRC Press, Boca Raton, Fla., 1990).

[14] A. Ranga, R. Mongrain, R. Mendes Galaz, Y. Biadillah and R. Cartier; *Journal of Medical Engineering and Technology* Vol.28 (2004) pp.95-103.

[15] J. De Hart, G.W. Peters, P.J. Schreurs, and F.P. Baaijens; *Journal of Biomechanics* Vol.36 (2003) pp.103-112.

[16] C.J. Carmody, G. Burriesi, I.C. Howard and E.A. Patterson; *Journal of Biomechanics* Vol.39 (2006) pp.158-169.

[17] W. Becker, J. Rowson, J. Oakley, A. Yoxall, G. Manson and K. Worden; *in Proceedings of 23rd International Conference on Noise and Vibration Engineering, Leuven* (2008) on CD..

[18] I. Vesely, D. Boughner and T. Song; *Annals of Thoracic Surgery*, Vol.46 (1988) pp.302-308.

[19] R.J. Brewer, J.D. Deck, B. Capati and S.P. Nolan; *Journal of Thoracic and Cardiovascular Surgery*, Vol.72 (1976) pp.413-417.

[20] M. Battipede and P.A. Gili; *Peculiar Performance of a New Lighter-Than-Air Platform for Monitoring*, 2004 Aviation Technology, Integration and Operation Forum, Chicago, IL – USA, September 2004.

[21] W.E. Becker, K. Worden, M. Battipede and C. Surace; *Submitted to AIAA Journal* (2010).

[22] A. Cappadona; *Analisi Strutturale su Dirigibile Non Convenzionale a Controllo Remoto: Configurazione Esoscheletro*, in *Facoltà di Ingegneria, Politecnico di Torino* (2008).

[23] Defense, U.S.D.o., *Polymer Matrix Composites Material Properties*, in *Composite Materials Handbook*. (2002).

[24] EASA, *Certification Specifications for Normal, Utility, Aerobatic and Commuter Category Aeroplanes CS23, Book 1*, E.A.S. Agency, Editor. (2009).

[25] J. Donea, A. Huerta, J.-Ph. Ponthot and A. Rodriguez-Ferran; *Arbitrary Lagrangian-Eulerian Methods*, in *Encyclopedia of Computational Mechanics*, E. Stein, R. De Borts, T.J.R Hughes (Eds.) (John Wiley & Sons, New York, 2004).

[26] J.O. Hallquist; *LS-Dyna Theory Manual* (Livermore Software Technology Corporation, 2006).

Session 1: Structural Health Monitoring

Applied Mechanics and Materials Vols. 24-25 (2010) pp 45-50
© *(2010) Trans Tech Publications, Switzerland*
doi:10.4028/www.scientific.net/AMM.24-25.45

Detection of Cracking in Gear Teeth Using Acoustic Emission

R. Pullin[1, a], A. Clarke[1,b], M. J. Eaton[1,c], K. M. Holford[1,d],

S. L. Evans[1,e] and J. P. McCory[1,f]

[1]Cardiff School of Engineering, Cardiff University, Queen's Buildings,

Cardiff, CF24 3AA, Wales, UK

[a]pullinr@cf.ac.uk, [b]clarkea7@cf.ac.uk, [c]eatonm@cf.ac.uk, [d]holford@cf.ac.uk,
[e]evanssl6@cardiff.ac.uk, [f]mccroryjp@cardiff.ac.uk

Keywords: Acoustic emission, gear teeth, structural health monitoring

Abstract. The detection of damage in gear teeth is paramount to any condition monitoring or structural health monitoring (SHM) tool for aerospace power transmissions such as those used in helicopters. Current inspection techniques include vibration analysis and time-inefficient visual inspection. Acoustic Emission (AE) is a very sensitive detection tool that has been successfully used in many SHM systems. Successful application of AE for damage detection in gear teeth will enable the optimisation of gear box design (and hence weight saving) in addition to safety improvements. This paper details a small aspect of a larger project designed to demonstrate automatic detection and location of common gear tooth defects. A novel test rig was designed to allow the fatigue loading of an individual gear tooth which was monitored using AE. The gear tooth was static in order to exclude the detection of AE signals arising from rotation; this allows initial development of the methodology prior to investigating rotating gears. Digital Image Correlation was used to determine the onset of cracking for comparison with the detected AE. Preliminary results of the investigation show that the developed methodology is appropriate for developing an automated gear health monitoring system and that future work should concentrate on the development of sensors and data acquisition methods associated with obtaining signals from rotating machinery.

Introduction

The development of detection and diagnostic methods for gear tooth faults in aerospace power transmission systems is an active research area. It is being driven largely by the interests of military organisations or large aerospace organisations such as NASA [1-3]. In certain situations, Health and Usage Monitoring (HUMS) systems are mandatory, such as helicopter operators servicing the North Sea oil and gas industry, and the majority of the commercial techniques are based on vibration methods [4,5]. In aerospace applications, the potential results of gear failure are serious, ranging from increased asset downtime and maintenance expenditure to, at worst, catastrophic failure with life-threatening consequences. Thus, new monitoring techniques which can identify early-stage failures are in demand.

Acoustic Emission (AE) monitoring is a technique which is used widely in other applications and which offers significant advantages in terms of early fault detection and diagnosis when compared with other techniques [6]. It is arguably the most sensitive NDT technique available, and relies upon the detection of stress waves generated by damage (cracks, plastic deformation etc) which propagate through the solid material as it undergoes strain. It is proposed that AE could offer significant advantages in gear monitoring applications. Recent advances at Cardiff University have led to novel techniques being developed to automatically to locate and identify damage in high noise environments [7,8,9] in metallic structures. When compared with the reasonably mature application of AE to structural monitoring, the use of AE to monitor rotating machinery in general, and gears in particular, is still at the developmental stage, particularly when applications such as high speed, heavily loaded aerospace transmissions are considered. Workers have investigated the

AE from spur gears [10,11,12,13] but there remains much confusion surrounding the sources of acoustic emissions in gear contacts. The RMS approach used in the reported work has shown some modicum of success but further work is required not only to understand the source mechanisms and propagation but also to provide an automated technique for identifying damage in these types of applications.

In terms of commercial usage, AE monitoring of gear and bearing systems has been limited to relatively slow-speed application. For example, the Aquilla AE-Pro system [14] developed by the steelmaking company Corus is in commercial use for bearing monitoring at rotational speeds below 80 rpm.

The work presented here is part of a wider project which ultimately aims to develop a monitoring system for aerospace gears which provides a measurement of the health of the gear system, and is capable of monitoring lubricant health, asperity contact levels, surface degradation and root fillet cracking. Given the inherent complexities of measuring AE signals from a rotating gear pair, the current work involves a non-rotating pair of gears which are cyclically loaded in order to allow initial investigation into the AE signals produced by the initiation and propagation of a root fillet crack.

Experimental Procedure

A novel test rig was designed and manufactured to allow the fatigue loading of an individual gear tooth (Fig. 1). This allows a full AE analysis of damage signatures in fatigue crack growth in gear teeth without the presence of constant background noise due to rotational and frictional sources. This is evidently a simplified version of actual gear systems but allows techniques to be trialed and tested quickly whilst allowing validation using techniques such as digital image correlation (DIC) or foil crack gauges. A number of teeth were notched to allow fatigue cracks to be initiated promptly.

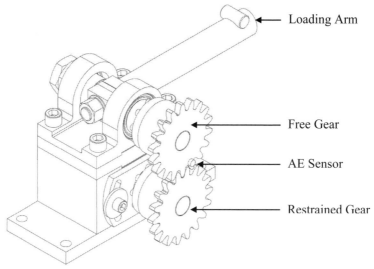

Loading Arm

Free Gear

AE Sensor

Restrained Gear

Fig. 1: Test Rig Design

A combination of AE and DIC was used to monitor the fatigue failure of a notched gear tooth. An initial static loading of an undamaged tooth was completed in order to validate an FEA model completed in PATRAN and to assist in determining the appropriate fatigue loading. To validate the AE results periodically during the investigation the test was stopped, peak load applied and DIC images captured. The gear tooth was loaded from 100 to 1400 N at 1 Hz.

During the fatigue investigation, a Physical Acoustics Limited (PAL) Pico (200-750 kHz) AE sensor was coupled and attached to the adjacent gear tooth using cyanoacrylate. Discrete AE data using a threshold of 45dB was captured using a PAL PCI2 system at a sample rate of 2 MHz for the duration of the investigation. Furthermore, after each stop in testing to capture DIC images, a wave-stream for 2 seconds (to ensure a complete load cycle was captured) at 2MHz was captured. Wave-streaming continuously records the output of the sensor and is independent of threshold.

Results and Discussion

A comparison of the DIC results at 1400 N (static load) with the FEA model results (at the same load) for strain in the Y-direction is shown in Fig. 2. When comparing the DIC with the FEA results, it must be noted that the DIC cannot analyse the entire tooth due to the method of averaging over a subset of pixels used by the image correlation software to improve accuracy. However, if strain is compared at corresponding positions, there is a broad agreement. On the edge of the tooth root fillets the DIC reports +/- 1200 microstrain whilst at the corresponding positions on the FEA, the strain is between 1100 and 1300 microstrain. It should also be noted that the DIC does not capture the strains due to the Hertzian contact which results in a significantly smaller maximum tensile and compressive strains to be reported. This will be explored in future investigations, and it is unclear currently whether this is due to the resolution of the DIC or some misalignment in the test rig.

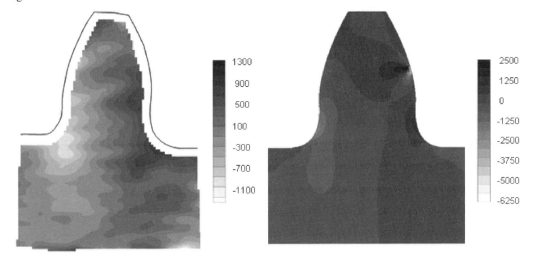

Fig 2. Comparison of y-strain from DIC (left) and FEA (right)

A fatigue crack of approximately 2 mm had propagated through the gear tooth after 23k cycles and the test was stopped. The DIC data was analysed after each block of fatigue testing whilst the AE data analysis was completed post test. Fig. 3 shows the DIC images captured periodically through the investigation. A key to the strain contours is provided but should not be considered to be a true representation of the actual magnitude of the strain values around the crack position due to the calculation method employed by the image correlation software. The measurements are calculated by examining the relative movements of adjacent pixels in the images; however whilst the DIC technique captures accurately the relatively large pixel displacements around the crack mouth, the resulting calculated strain is an over-estimation as these displacements are largely caused by crack opening as opposed to elastic strain.

The strain contours clearly show the growth of a high stress region propagating from the notch in the tooth as the test progresses due to increasing crack length and furthermore a subtle difference can be noted after 1k cycles suggesting the crack had already started propagating at that point. A further method of identifying crack growth based on the DIC measurements was employed. The relative movement of two pixels either side of the notch were used to provide a crack mouth opening displacement (CMOD) measurement in order to validate and support the understanding of the collected AE data.

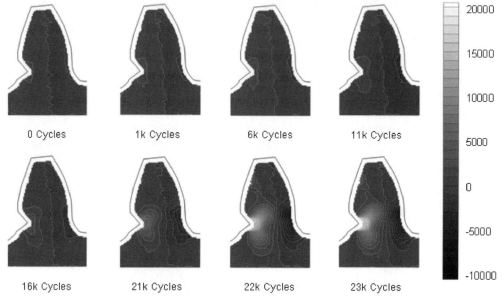

Fig 3. Development of microstrain during fatigue loading

Fig. 4 shows the CMOD measurement determined using the DIC system. Using an optical measurement offers distinct advantages over a foil crack gauge or an actual CMOD measurement. A non-contact measurement technique does not introduce sources of AE due to friction or glue cracking which would occur in the crack region, thus providing false or misleading results. The cumulative acoustic energy detected by the AE sensor is also shown in Fig. 4. The plot demonstrates that the AE is detecting the crack growth and that at the point of a large increase in crack length after 21k cycles, there is a corresponding large increase in detected energy. Furthermore, at the very onset of the test there is little AE detected until approximately 200 cycles suggesting that this is the point of crack initiation.

A simple measurement of energy will be insufficient for detecting fractures in a high noise environment associated with rotating machinery due its lack of sensitivity but it clearly demonstrates that AE can detect cracking of gear teeth without the presence of rotating noise. Subsequent tests with a sensor mounted on bearing housing have shown the AE method to still be capable of detecting tooth cracking. This work will be the subject of a further publication.

Fig. 5 shows the AE wave-streams captured at discrete points throughout the investigation with the load cycle overlaid. The initial wave-stream prior to cracking in Fig 5a shows only low level background noise, however as soon as cracking has occurred, based on the DIC and traditional AE data, the wave-streams exhibit clear bursts of activity that occur at peak loads - further demonstrating that the AE is detecting the crack growth in the gear. This again is probably insufficient for detecting fractures in a high noise environment and offers very little scope for providing an automated system.

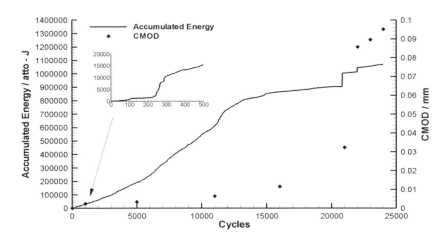

Fig 4 Recorded AE data compared with crack mouth opening measurements

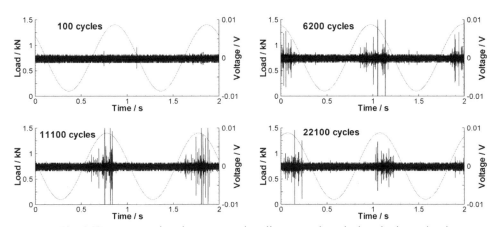

Fig. 5 Wave streaming data captured at discreet points during the investigation

A further method of exploring the data is shown in Fig. 6. A wavelet transform is shown of a 1 second sample which contains the peak load, prior to cracking and at 11.1k cycles. Wavelet transforms offer significant benefit in analysing the collected AE data. The wavelet transform analyses a signal in the time domain for its frequency content thus providing information of the content and temporal position at frequencies within a signal. This is clearly advantageous as it is easier to identify differing sources within the dataset. Regions of cracking will have different frequencies to that of rotating noise.

This offers a great possibility for developing an automated system for identifying fractures in gear teeth. By using either a novelty or cross correlation technique of a wavelet transform of a wave-stream recorded on one complete rotation, new sources can be identified and flagged as suspect cracks. This could be further enhanced by identifying frequency bands for cracks and use them as a further crack indicator. The techniques discussed will be the focus of future tests and research publications.

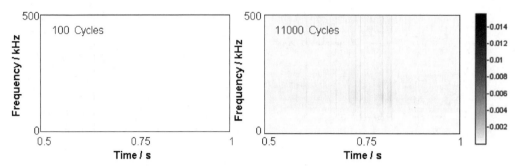

Fig. 6 Wavelet transforms of wave-streams centred around peak load (plots have same peak scale)

Conclusions

The completed work has demonstrated the capabilities of AE for detecting fatigue fractures in gear teeth. Furthermore techniques for developing an automated system have been explored and demonstrate real potential for designing an automated gear health monitoring system

Acknowledgements

The authors would like to acknowledge the support and advice provided by the technical staff of Cardiff School of Engineering with particular thanks to Paul Malpas who manufactured the test rig.

References

[1] Wong AK, *Vibration-based helicopter health monitoring,* (DSTO, 2001)

[2] Decker HJ, *Gear crack detection using tooth analysis*, (NASA-TM2002-211491, 2002)

[3] Decker HJ, *Crack detection of aerospace quality gears*, (NASA-TM2001-211492, 2002)

[4] Larder BD, *Helicopter HUM/FDR: Benefits and Developments*, (American Helicopter Society, Montreal, Canada, 1999)

[5] McColl J, *Overview of transmissions HUM performance in UK, North Sea Helicopter Operations*, (Institution of Mechanical Engineers, London n.d.)

[6] Loutas TH *et al.*, Applied Acoustics Vol. 70 (2009), p. 1148

[7] Holford K M *et al.*, Proc IMechE Part G, Journal of Aerospace Engineering Vol 223 (2009), p. 525

[8] Pullin R *et al.*, Journal of Acoustic Emission Vol 25 (2007), p. 215

[9] Hensman J J *et al.*, Applied Mechanics and Materials Vols 13 – 14 (2008), p. 251

[10] Toutountzakis T, Tan CK, Mba D, NDT E Int Vol 38(1) (2005), p.27

[11] Tan CK and Mba D, Tribol. Int. Vol 38(5) (2005), p. 469

[12] Tan CK and Mba D, Proc Inst Mech Eng Part J J. Eng Tribol Vol 219 (2005), p. 401

[13] Loutas TH *et al.*, Trans ASME J Vibration and Acoustics Vol 130 (2008)

[14] Corus Northern Engineering Services, PCM Brochure 001 "Aquilla AE Pro", (2009)

Applied Mechanics and Materials Vols. 24-25 (2010) pp 51-56
© *(2010) Trans Tech Publications, Switzerland*
doi:10.4028/www.scientific.net/AMM.24-25.51

Active sensor arrays for damage detection

P. H. Malinowski[1,a], T. Wandowski[1,b] and W. M. Ostachowicz[1,2,c]

[1]Institute of Fluid-Flow Machinery, Polish Academy of Sciences

Fiszera 14, 80-952, Gdansk, Poland

[2]Gdynia Maritime University, Faculty of Navigation

Aleja Jana Pawla II 3, 81-345, Gdynia, Poland

[a]pmalinowski@imp.gda.pl, [b]tomaszw@imp.gda.pl, [c]wieslaw@imp.gda.pl

Keywords: damage detection, piezoelectric transducers, Structural Health Monitoring, SHM, Lamb waves

Abstract. In presented research a problem that belongs to the Structural Health Monitoring (SHM) topic was investigated. Special arrays of active sensors were used for damage detection. These sensors were piezoelectric transducers. They were attached to specimen under investigation and used to excite and sense guided elastic waves – Lamb waves. Each array comprised of uniquely placed transducers. The total number of transducers was the same for all considered arrays. This ensured that the same number of signals was used to obtain damage information. A numerical algorithm was proposed to process these signals. It was designed to be independent of sensor arrangement so it could be used for all considered arrays. The principal idea behind the algorithm is that obstacles on a wave path cause wave reflection. These reflections are represented in the time signals. The algorithm was used to associate energy of these reflections with a particular area of the investigated specimen. The value of the energy was extracted from all the signals and projected to coordinate system associated with the specimen edges. In order to test and compare proposed arrays artificial defects were introduced to the specimen to model damaged structure. Because the specimen with defect and signal processing algorithm were the same, the only variable that could influence damage detection was the type of the array.

In the investigation damage detection results were obtained for considered arrays. Although the number of sensors were invariable, differences in damage indication exist. This suggest that the type of sensor array should be precisely chosen for a particular application. Even simplest linear array may be sufficient but it depends where we want to apply it.

Introduction

In this paper a research on the SHM topic is presented. A system with sensors that interrogate a structure on-line would increase the safety and reduce the cost connected with unexpected failure. One of the promising methods for such system is based on elastic wave propagation phenomena. The interaction of these waves with defects can bring the diagnostic information. In case of thin walled structures such as aircraft panels these wave are guided by the two parallel surfaces therefore they are called guided waves [1]. However if isotropic material is considered they are called Lamb waves to highlight the Horace Lamb input to the subject [2]. One of the most important features of these waves is that they propagate as symmetric (S_0, S_1, S_2,...) and antisymmetric (A_0, A_1, A_2,...) modes [3]. The number of both modes is infinite and they are highly dispersive. The dependence of phase velocity on frequency is given be the Rayleigh-Lamb equations [3]:

$$\frac{\tan(ad)}{\tan(bd)} = -\left[\frac{4k^2ab}{(a^2-k^2)^2}\right]^{\pm 1},\qquad(1)$$

where d is the half of the plate thickness, k is the wave vector length and ω is an angular frequency. a and b are defined by

$$a^2 = \left(\frac{\omega}{c_T}\right)^2 - k^2, \ b^2 = \left(\frac{\omega}{c_L}\right)^2 - k^2 \tag{2}$$

The exponent +1 is for the symmetric modes, while -1 is for the antisymmetric modes. c_T and c_L denote the transverse and longitudinal wave velocities, respectively.

There are many way to excite Lamb wave in a structure. Such methods can be listed as angle beam transducers [3], EMATs [4], laser sources [5,6] or piezoelectric elements [7]. By the same means registration can be performed. Nevertheless for an on-line system it is necessary to use small and light sensors that can be permanently attached on or embedded in the structure. Piezoelectric transducers can meet these demands.

Damage detection approaches can be divided in two general groups. One based on concentrated and the second one on distributed arrays. In the first approach such solutions, among others, as phased arrays [7], clock-like array [8] and star-shaped array [9] can be pointed out. As far as distributed arrays are concerned a triangular array [10] and rectangular one [11] are worth mentioning.

In this article an experimental investigation was conducted. Small piezoelectric sensors were used to detect damage in an aluminium alloy panel. These transducers were arranged into three different concentrated arrays and their performance on simulated damage case was tested.

Investigated arrays

Research reported in this article was focused on three arrays of active sensors (linear, cross and square - Figure 1). Each of considered arrays consisted of nine sensors. Spacing between centres of these elements was equal to 5 mm. Middle sensor of the array was used for Lamb wave excitation. Remaining eight was receiving the reflected waves. The dimensions of the sensors (manufactured by Noliac) were following: $3 \times 3 \times 2$ mm^3. The goal of the investigation was to employ these arrays for damage detection. In order to do this an object of investigation in the form of an aluminium alloy AA5754 panel was chosen. It's dimensions were following: $1000 \times 1000 \times 1$ mm^3.

Damage was simulated in the panel by drilling two through thickness holes (Ø6) and introducing two shallow notches, one 10 mm long and second 184 mm long. The notches were 0.2 mm wide. Considered arrays were attached to the panel in such way that the middle sensor of each array was at the centre of the panel. The panel with one of the arrays is presented in Figure 2a. A wax for accelerometer mounting was used to install the sensors, therefore it was possible to use three considered array on the same panel.

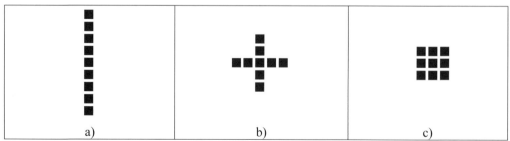

a)   b)   c)

Figure 1. Considered arrays

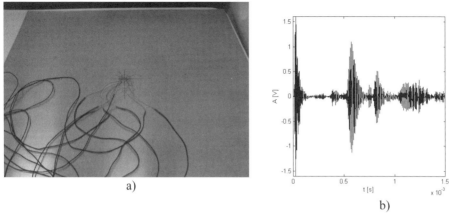

Figure 2. a) Aluminium alloy panel with simulated damage and an array of transducers; b) a part of one of the measured signals

In order to ensure high sensitivity to damage a relatively high frequency waves were used for damage detection. The excitation was chosen in the form of five cycles of sine with Hanning window modulation. The sine frequency was 220 kHz. This signal was applied to the middle sensor and registration was performed in remaining eight sensors. Registered signals were processed with special signal processing algorithms in order to obtain information about the panel condition.

Signal processing

Conducted measurements resulted in eight signals for each considered array. A part of an typical registered signal is depicted in Figure 2b. To facilitate the process of obtaining damage information a signal refining procedures were applied. Firstly, signals were shifted by

$$\Delta t = 0.5\frac{N}{f},$$
(3)

where N is the number of cycles and f is the excitation frequency. Value of Δt corresponds to the half of the excitation signal duration, therefore this shifting ensures that in the initial instant t=0 the maximum of the excitation is placed. In this research N=9 and f=220 kHz. Secondly, all registered signals were measured several times and averaged to increase signal to noise ratio. Thirdly, due to imperfect electromechanical shielding a crosstalk occurred resulting in excitation signal present at the beginning of each registered signal. Additionally, because of the fact that the sensors were placed so close to each other wave reflections between them occurred and were registered in the signals. Crosstalk and sensors reflections overlapped in the initial part of the signal making this part useless for damage detection purposes (Figure 2b). In order to remove this part from further analysis the initial parts of the signals lasting 4×Δt (two times the length of excitation) was set to zero. Fourthly, because the distance from the array to the edges is approximately 0.5 m, the part of the signals representing the response from more than 0.45 m distance was removed. This ensured that edge reflections do not disturb the damage detection procedure. Lastly, the signals refined in previous steps were normalised by the maximal values. This guaranteed that each of them had an equal input in the damage detection procedure.

Imaging for damage detection

Signals, that were prepared according to description in the previous section, were used for damage detection. Due to the fact that waves excited in the middle sensor propagate omni-directionally guided by the panel a special mesh of points was introduced. The mesh consisted of concentric circles of points (Figure 3). These points were equidistant in radial direction $\Delta r = 10$ mm. Also a constant angle step was chosen $\Delta \alpha = 5°$. The start radius was zero and end radius was equal to 0.5 m (distance from the middle sensor to the panel edges). The task for the numerical algorithm was to assign to the point of the mesh a numerical value extracted from measured signals. As it can be seen in registered signals (Figure 2b) various wave reflections were registered and the goal is to determine to which simulated defect they correspond.

In order to perform damage detection distances $|GP_i|$ and $|P_iR_k|$ were calculated. G is the point where the generating sensor is placed, R_k is the point where one of the receiving sensors is placed (k=1,2,...,8) and P_i is one of the mesh points. These distances were used to cut out a part of the signal registered in R_k. The cut out part, F_n, has a length of $l=\Delta r/c_g$ and starts at

$$t_{ik} = \frac{|GP_i| + |P_iR_k|}{c_g},$$
(4)

where c_g is the wave group velocity obtained in separate measurement. In considered case $c_g=2315$ m/s. It corresponds to the Lamb wave A_0 mode velocity. F_n is discrete so the index takes values n=1,2,...,N, N depends on the length l. All the eight registered signals were mapped into point P_i by summing squared signal from all the receiving sensors (R_k):

$$M(P_i) = \sum_{k=1}^{K} \sum_{n=1}^{N} F_n^2$$
(5)

This procedure is repeated for all point P_i in considered mesh. Such signal processing approach caused that the $M(P_i)$ lies on an ellipsis with loci at T and R_k [12].

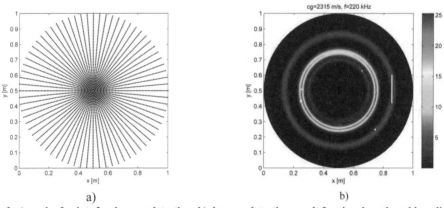

a) b)

Figure 3. a) mesh of points for damage detection; b) damage detection result for signals gathered by a linear array; white marks indicate the damage position

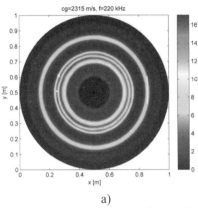
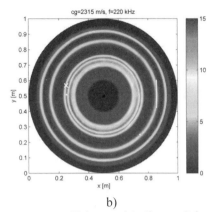

a)   b)

Figure 4. a) damage detection result for signals gathered by a cross array; b) damage detection result for signals gathered by a square array; white marks indicate the damage position

Proposed algorithm returns a 3D image as a result. The values M for P_i points are obtained directly by calculation, while for the intermediate points they are interpolated by the MATLAB plotting function. Figure 3b, Figure 4a and Figure 4b present the damage detection results for linear, cross and square arrays, respectively. They are investigated thoroughly in subsequent section.

Results and comparison

Proposed damage detection algorithm was applied to signals gathered by three sensor arrays. Difference between arrays was in individual sensors placement. The individual signals were normalized, therefore the $M(P_i)=1$ was a threshold value. If the presence of a damage is detected value of $M(P_i)$ should be higher than this threshold. The results obtained for all the arrays (Fig.3b, 4a and 4b) indicate that the amplifications match the damage positions. One can notice high amplitude circles which radii correspond to simulated damage (short notch, long notch and two drilled holes). One circle is noticeable for the long notch and the drilled hole in right lower corner. The reason for this is the fact that the distance from the arrays to this two simulated defects is roughly the same. The linear and cross array detected the simulated defect and nothing else, while the square array results showed an amplification circle with radius larger than for the rest of the defects. Most probably the cause of this additional indication were wires that was laid on the specimen.

Inspecting the values of $M(P_i)$ one can notice that for all three arrays the highest amplification is achieved for the hole on the left hand side. However comparing the amplification value only for this defect, the strongest response (above 25, compare the colorbar) was given by the linear array, lower value is for the cross array (above 17) and the lowest (approx. 15) for the square array. The comparison of the amplification levels for all the arrays is gathered in Table 1. One can notice that the cross and square arrays have the same hierarchy of sensitivity (1. left hand side hole, 2. short notch, 3. long notch with right hand hole) while for the linear array short notch gives lower response than the long notch with right hand hole. This suggest that the linear array is more sensitive to damage that cause wave reflection perpendicularly to the array.

Table 1. Amplification levels for considered arrays and damage types

array damage type	linear	cross	square
left hand side hole	25.27	17.10	15.01
short notch	3.26	11.33	9.40
long notch + hole	5.86	10.18	6.67

Results obtained for the three arrays (Fig.3b, 4a and 4b) showed that the information about damage presence can be obtained. Moreover partial localisation was achieved because the radii on which the defects lie were found. In order to obtain complete localisation an angular prediction was necessary. To obtain it the areas around the found radii were investigated. The maximal values of M in this areas were calculated. For left hand side hole the maximal value is indicated by white × in Fig. 3b, 4a and 4b. Unfortunately the position of found maxima differ considerably from the true damage location by 73, 100 and 57 mm for linear, cross and square array, respectively.

The attempt to obtain angular prediction of the defects failed, therefore only partial information about damage position was obtained. A solution for this could be a phased array processing algorithm [7]. However it needs a great care in selection of sensors spacing and the frequency of excitation.

Summary

In this paper an experimental study was performed. Its main goal was to compare the performance of the damage detection with three concentrated arrays of sensors. These arrays consisted of nine sensors. The only difference was in the arrangement of the sensors – linear, cross and square. A special numerical algorithm was used to process signals gathered by the sensors and to perform damage detection. A new type of mesh of points was introduced and used with this algorithm. Results of the detection algorithm were compared. In particular it was shown that a linear array is most sensitive to damage that cause wave reflection perpendicular to the array. All the arrays had the highest sensitivity to left hand side hole. However only in the case of a linear array the amplification connected with this damage was so dominant in relation to remaining defects. It was also shown that with the proposed algorithm damage localization is impossible. Phased array algorithm should help to overcome this drawback.

Acknowledgements

The authors acknowledge the support provided by project MONIT. Pawel Malinowski was also supported by the Polish Ministry of Science and Higher Education (ref. no. N501 335 134).

References

[1] Z. Su, L. Ye and Y. Lu: Journal of Sound and Vibration Vol. 295 (2006), pp. 753-80.

[2] H. Lamb: Proc. R. Soc. Lond. A Vol. 93 (1917), pp. 114-28.

[3] J. L. Rose: *Ultrasonic waves in solid media* (Cambridge University Press Cambridge,1999).

[4] P. Wilcox: IEEE T. Ultrason. Ferr. 50 (2003), pp. 699-709.

[5] K. Hongjoon, J. Kyungyoung, S. Minjea and K. Jaeyeol: NDT&E Int. 39 (2006), pp. 312-19.

[6] K. Hongjoon, J. Kyungyoung, S. Minjea and K. Jaeyeol: Ultrasonics 44 (2006), pp. 1265-68.

[7] V. Giurgiutiu: *Structural Health Monitoring with Piezoelectric Wafer Active Sensors* (Academic Press, Amsterdam 2008).

[8] W. Ostachowicz, P. Kudela, P. Malinowski and T. Wandowski: Mech. Syst. Signal Pr. 23 (2009), pp. 1805-29.

[9] P. Malinowski, T. Wandowski, I. Trendafilova and W. Ostachowicz W: Struct. Health Monit. 8 (2009), pp. 5-15.

[10] T. Wandowski, P. Malinowski, P. Kudela and W Ostachowicz: Proc.7th Inter. Workshop on Structural Health Monitoring (2009), pp. 2315-22.

[11] W. Qiang and Y. Shenfang : J. Intel. Mat. Syst. Str. 20 (2009), pp. 1663-73.

[12] Y. Lu, L. Ye and Z. Su: Smart Materials and Structures Vol. 15 (2006), pp. 839-49.

Applied Mechanics and Materials Vols. 24-25 (2010) pp 57-62
© (2010) Trans Tech Publications, Switzerland
doi:10.4028/www.scientific.net/AMM.24-25.57

Towards a hybrid infrared approach for damage assessment

R. K. Fruehmann[a], J. M. Dulieu-Barton[b], S. Quinn[c]

University of Southampton, School of Engineering Sciences
Southampton, SO17 1BJ, UK

[a]rkf@soton.ac.uk, [b]janice@soton.ac.uk, [c]s.quinn@soton.ac.uk

Keywords: Thermoelastic stress analysis (TSA), vibro-thermography (VT), pulsed phase thermography (PPT), non-destructive evaluation (NDE)

Abstract. The thermoelastic response obtained from an infra-red (IR) detector contains two components: the magnitude of the small stress induced temperature change caused by the thermoelastic effect and the phase angle of the temperature change relative to a reference signal generated by an application of a stress change. The phase angle is related to nonlinearity in the thermoelastic response and departures from the simple linear relationship that underpins thermoelastic stress analysis (TSA). The phase data could be used to make an assessment of temperature evolutions caused by viscoelastic behaviour resulting from damage and provide a basis for its evaluation. In the current paper the physics of other infra-red techniques used for non-destructive evaluation is used to better understand the nature of the thermoelastic response. The objective is to provide better exploitation of TSA by alternative processing of the IR measurements. Three case studies are presented that demonstrate the potential of the alternative processing for evaluating damage.

Introduction

The detection and qualification of damage in engineering materials and structures presents an ongoing challenge. Accordingly there are many tools and many different approaches for assessing damage evolution. One such technique is thermoelastic stress analysis (TSA). The potential for better exploitation of the technique in industrial applications has been demonstrated [1] by modifying the approach to data collection. Placed in the wider context of infra-red (IR) thermography techniques currently used in non-destructive evaluation (NDE), the ultimate aim is to combine pulsed phase thermography (PPT) and acoustic or vibro-thermography (VT) with TSA to enhance NDE procedures. Some initial background studies have been carried out with this in mind [2].

In TSA, PPT and VT, the material or component under inspection is subjected to some form of thermal stimulus, either *via* an external heat source or *via* a mechanical input. The resulting evolution of the surface temperature field is analysed to ascertain the location and size of defects within the material. The unique feature of TSA is that it can be used to identify and quantify changes in the stress field due to damage in virtually real time, as the material undergoes cyclic fatigue [1, 3]. A further attractive feature is that, as the technique requires a stress change, measurements can be taken during dynamic cyclic loading, without interrupting the fatigue loading. However, applications to full-scale engineering components are few, e.g. for a wind turbine blade [4]. One reason for the lack of applications in the field is the challenge of applying the technique without the application of a cyclic load. This topic has been examined in [5] and more recently in [1]. To make TSA more effective away from the laboratory, the measured IR signal must be optimised by using alternative means of remote loading (i.e. imparting the temperature change). Techniques for achieving the loading could be drawn from those used with other NDE techniques.

Recent work on fatigue in composite materials has indicated that nonlinearities in the thermoelastic response provide a useful means of identifying damage [6]. As the most notable cause of nonlinearity in the thermoelastic response results from heat transfer, a means of separating the adiabatic and non-adiabatic processes is necessary. This may be achieved by utilising the phase of the thermoelastic response. The purpose of the present paper is to provide further insight into the potential for using TSA as a means of quantitative damage assessment. The influence of heat transfer on the magnitude and phase angle of the thermoelastic response is the primary consideration. However, insight into the benefits of a hybrid approach to IR sensing for NDE is also discussed through three case studies. Here TSA data is examined with a focus on considering the non-adiabatic components of the temperature changes. The first two examples focus on the phase data. The third example examines the absolute temperature field as damage evolves under low cycle fatigue.

IR thermography techniques

In TSA changes in the temperature of a dynamically loaded component are correlated to the changes in the elastic stress at the surface [7]. The technique utilises the reversible change in temperature resulting from an elastic strain. The amplitude of the change in temperature is directly proportional to the amplitude of the change in stress. Typically, a sinusoidal load is applied, producing a corresponding sinusoidal response in the surface temperature that can be filtered from the IR signal and the amplitude measured with a precision to the order of mK. The IR signal is processed to extract the reversible components of the temperature change at the surface of the material. Two additional data-sets are also produced: the mean temperature field and the phase angle of the temperature change relative to a reference signal generated by a stress change. If the phase angle is zero (or 180°) then the thermoelastic temperature change is occurring at the same instance as the stress change, i.e. in phase and the response is isentropic. If it is not then the response is nonlinear as a result of a combination of effects, the most likely being heat transfer. Wang *et al* [8] provide an illustration of the effect of non adiabatic behaviour. The mean temperature field can also be used to assess irreversible heat generation, for example, due to frictional or viscoelastic heating. It is in the interpretation of these two data sets that concepts from PPT and VT can be exploited.

In PPT an external heat pulse is used to introduce a temperature change at the surface of a material. The pulse is transmitted through the material as a wave containing many frequency components. When the wave meets a discontinuity, for example a region of damage, the wave is reflected causing a temperature change at the surface. The depth of the defect is obtained using the phase angle of the reflected wave, while the size of the defect can be measured by its spatial extent of data with the same phase angle [9,10]. Depending on the conductivity of the material and the magnitude and frequency of the thermal pulse, individual frequency components are propagated and reflected in different ways. Lower frequency components can propagate further into the material, while higher frequency components are attenuated more quickly but allow smaller defects to be identified.

In VT, an ultrasonic wave is used to excite the material. In regions of damage, frictional and viscoelastic effects cause local heating. The generated heat is transmitted to the surface of the material where defects result in regions of elevated temperature. The depth of defect that can be detected depends on the magnitude of the energy release rate at the defect, the duration of the excitation and the conductivity of the material [11]. Successful application of the technique is crucially dependent on the ability to introduce an ultrasonic wave in such a manner that existing damage is not propagated further and that new damage is not introduced at the point of excitation. This must be balanced with the need for an excitation of sufficient magnitude to enable the existing damage to be identified.

In TSA the material can be treated as a periodic heat source. Non-uniformities result in a gradient in the amplitude of the local heat sources, and hence lead to heat transfer effects, similar to those exploited in PPT and VT. The two important concepts from PPT and VT that may be of interest in TSA are periodic temperature changes being transmitted through the material as a thermal wave, and the use of localised heating as a means of locating damage.

Case studies

Figure 1 shows the full set of TSA data taken from the neighbourhood of a 30° slot in a thin aluminium plate. The plate is undergoing tension-tension cyclic loading and as a result a cracks are propagating from each slot tip transversely to the loading direction. In this data the load amplitude was maintained at a low level to avoid further growth of the crack during the data collection. From the mean temperature across the plate (Figure 1a), it is evident that the mean temperature field is uniform indicating that the damage is not leading to localised heating. The TSA data, ΔT (Figure 1b) clearly shows two large stress concentrations either side of the slot. However, the exact position of the crack tip, and the exact magnitude of the stress concentration are difficult to identify from the ΔT data. This is due to heat transfer near the crack tip where the stress induced temperature gradients are large. The effect of heat transfer can be seen clearly in the phase data (Figure 1c). Taking a line of data through the upper stress concentration (line y = 108) allows the deviations in the phase angle to be correlated to the ΔT data (Figure 1d). The dotted line (left y-axis) shows the TSA data; the crosses (right hand y-axis) show the phase angle. The position of the

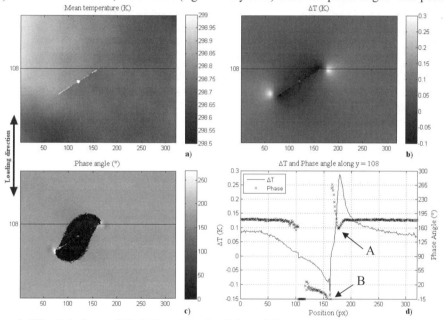

Figure 1: TSA data from a 30° slot in a thin aluminium plate

local phase minimum (point A in Figure 1d) was used by Zanganeh et al [12] to identify the location of the crack tip. This minimum occurs due to heat conduction away from the region of highest ΔT and is best understood by considering the stress concentration as a point source generating a thermal wave. As the wave propagates away from the source, it results in a thermal oscillation in the adjacent region with a time delay dependent on the conductivity of the material, which occurs over about 10 pixels (i.e. 5 mm). The phase angle departs from 180° indicating that heat transfer is taking place as a result of the large stress/thermal gradient adjacent to the crack tip. The transition from tensile to compressive stress to the left of point A shows first an increase in phase to around 260° before a step change to 0° to the left of point B. The stress is close to zero in the region between points A and B where the crack line is crossed just above the tip. Here the ΔT readings result *only* as a consequence of heat transfer.

Very similar phenomena can be seen in woven composite materials undergoing high cycle fatigue loading. The data shown in Figure 2 is taken from a 2 x 2 twill woven E-glass / epoxy, single ply composite specimen loaded in tension. The weave pattern is superimposed on the images, with the warp yarns running parallel to the loading direction. A transverse crack (circled in

Figure 2a, b and c) is present in one weft yarn. Again, the TSA and phase data from a line (y = 40) drawn through the damaged region has been plotted in the bottom right image in Figure 2. The phase data in the vicinity of the damage in the woven composite material shows similar trends to those from the aluminium plate, despite the non-uniform ΔT field; there is a reduction in the phase angle close to the crack tip because of the large stress gradients in the damaged region. In [6] it was demonstrated that the phase data thereby enables damage to be detected within a noisy / non-uniform ΔT field.

An additional feature seen in the TSA data in Figure 2, is a 15° difference in phase between the tows running parallel and transverse to the loading direction. The warp yarns exhibit the positive phase shift indicating heat transfer to the warp yarns. In the TSA data it can be seen that the weft yarns have the larger thermoelastic response, despite the majority of the stress being carried by the warp yarns [13]. In this case, a stress gradient is occurring through the thickness of a material, rather than in the plane. The surface phase shift therefore identifies a subsurface stress

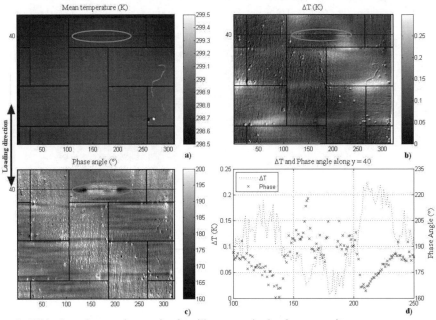

Figure 2: TSA phase images from a 2 x 2 twill woven, single ply composite

concentration, the magnitude of the phase shift depending on the thermal conductivity of the material and the loading frequency. This idea has been used in [14] and links in with ideas from PPT, where a thermal pulse consisting of a summation of thermal waves of different frequencies and magnitudes is introduced into the material. By decomposing the thermal wave as it reaches the surface it is possible to use the phase of the different frequency components to obtain information regarding a range of sizes and depths of defects. This idea could equally be applied to TSA data. Instead of using only a single excitation frequency (the loading frequency), random loading, or loading with multiple simultaneous frequencies may provide a more useful thermoelastic response to maximise the information obtained from the phase data. This idea could be applied to the study of components during in-service loading. Alternatively, a single transient step load could be applied, generating a square wave which comprises multiple frequencies, as in PPT.

Figure 3 shows 250 frames from a 1050 frame recording taken at 101 Hz of the temperature-time history from three areas on an E-glass epoxy composite strip specimen with a cross ply [90, 0]$_{2S}$ layup. A 2.2 mm long, 0.7 mm wide notch has been cut into its right hand edge and cyclic tensile loading applied. Figure 3 also shows the corresponding TSA data, i.e. the amplitude of the temperature shown in the graph, but calculated using only the data collected after frame 300.

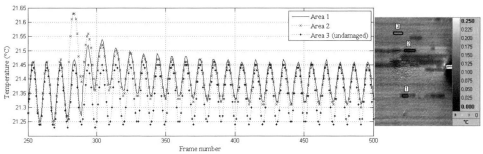

Figure 3: Time history of the temperature in three regions of a cross-ply tensile specimen

Damage growth can be observed in the 90° surface ply, away from the notch region in areas 1 and 2. Area 3 shows a region that does not sustain damage during the measurement period. The significant feature in this data is the sharp increase in temperature at the instant the damage occurs which can be linked to the energy released as a crack forms. The increase in temperature quickly dissipates, and the amplitude of the temperature signal is reduced as the stress is redistributed to the 0° ply below. (The small decrease in the mean temperature between frames 220 and 230 can be discounted as a background reflection.)

Filtering out the 10 Hz frequency shows the non-adiabatic temperature changes occurring in the material, shown in Figure 4. A significant difference between the temperature profiles of area 1 and 2 can be seen in the rise time of the peak at frame 280. While in area 2 there is an almost instantaneous rise to the peak temperature, in area 1 the rise takes place over the course of approximately twenty frames (0.2 seconds). Two explanations are proposed for the difference in

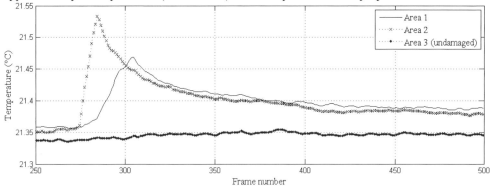

Figure 4: Mean temperature of areas 1, 2 and 3 from Figure 3

the response in the two areas. Firstly, area 1 could contain a subsurface crack and therefore a time lag would be expected as the heat travels to the surface or, secondly, fibre bridging might be occurring as the crack forms, slowing the rate of growth and dissipating energy over a longer time. The area under the curves could be used to estimate the energy released during the damage event.

Conclusions

It is has been shown that the phase of the thermoelastic response can be exploited to derive more detailed information regarding the depth of defects and the nature of subsurface damage. It has been suggested that similar ideas to those used in PPT would enable the application of the phase data in a quantitative manner. The most interesting aspect of this is the potential for evaluating components under in-service loading, where multiple frequency loading could be used as an advantage over conventional sinusoidal loading used in laboratory work. In this context, alternative loading scenarios for laboratory testing of materials and components may also be developed.

It has also been shown that the change in temperature as energy is released during the formation and growth of cracks can be obtained from the IR measurements. To fully exploit this aspect, suitable algorithms for identifying the energy release event are required. These could be used to trigger a TSA measurement from which the influence of the damage on the stress field can be evaluated.

The paper shows how a holistic approach towards IR measurement has the potential to provide a powerful tool for strain based NDE of engineering materials and structures.

References

[1] R.K. Fruehmann, J.M. Dulieu-Barton and S. Quinn: Experimental Mechanics, (2009), DOI: 10.1007/s11340-009-9295-9

[2] T.R. Emery and J.M. Dulieu-Barton: Key Engineering Materials Vol. 347 (2007), p. 621

[3] T.R. Emery and J.M. Dulieu-Barton: Composites Part A (2009), DOI: 10.1016/j.compositesa.2009.08.015

[4] R. J. H. Paynter and A.G. Dutton: Strain Vol 39 (2003), p. 73

[5] U. Galietti, D. Modugno and L. Spagnolo: Measurement Science and Technology Vol 16 (2005), p. 2251

[6] R.K. Fruehmann, J.M. Dulieu-Barton and S. Quinn in: *Proceedings of ICCM 17* (2009), on cd

[7] J. M. Dulieu-Barton and P. Stanley: Journal of Strain Analysis Vol. 33 (1998), p. 93

[8] W.J. Wang, J.M. Dulieu-Barton and Q. Li : Experimental Mechanics (2009), Vol. 50, p 449. DOI: 10.1007/s11340-009-9249-2

[9] X. P. Maldague and S. Marinetti: Journal of Applied Physics Vol 79(1996), p. 2694

[10] G. Busse, D. Wu and W. Karpen: Journal of Applied Physics Vol 71 (1992), p. 3962

[11] T.J. Barden, D.P. Almond, S.G. Pickering, M. Morbidini and P. Cawley: Nondestructive Testing and Evaluation Vol 22 (2007), p. 71

[12] M. Zanganeh, R.A. Tomlinson and J.R. Yates: Journal of Strain Analysis Vol: 43 (2008), p. 529

[13] R.K. Fruehmann, J.M. Dulieu-Barton and S. Quinn: Journal of Strain Analysis Vol: 43 (2008), p. 435

[14] N. Sathon and J.M. Dulieu-Barton: Applied Mechanics and Materials Vol: 7-8 (2007), p. 153

Session 2: Vibration

Applied Mechanics and Materials Vols. 24-25 (2010) pp 65-70
© (2010) Trans Tech Publications, Switzerland
doi:10.4028/www.scientific.net/AMM.24-25.65

Designing a Hollow Langevin Transducer for Ultrasonic Coring

P. Harkness[1,a], A. Cardoni[1,b], J. Russell[1,c] and M. Lucas[1,d]

[1]School of Engineering, University of Glasgow, Glasgow, G12 8QQ, UK

[a]p.harkness@eng.gla.ac.uk, [b]a.cardoni@eng.gla.ac.uk

[c]0501885R@student.gla.ac.uk, [d]m.lucas@eng.gla.ac.uk

Keywords: Bolted Langevin Transducer, Effect of Preload, Response Functions

Abstract. A number of architectures for a hollow Langevin ultrasonic transducer are proposed and evaluated. One of these is optimised by finite element modelling and is then manufactured and analysed experimentally. The preload on the transducer ceramics is increased and the effect on the performance is measured. At maximum preload the results of an experimental modal analysis are used to determine the natural frequency and response of both the operating longitudinal mode and unwanted bending modes. The performance of the hollow transducer is compared to a solid commercial transducer containing the same volume of piezoceramic material. The efficiency is shown to be comparable. Higher ultrasonic displacement amplitudes are achieved with the hollow transducer although a lower Q-factor is found.

Introduction

Bolted Langevin transducers are solid structures, with a central bolt running through the centre of a stack of piezoceramic rings and securing the end-masses. The bolt delivers the prestress required to keep the piezoceramics in a state of compression throughout the vibrational cycle. However, for some applications such as rock coring, or when the transducer requires extensive cooling, access through the centre of the transducer may be required.

For example, ultrasonic tools have been suggested as a mechanism by which rock cores can be taken from the subsurface by planetary landers [1]. However, these designs all suffer from the requirement that the ultrasonic apparatus be removed to extract the sample. A hollow transducer, horn and cutting assembly could obviate this need [2]. Additionally, heavy-duty ultrasonic systems often require fan cooling systems [3]. Permitting passage of the coolant through the centre of the stack would increase the surface area available for heat transfer.

Architecture of the Hollow Transducer

Design procedure. The traditional bolted Langevin transducer (BLT) has a solid central bolt which must be replaced by a hollow structure. There are two options: either the bolt can be moved to the outside of the structure, as a cylinder or as a ring of peripheral columns, or the bolt can be retained inside the transducer but made hollow. Two alternative architectures, a cage and a hollow-bolt, are therefore explored and compared by considerations such as application of prestress, part-count, cost and ease of final assembly. The hollow-bolt architecture is shown to be superior, although the problems associated with an inherently low-stiffness device are recognised. Using different materials for the front and end-masses (steel and aluminium) to maximise unidirectional vibration output, and with reference to the fatigue life of the thin-walled titanium bolt, a prototype of this device is designed using finite element (FE) analysis.

The transducer is manufactured, an experimental modal analysis is conducted using 3D laser Doppler vibrometry, and the extracted modes are compared (in terms of mode shape and natural frequency) to the numerical predictions. Impedance analysis is also carried out and, as the preload in the system is progressively increased, the frequency of the response of the operating longitudinal mode tends towards the modelled predictions. When sufficient preload to stabilise the operating frequency is reached a response spectrum is measured and the Q-factor of the transducer is calculated. Further addition of preload is then shown to increase the magnitude and quality factor of the response at the operating frequency. Finally, the hollow transducer is contrasted with a traditional transducer of similar size, to determine its performance with reference to its solid competitors.

External bolting. An external cylinder is unlikely to be practical because it would impede access to the piezoceramics. It would also have to be, effectively, a thin-walled shell. Thin-walled structures, which are inherently non-stiff, are liable to exhibit detrimental transverse modes at ultrasonic frequencies. Furthermore, the structure would preclude the use of all but the thinnest screwthreads for prestressing purposes, and would be extremely vulnerable to fatigue.

The alternative external-bolting approach, consisting of a series of columns around the circumference of the transducer, permits access to the ceramics but would require that the ceramics themselves, which are commercially produced as rings, be shaped to fit around the columns. The number of columns and non-standard ceramics would be difficult, slow and expensive to both manufacture and assemble. The stress applied through each bolt would also have to be equalized to ensure even compressive load on the ceramic stack, as uneven prestress is known to be associated with performance inefficiencies [4].

Internal bolting. Modifying the central bolt, such that it becomes a hollow structure, appears to be a more attractive approach because the walls of the bolt remain sufficiently thick to accommodate screwthreads for assembly purposes. It also permits the use of standard piezoceramic discs and facilitates the even distribution of prestress across the piezoceramic stack. In the absence of other factors, this architecture is preferred for the targeted coring application.

Design of the Hollow Transducer

In common with several conventional transducers, end masses of steel and aluminium are used in this design to maximise longitudinal vibration output by taking advantage of their different stiffness-to-density ratios. A 90Ti/6V/4Al bolt is used because of its high quality factor and fatigue resistance, and PZT-8 ceramics driven by copper electrodes provide the vibration excitation.

The thickness of the bolt wall is set at 3 mm, and the volume of PZT-8 is required to be close to 5×10^{-5} m^3 to match the volume of ceramic contained in a commercial transducer known to be effective for rock-coring operations. A separation of 2 mm is maintained between the central bolt and the electrodes and the interstitial space is filled with dielectric material to prevent internal shorts. A preload screw is provided at one end for final assembly and the other is threaded to accept a half-inch UNF stud. Although a system utilising a hollow transducer would be hollow throughout, the transducer is to be further tested using a hollow horn designed to connect to the commercial transducer and so the central stud connection is required.

The design frequency of the hollow transducer is 20 kHz for operation in the first longitudinal mode. To achieve this, an FE model of the hollow transducer is created in Abaqus, with reference to the fixed design parameters described above. The non-fixed parameters, such as the lengths of the end-masses, are then adjusted to tune the operating frequency. Finally the oscillatory stress in the central bolt is evaluated, assuming a peak-to-peak displacement of the transducer of 10 μm and is found to be well below the fatigue limit. The resulting design is presented in Fig. 1.

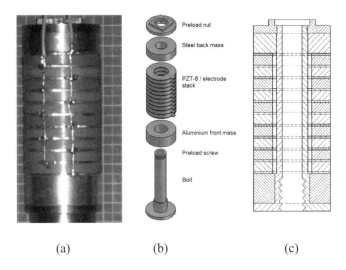

(a)   (b)   (c)

Figure 1. The hollow transducer, (a) as manufactured, (b) in exploded and (c) in assembled form. The grid spacing in the photograph is 5 mm and the external wiring to the electrodes is apparent.

Application of Preload

Clamps are offered up to flats ground on the bolt, front mass and back mass (see Fig. 2(a)), whilst the preload nut is tightened to increase the preload on the stack. This allows preload to be applied to the stack without any torsion which could damage the electrodes. Whilst the actual preload is almost impossible to predict due to the complexities of the transducer structure [5,6], it can be found in practice by measuring the charge on the piezoceramics during the prestressing process [7]. It is known that increasing the preload will tend to increase the first longitudinal resonance frequency of a bolted Langevin transducer as well as the associated Q-factor, and it is reasonable to expect that the same will be true for hollow transducers.

The preload is increased from the minimum required for solid assembly (bias 1) to the point at which no further changes in the frequency of the first longitudinal mode appear to be achievable (bias 4). To measure the prestress applied in this condition, a sample of PZT-8 was subjected to compression testing. After ten seconds at a range of applied pressures, the voltage across the faces was recorded to form a look-up table which was then used to deduce the preload in the transducer after tightening. The minimum preload required to stabilise the frequency behaviour of the device was found to be 10 MPa.

Broadband Results

Fig. 2(b) shows the impedance analysis, and the progression from bias 1 to bias 4 shows how the results of the repeated impedance analyses are affected by increasing preload.

An experimental modal analysis of the transducer is performed at bias 1 using a 3D laser Doppler vibrometer, where the responsive modes, marked A, B, C and D in Fig. 3(a), are extracted during excitation by a white noise signal between 0 and 40 kHz. The frequencies of these modes, which are respectively first bending, second bending coupled with first longitudinal, third bending and first radial, are indicated on the frequency response function (FRF) shown in Fig. 3(a). Fig. 3(b) shows a result of the second experimental modal analysis, where a much higher preload (bias 4) has been applied. The responsive modes, which are first bending, second bending, first longitudinal and third bending and are marked E, F, G and H respectively in the figure, are extracted.

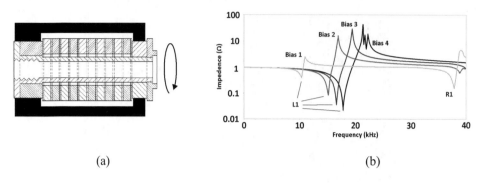

(a)   (b)

Figure 2. Clamps are applied (a) such that torque does not reach the PZT as preload is increased, and (b), the measured impedance of the transducer as the preload is increased from bias 1 to bias 4. L1 peaks refer to the first longitudinal mode and R1 peaks to the first radial mode.

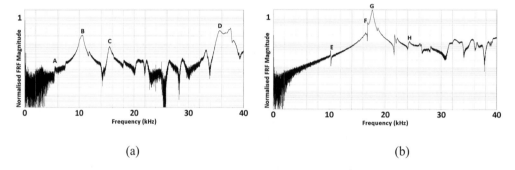

(a)   (b)

Figure 3. Normalised longitudinal FRF magnitude at (a) bias 1 and (b) bias 4.

Initially, the operating mode (first longitudinal) of the transducer is excited at 10.3 kHz and is coupled with the second bending mode. Increasing the preload causes the operating mode to decouple from the bending mode, increase in strength by almost an order of magnitude and move towards the 20 kHz value predicted by the finite element model. This final trend is consistent with the behaviour of conventional bolted Langevin transducers [8].

	Modal Frequencies [kHz]				
	Bias 1	Bias 2	Bias 3	Bias 4	FE Prediction
1st Longitudinal	10.3 (52%)	15.2 (77%)	16.6 (84%)	17.8 (90%)	19.8
1st Bending	5.6 (51%)	-	-	10.2 (94%)	10.9
2nd Bending	10.3 (56%)	-	-	16.8 (92%)	18.3
3rd Bending	15.5 (57%)	-	-	24.2 (90%)	27.2

Table 1. Modal frequencies measured at different bias levels. Values in parentheses are percentages with respect to the predicted frequency.

Increasing prestress also causes the frequencies of the radial and bending modes to increase, due to the stiffening of the structure as microscopic voids between the elements are progressively removed. This process can take place due to cold flow of the soft copper electrodes into surface imperfections [9]. As prestress increases from bias 1 to bias 4, the modal frequencies approach the values predicted by the finite element model, as shown in Table 1. However, increasing prestress beyond 10 MPa has a negligible effect on the frequencies of the modes studied.

Modal Analysis

The mode shapes, as predicted by FE and measured by EMA are presented in Table. The transducer was self-excited by a random signal.

Mode	EMA	FE	Mode	EMA	FE
1st Longitudinal			2nd Bending		
1st Bending			3rd Bending		

Table 2. Mode shapes measured by EMA and predicted by FE analysis.

Improvements to the Design

The process of preloading the transducer demonstrated that the application of significant preload requires very considerable torque. The 10 MPa loading needed to stabilise the transducer required approximately 50 Nm and so the preload nut was redesigned to facilitate the use of a hexagonal socket.

FE analysis indicated that this new nut has a negligible influence on the natural frequency of the transducer, but it permitted the preload on the piezoceramic stack to be increased to 20 MPa. Although some sources recommend preloads as high as 40 MPa [10], the thin-walled structure of the central bolt makes high loads difficult to apply safely.

Narrowband (Operating Mode) Results

Fig. 4(a) shows the response of the longitudinal operating mode (+/- 2 kHz range) in bias condition 4, compared to the response of a solid commercial transducer containing the same volume of piezoceramic. In both cases a sinusoidal excitation of 40 V (peak-to-peak) has been applied. Fig. 4(b) shows the same measurements after the new nut has permitted doubling of the applied preload.

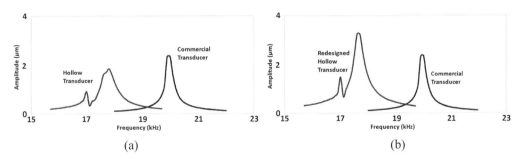

(a)   (b)

Figure 4. The response of the hollow transducer (a) before and (b) after the prestress-enhancing modifications, compared to that of a commercial transducer containing the same volume of PZT.

The transducer resonance is 10% below the design frequency and the presence of a bending mode near the tuned frequency requires careful driving for effective use. However, the additional preload beyond the point of resonance stabilisation increased the amplitude response of the hollow transducer from 1.8 to 3.2 μm and increased the Q-factor to a level comparable to that of the commercial transducer.

These disparities are to a large extent the result of high tolerance requirements for the manufacture of the hollow transducer. The wall-thickness of the central bolt is just 3 mm, and so a small manufacturing error can have a significant effect on the stiffness of the structure, whereas the same error on a solid bolt has a negligible effect. The hollow transducer is effective in delivering the required ultrasonic amplitude to a coring device and will form the basis of the ongoing design of a completely hollow ultrasonic coring system.

Conclusions

The design process leading to a prototype of a hollow Langevin transducer has found that the frequency of the operating longitudinal mode increases and tends towards the FE predictions as the mechanical preload is increased, until frequency stabilisation occurs. Thereafter the results indicate that increased preload continues to improve the mechanical quality factor of the transducer and therefore its ultrasonic vibration amplitude.

In terms of mechanical efficiency, the performance of the prototype hollow transducer was comparable to that of a commercial transducer containing the same volume of piezoceramic. The peak output at 40 V peak-to-peak excitation was over 3 μm compared to just over 2 μm for the commercial model. However, the Q-factor was significantly lower and an unwanted bending mode close to the main resonance remains a cause for concern.

Technical issues highlighted by the investigation centre on the design of the central bolt. Extremely tight manufacturing tolerances are required to ensure that its mechanical stiffness is correct, and the thread which receives the preload nut should be shallow yet extend over the greatest possible axial distance.

The hollow transducer also requires a long stack of piezoelectric elements if a significant volume of ceramic is to be incorporated. This, coupled with the low stiffness of the hollow bolt, places a constraint on the maximum operating frequency. Unwanted bending modes also appear to be a more significant problem than is the case for traditional solid bolted Langevin transducers. However, the long stack does lend itself to higher ultrasonic vibration amplitudes because equivalent percentage strains translate to higher absolute displacement.

References

[1] X. Bao, Y. Bar-Cohen, Z. Chang, B. Dolgin, S. Sherrit, D. Pal, S. Du and T. Peterson: IEEE Transactions of UFFC. Vol. 50 (2003), p. 1147

[2] P. Harkness, A. Cardoni and M. Lucas: AIAA 09-6507 (2009)

[3] J. Gallego-Juarez, G. Rodriguez-Corral and L. Gaete-Garreton: Ultrasonics Vol. 16 (1978), p. 267

[4] A. Abdullah, M. Shahini and A. Pak: J. Electroceram. Vol. 22 (2009), p. 369

[5] K. Adachi, M. Tsuji and H. Kato: J. Acoust. Soc. Am. Vol. 105 (1999), p. 1656

[6] T. Takahashi and K. Adachi: Jpn. J. Appl. Phys. Vol. 47 (2008), p. 4736

[7] F. Arnold and S. Muhlen: Ultrasonics Vol. 41 (2003), p. 191

[8] B. Fu, T. Li and Y. Xie: *Model-Based Diagnosis for Pre-Stress in Langevin Transducers* IEEE Circuits and Systems Technical Conference (2009).

[9] G. Bromfield, U.S. Patent 7627936. (2009).

[10] B. Dubus, G. Haw, C. Granger and O. Ledez: Ultrasonics Vol. 40 (2002), p. 903

Applied Mechanics and Materials Vols. 24-25 (2010) pp 71-76
© *(2010) Trans Tech Publications, Switzerland*
doi:10.4028/www.scientific.net/AMM.24-25.71

Experimental Modal Analysis of an Automotive Powertrain

C. Delprete[1,a], A. Galeazzi[1,b] and F. Pregno[1,c]

[1]Politecnico di Torino, C.so Duca degli Abruzzi, 24 – 10129 Torino
[a]cristiana.delprete@polito.it, [b]g.alex86@tiscali.it, [c]fabio.pregno@gmail.com

Keywords: Experimental, modal analysis, powertrain, engine, vibration.

Abstract

The work is devoted to study the dynamic properties of a powertrain, performing an experimental modal analysis (EMA). The aim is to determinate structural modes and frequency response function (FRF) using an experimental approach. Two types of excitation mechanism are applied and compared for the EMA: a modal exciter (electromagnetic shaker system) and an impact hammer. Both the analyses with modal shaker and with impact hammer are carried on measuring the acceleration of the structure in the same set of eighty-one points. In both cases, the excitation is performed along three directions (vertical, lateral and longitudinal with respect to the structure). The two different modal analysis methodologies are described, and results (modal parameters such as natural frequencies, damping ratios and modal shapes are identified with commercial software) are compared. The comparison is made in term of result accuracy, reliability and testing time required.

Introduction

The noise, vibration, and harshness (NVH) performances are one of the major challenges in the automotive design field. In a vehicle, there are many sources of noise and vibration: engine, driveline, brakes, tire contact patch and road surface. Designers have to consider NVH during the development of a new internal combustion engine. Noise and vibration reduction is fundamental for the comfort of the passengers. Another design milestone is the weight reduction in order to decrease fuel consumption and pollutant emissions. Weight reduction has not to compromise the structural performance since a lighter structure could not match NVH characteristic requested for passengers' comfort. Experimental modal analysis (EMA) can help designers to study NVH characteristic of an existing engine in order to set a series of goals to improve a new design project. Moreover, NVH performances of new engines can be evaluated. EMA is based on vibration measurements in order to obtain frequency response function relating output vibration to input excitation.

The EMA is performed on a three cylinder spark ignition engine with gearbox in order to evaluate natural frequencies, damping ratios and mode shapes. It is a multi-input multi-output analysis (MIMO): output responses are accelerations and they are measured with piezoelectric transducers (accelerometers), input signals are obtained using two different excitation instruments. The final critical analysis of the experimental results allows understanding which excitation technique is more suitable.

Instrumentation

The test-bench is set up by:
- 1 Impact Hammer *PCB Piezotronics*[INC.] model T086C03;
- 1 Modal Exciter *Dongling Vibration* model ESD-045 with PA-1200 power amplifier;
- 3 accelerometers *PCB Piezotronics*[INC.] model TLB356A12;
- 1 load cell *PCB Piezotronics*[INC.] model 208C03;
- 1 portable front-end *LMS* SCADAS III model 305;
- 1 Hydraulic Floor Crane;
- Suspension equipment;
- 1 personal computer;
- *LMS* Test.lab software.

Experimental methodology

Eighty-one points of measure were identified in order to obtain a complete model of the powertrain, which gives a good representation of the geometry of the whole structure and its components (as illustrated in [3, 4]). The tridimensional model is created with *LMS TestLab* software using the Cartesian coordinates of these points.

The experimental modal analysis was performed using two different excitation instruments: impact hammer and modal exciter (electromagnetic shaker).

The powertrain was suspended by means of two springs in both cases, in order to reproduce free-free conditions.

The analysis of the experimental data allows the comparison of the two different excitation methods.

Modal exciter

As regards to the use of modal exciter, the excitation was applied into two points of connection between the instrument and the structure. In this way it was possible to excite the structure along its three main directions: longitudinal (x in Fig 3), transversal (y in Fig 3) and vertical (z in Fig 3).

Firstly the shaker was connected to a point on the bedplate, skewed so that the transmitted force could be split into two components in order to excite the powertrain along two directions: transversal and vertical (see Fig. 1). Then, the modal exciter was connected horizontally to a point on the engine lateral carter, in order to apply the excitation force along the longitudinal direction (see Fig. 2).

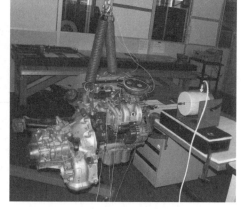

Figure 1 – First excitation point.   Figure 2 – Second excitation point.

Figure 3 shows the powertrain model for modal exciter analysis: the point 37 is for the transversal and vertical excitation while the point 15 is for the longitudinal excitation.

The force transmitted to the structure was measured with a load cell, interposed between the shaker and the point of excitation on the structure. The connection with the engine was made by means of a threaded link.

It was used a Sine Sweep excitation signal, applied with a frequency bandwidth of 1600 Hz. It was necessary to split the excitation bandwidth into four steps because the output signal obtained from the excitation of the whole bandwidth was completely influenced by noise, as Figure 4 shows. This phenomenon is due to the fact that the system is excited in a too wide frequency band and as a consequence the response is obtained by the superimposition of too many modes, which are too lightly-damped.

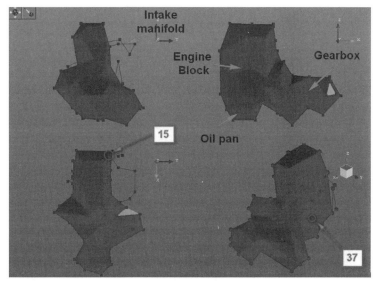

Figure 3 – Powertrain model.

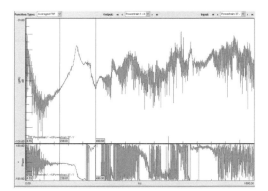

Figure 4 – FRF obtained from the whole bandwidth excitation.

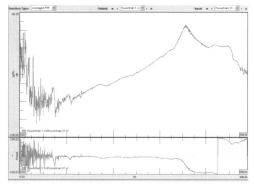

Figure 5 – FRF obtained from the first bandwidth step excitation.

Figure 5 shows the FRF for the first bandwidth step.

Figures 4 and 5 present the frequency [Hz] on the horizontal axis and the accelerance [g/N] on the vertical. The four frequency band steps (Table 1) in which the excitation bandwidth is splitted allow obtaining FRFs with not relevant noise.

Step	Frequency [Hz]
1	0-490
2	390-1100
3	690-1300
4	1100-1600

Table 1 – Frequency band steps.

Impact hammer

As regards to the use of impact hammer, the excitation was applied into three points of the engine, in order to excite the structure along its three main directions: longitudinal, transversal and vertical.

Figure 3 shows the powertrain model for impact hammer analysis and it is the same model used for the analysis with the modal exciter.

Excitation along the transversal direction was applied on a point of the bedplate (point 37 of Fig. 6), for the longitudinal excitation was used a point of the engine lateral carter (point 16 of Fig. 6). Lastly, the excitation along vertical direction was performed on a point of the engine head cover (point 4 of Fig. 6).

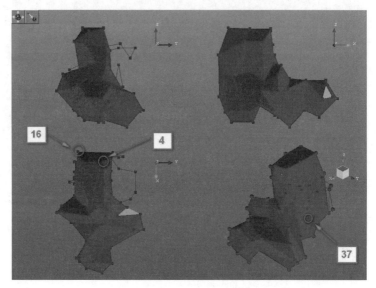

Figure 6 – Powertrain model.

Results

In both excitation cases, input and output signals were conditioned with Hanning window in order to reduce leakage error (see [5] for details) and each FRF was computed as the mean of five measurements and with the so called H_v estimation method [1].

The experimental data were processed with *Polymax modal estimation method*, which is based on the *Least squares frequency domain* (LSFD) method [2]. This method allows identifying stable modes of the structure, estimating the corresponding natural frequencies, damping ratios and modal shapes. Modal shapes are shown with an animation clip of the modal deformations.

Figure 7 shows the modal shape of the first mode evaluated with the modal exciter. The red lines represent the undeformed model.

Figure 7 – Modal shape of the first mode (modal exciter).

Tables 2 and 3 illustrate the results obtained with the modal exciter and the impact hammer.

Mode	Frequency [Hz]	Damping ratio [%]	Modal shape		Mode	Frequency [Hz]	Damping ratio [%]	Modal shape
1	326	3.14	Bending		1	331	2.96	Bending
2	337	1.30	Bending		2	355	2.35	Bending
3	356	2.05	Bending		3	360	1.66	
4	361	1.72	Intake only		4	368	0.18	Bending
5	414	1.66	Torsional		5	415	2.19	Bending
6	453	1.39			6	455	2.00	
7	464	1.37			7	474	0.21	
8	510	1.52			8	514	0.30	
9	522	0.14	Oil pan only		9	567	1.09	
10	610	0.14	Torsional		10	605	1.21	
11	632	2.02			11	636	3.75	Torsional
12	704	1.63			12	701	1.19	
13	734	0.04	Engine only		13	757	1.33	
14	745	0.01	Alternator		14	864	0.88	
15	753	0.16			15	886	1.44	
16	759	1.89			16	1015	0.66	
17	830	0.69			17	1133	0.52	
18	897	1.01			18	1167	0.52	
19	1007	0.28			19	1348	1.08	
20	1084	0.35			20	1412	0.18	
21	1132	0.71			21	1540	1.24	
22	1154	1.01						
23	1189	0.28						
24	1328	0.87						
25	1391	1.64	Engine head					
26	1468	2.31						

　　　Table 2 – Modal exciter results　　　　　　　　　　Table 3 – Impact hammer results

The comparison between the data presented in Tables 2 and 3 shows that it was possible to find a major number of modes using the modal exciter instead of the impact hammer. The difference between the values of natural frequencies of the corresponding modes identified by both excitation strategies is limited to a range of 5 Hz.

Moreover, with the modal exciter it was possible to find a quantity of modes which modal deformation involves only single components (i.e. mode #9 involves only the oil pan); the impact hammer allows identifying only modes which modal deformation involves the whole structure.

Another term of comparison between the two different excitation techniques is offered by the analysis of the identification method parameter called *Model size*. The *Polymax modal estimation method* requires that the following relation is respected:

$$r \cdot N_i \geq 2 \cdot N \tag{1}$$

where r indicates the model size, N_i the number of input points and N the number of modes of the structure [2].

It was necessary to set the model size to the value of 60 to apply the identification method to the FRFs obtained from the impact hammer excitation, while it was set to 40 to analyse the data deriving from the modal shaker excitation (see Fig. 8 and 9). Both values respect Equation 1, but modal exciter permitted to obtain more modes with a smaller *model size.* Figures 8 and 9 present the frequency [Hz] on the horizontal axis and the accelerance [g/N] on the vertical axis.

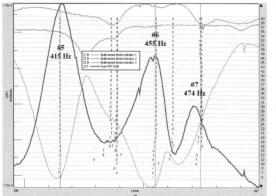
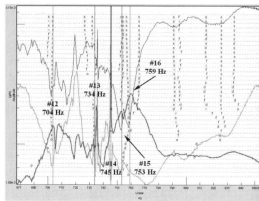

Figure 8 – Stabilization diagram (impact hammer).   **Figure 9 – Stabilization diagram (modal exciter).**

Conclusions

Experimental modal analysis on powertrain can be performed with two different excitation instruments: impact hammer and modal exciter.

Modal exciter seems to be more suitable for this experimental investigation because it permitted to identify a major number of modes with a smaller model size.

Moreover, the modal exciter analysis allows finding structural modes which modal shapes deformation involves only single engine components. It is clear that with the modal exciter it is possible to transfer a major amount of excitation energy to the structure with respect to the impact hammer.

Furthermore, impact hammer is more suitable to study fairly linear structures, not too lightly damped and not too heavily damped. Modal exciter is more suitable to study barely linear structure.

The impact hammer analysis has the advantage of quick set-up and execution times: the analysis with the impact hammer lasts one day while that with the modal exciter lasts three days.

References

[1] D. J. Ewins: *Modal Testing*, Research Study Press Ltd, Baldock, Hertfordshire (2000)

[2] LMS, *The LMS Test.LAB Impact Testing Manual*, supplied with LMS Test.LAB rev 9A, (2008)

[3] Y. Honda, K. Wakabayashi and T. Kodama: A Basic Study on Reduction of Cylinder Block Vibrations for Small Diesel Cars, SAE Technical Paper 2000-01-0527 (2000)

[4] H. Okamura, Susumu Arai: Experimental Modal Analysis for Cylinder Block-Crankshaft Substructure Systems of Six-cylinder In-line Diesel Engines, SAE Technical Paper 2001-01-1421 (2001)

[5] F. G. Ferraz, A. L. Cherman, D. Silveira de Abreu, R. Soares: Experimental Modal Analysis on Automotive Development, SAE Technical Paper 2003-01-3610 (2003)

Applied Mechanics and Materials Vols. 24-25 (2010) pp 77-82
© *(2010) Trans Tech Publications, Switzerland*
doi:10.4028/www.scientific.net/AMM.24-25.77

Investigation into the damping and stiffness characteristics of an elevator car system

I. Herrera[1, a], H. Su[2,b] and S. Kaczmarczyk[2,c]

[1]Escuela de Ingenierías Industriales, Universidad de Extremadura, Avda. de Elvas s/n, 06071 Badajoz, Spain

[2]School of Applied Sciences, University of Northampton, St. George's Avenue, Northampton, NN2 6JD, UK

[a]iherrera@unex.es, [b]huijuan.su@northampton.ac.uk, [c]stefan.kaczmarczyk@northampton.ac.uk

Keywords: Stiffness coefficient, damping coefficient, elevator car system

Abstract. Modelling the dynamic performance of an elevator car system represents a complex task and forms an important step in the elevator system design procedure. The need to consider the behaviour of passengers travelling in the car complicates the procedure further. This paper presents an original approach to identify the stiffness and damping characteristics of an elevator car system. A simplified model is developed and the experimental rig with a rectangular elevator platform fixed on the top of four silent blocks attached to a shaker is setup. The transmissibility measurements are carried out with a harmonic excitation applied first to a platform with no passenger load and then to the platform with one passenger within the frequency range of 1 – 20 Hz. A single person standing on the platform is employed in order to assess the passenger's contribution to the dynamic behaviour of the elevator car system. The curve fitting technique implemented in MATLAB is used to determine the damping and stiffness coefficients both for the empty car system and the car-passenger system. Investigation on the tolerances for both parameters is carried out. An approach to simplify the experimental procedure and to reduce the number of individual tests is proposed.

Introduction

The dynamic characteristics of the elevator car system play a very important role in elevator engineering. Nowadays, high ride quality of elevators is demanded. In particular low vibration and noise levels are related to high ride quality. These are usually checked by elevator companies before launching a new product into market by experiments and/or modelling. Modelling the dynamic performance of an elevator car system is a complex task and forms a significant step in the elevator system design procedure. As stated in the ISO standard [1] for elevator ride quality measurements, the components affecting the acceleration measurements in the car system are the instrument, persons and floor covering. Much effort has been made to predict passenger's behaviour and characteristics on the dynamic response of an elevator car [2,3,4]. In the previous investigations [2] passengers were modelled as rigid bodies with different types of shoes and it was concluded that the dynamic response of the elevator car could be substantially influenced by the passenger load. The need to consider the dynamics of passengers riding in the car complicates the procedure further [5,6].

In this paper the results of a study to assess the influence of the dynamic characteristics of a passenger on the lift car response are presented. The transmissibility measurements have been carried out. The curve fitting toolbox implemented in MATLAB [7] is then used in order to identify the stiffness and damping characteristics of the elevator car–passenger system. An approach is proposed to simplify the experimental procedure by reducing the number of transmissibility measurements by focusing on the resonance frequency region rather than using the whole range of the test frequencies.

Mechanical Model of the Car System

A typical elevator car system consists of the cabin (car enclosure) and the car frame (see Fig. 1). The cabin carries passengers and the car frame is connected to the means of suspension and to the platform (the floor) which directly supports the passengers. Considering vertical only, a simplified two degree of freedom model as shown in Fig. 2 is used to represent the car-passenger system. The model includes a single passenger represented by a rigid mass m_{PA}. The passenger is coupled to the platform via a spring-damper element of coefficient of stiffness k_{PA} and of coefficient of viscous damping c_{PA}, respectively. The car enclosure is modelled as a rigid mass m_{CA}. The platform is isolated from the frame by four isolation blocks represented by a spring-damper element of stiffness coefficient k_{CA} and of viscous damping coefficient c_{CA}, respectively. The car frame is subjected to a harmonic base motion excitation y_{FR}. The vertical displacements of the passenger and the car enclosure are denoted as y_{PA} and y_{CA}, respectively.

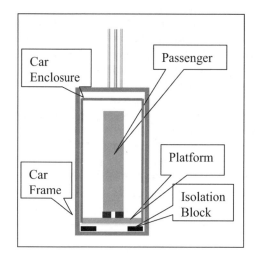 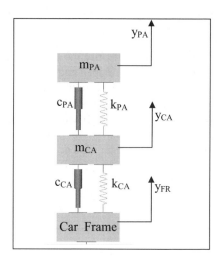

Figure 1. The car-passenger subsystem   **Figure 2.** A model of the car-passenger system

The coefficients k_{CA} and c_{CA} are determined as [2],

$$k_{CA} = nk_{sp}; \quad c_{CA} = nc_{sp} \tag{1}$$

where k_{sp} and c_{sp} are the coefficients of stiffness and damping of a single isolation block, respectively, and n is the number of isolation blocks used.

The coefficients k_{PA} and c_{PA} are then defined as

$$\frac{1}{k_{PA}} = \frac{1}{k_{pa}} + \frac{1}{k_{sh}} + \frac{1}{k_{fl}}; \quad \frac{1}{c_{PA}} = \frac{1}{c_{pa}} + \frac{1}{c_{sh}} + \frac{1}{c_{fl}} \tag{2}$$

where k_{pa} and c_{ca} denote the stiffness and damping coefficients of a human body in the vertical direction, respectively, k_{sh} and c_{sh} are the vertical stiffness and damping coefficients of the passenger's shoes, and k_{fl} and c_{fl} represent the vertical stiffness and damping coefficients of the platform/floor covering.

The differential equation of motion of the car-passenger system in matrix form is given as:

$$\begin{bmatrix} m_{CA} & 0 \\ 0 & m_{PA} \end{bmatrix}\begin{bmatrix} \ddot{y}_{CA} \\ \ddot{y}_{PA} \end{bmatrix}+\begin{bmatrix} c_{CA}+c_{PA} & -c_{PA} \\ -c_{PA} & c_{PA} \end{bmatrix}\begin{bmatrix} \dot{y}_{CA} \\ \dot{y}_{PA} \end{bmatrix}+\begin{bmatrix} k_{CA}+k_{PA} & -k_{PA} \\ -k_{PA} & k_{PA} \end{bmatrix}\begin{bmatrix} y_{CA} \\ y_{PA} \end{bmatrix}=\begin{bmatrix} c_{CA}\dot{y}_{FR}+k_{CA}y_{FR} \\ 0 \end{bmatrix} \tag{3}$$

In order to identify the resonance frequency, a frequency sweep is applied to investigate the forced response of the system. The sinusoidal frequency sweep is applied as

$$y_{FR}(\tau)=y_0 \sin(\Omega \tau) \tag{4}$$

where y_0 is the amplitude and

$$\Omega=\omega/\omega_{nCA} \tag{5}$$

where ω is the angular frequency of the shaker excitation and τ represents the non-dimensional time defined as $\tau=\omega_{nCA}t$.

The non-dimensional parameters are defined as

$$\omega_{nCA}^2=\frac{k_{CA}}{m_{CA}};\ \omega_{nPA}^2=\frac{k_{PA}}{m_{PA}};\ m_r=\frac{m_{PA}}{m_{CA}};\ \omega_r=\frac{\omega_{nPA}}{\omega_{nCA}};\ 2\xi_{CA}=\frac{c_{CA}}{m_{CA}\omega_{nCA}};\ 2\xi_{PA}=\frac{c_{PA}}{m_{PA}\omega_{nPA}} \tag{6}$$

Introducing the non-dimensional parameters and time τ into Eq. (3), computing the Laplace transform and transposing yields the transfer function (the displacement transmissibility)

$$G(\Omega)=\left|\frac{Y_{CA}(\Omega j)}{Y_{FR}(\Omega j)}\right|$$

$$=\left|\frac{(1+2j\xi_{CA}\Omega)(\omega_r^2-\Omega^2+2j\xi_{PA}\omega_r\Omega)}{(\Omega^4-[(\omega_r^2+2j\xi_{PA}\omega_r\Omega)(1+m_r)+(1+2j\xi_{CA}\Omega)]\Omega^2+(1+2j\xi_{CA}\Omega)(\omega_r^2+2j\xi_{PA}\omega_r\Omega)}\right| \tag{7}$$

In the case of an empty car system, Eq. (7) is reduced to a single degree of freedom model and the displacement transmissibility assumes the following form

$$G^1(\Omega)=\left|\frac{1+2j\xi_{CA}\Omega}{1-\Omega^2+2j\xi_{CA}\Omega}\right| \tag{8}$$

Experimental Setup

The experimental rig has been developed to carry out the testing programme. The rig employed a rectangular 110cm by 106cm platform of mass 202 kg fixed on the top of four silent blocks. The platform was then excited by a shaker actuator. The CSI 2130 portable analyzer was used to carry out data acquisition and signal analysis. Two IMI-ICP 626A04 piezoelectric accelerometers, designed for low frequency usage, were fixed to the actuator table and to the rigid base of the car platform. First, the acceleration transmissibility was measured using a harmonic excitation applied to the platform with no passenger load within the frequency range from 1 to 20 Hz. This facilitated the identification of the resonance frequency (denoted as f_r, see Fig. 3) and the damping ratio ξ_{CA} using Eqs. (5) and (8) in order to determine the coefficients of stiffness k_{CA} and damping c_{CA}, respectively. Next, the acceleration transmissibility corresponding to the platform carrying a passenger of mass m_{PA} = 80kg was measured and the experimental data were used to determine the effective coefficient of stiffness k_{PA} and the effective damping coefficient c_{PA} of the passenger by using the MATLAB curve fitting toolbox [7].

Results and Discussions

The curve fitting technique employing the nonlinear least square method together with the Trust-region algorithm is applied to determine the damping and stiffness coefficients both corresponding to the empty car system (no passenger load) and to the car-passenger system with one passenger load. First the method is used to determine the resonance frequency (f_r) and the damping ratio (ξ_{CA}) of the empty car system. The experimental transmissibility measurements for the car system with no passenger load are shown in cross points in Fig. 3. The data is fitted using the displacement transmissibility Eq. (8) together with Eq. (5) for the two unknowns, f_r and ξ_{CA} respectively. The fitted curve is shown in Fig. 3 as a solid line together with two thinner lines representing the fitted curves with 95% observation prediction bounds, which define the lower and upper values of the width of the interval. In other words, this indicates that there is a 95% chance that the observation is actually contained within the lower and upper prediction bounds. In order to determine whether the best fit is achieved, the numerical fits are required to be examined. The curve fitting toolbox in MATLAB supports the goodness-of-fit statistics for parametric models. They are the sum of squares due to error (SSE), R-square and root mean squared error (RMSE). These values for the empty car system are obtained as 0.666, 0.9954 and 0.2961, respectively. The value of 0.9954 of R-square means that the fit explains 99.54% of the total variation in the data about the average. Considering all three values of statistics, it indicates a good fit achieved.

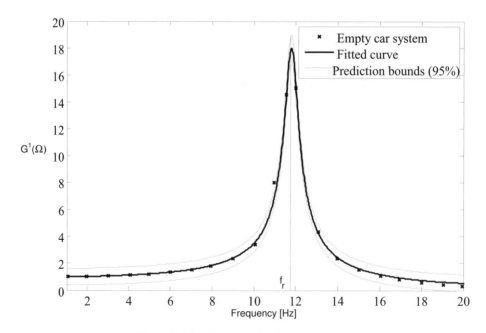

Figure 3. Fitted curves for the empty car system

As the confidence intervals on the results determine the accuracy, the two coefficients are obtained with 95% confidence bounds. The resonance frequency (f_r) is identified as 11.79Hz having the lower and upper bounds 11.77Hz and 11.81Hz with the interval width 0.04. The damping ratio

(ξ_{CA}) is found to be 0.02782 having the lower upper bounds 0.02661 and 0.02903, respectively, with the interval width 0.00242. Having obtained the parameters f_r and ξ_{CA}, the coefficients of stiffness (k_{CA}) and damping (c_{CA}) of the car system are calculated from Eq. (6) as 1108.5kN/m and 832.59kg/s respectively.

The tests are then carried out with the single passenger load on the platform within the frequency range of 1-20Hz. The acceleration transmissibility is measured and the curve fitting technique is applied to the experimental data. Note the same procedure for the curve fitting is employed but Eq. (7) representing the two degree of freedom model taking into account the contribution of the passenger is now applied. The measured values of the transmissibility against the frequency range are shown in cross points in Fig. 4. Having identified the coefficients ξ_{CA} and k_{CA} of the platform from the single degree of freedom model, the damping ratio ξ_{PA} and the stiffness coefficient k_{PA} representing damping and stiffness characteristics of the passenger are determined using the curve fitting toolbox. The coefficients obtained using the data from the full frequency range are shown in Table 1. With the 95% confidence bounds, the stiffness coefficient k_{PA} is identified as 115.6 kN/m with the lower and upper bounds of 105.6 kN/m and 125.5 kN/m, respectively. The error difference is calculated as ±11.24%. The damping ratio is 0.421 with the lower and upper bounds 0.3736 and 0.4684, respectively. The error difference is ±11.26%. The damping coefficient (c_{PA}) of the passenger is calculated using Eq. (6) as 2560.6kg/s.

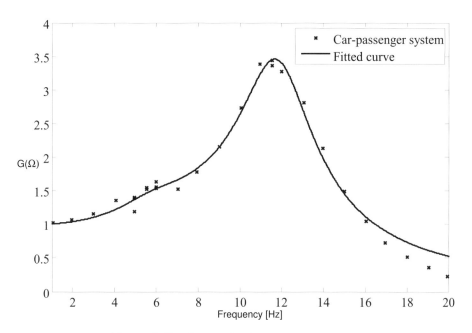

Figure 4. Fitted curve for the car-passenger system

Using reduced experimental data concentrated on the frequency range from 8-14 Hz, i.e. around the second resonance frequency, the two coefficients of the passenger are determined using the curve fitting technique. The results together with the statistics are shown in the third column in Table 1. With the 95% confidence bounds, the stiffness coefficient k_{PA} is 115.9 kN/m with the lower and upper bounds 106.3 kN/m and 125.5 kN/m. The error difference is ±8.28%. The damping ratio ξ_{CA} is determined as 0.4145 with the lower and upper bounds of 0.3692 and 0.4597, respectively. The error difference is $\pm10.93\%$. The damping coefficient (c_{PA}) of the passenger is calculated using Eq. (6) as 2524.3kg/s.

The error difference of the coefficients using the two sets of data is shown in the last column in Table 1, it is evident that the differences are small. Comparing the statistics results of SSE, R-square and RMSE in Table 1, an improved fitted curve is achieved using the reduced frequency range data than the one using the full frequency range. Thus, the parameter identification procedure using the curve fitting based on experimental data within the frequency range reduced to the vicinity of the resonance peak, instead of carrying out the test for the whole frequency range, can be proposed. The tests conducted show that such a reduction of the number of experimental data without increasing the standard deviation considerably lead to better curve fitting results.

Table 1. The curve fitting results of coefficients with the statistics for the car- passenger system

Coefficients	Full frequency range	Reduced frequency range	Difference (%)
k_{PA} [kN/m]	115.6 (105.6, 125.5) ($\pm11.24\%$)	115.9 (106.3, 125.5) ($\pm8.28\%$)	0.2
ξ_{PA}	0.421 (0.3736, 0.4684) ($\pm11.26\%$)	0.4145 (0.3692, 0.4597) ($\pm10.93\%$)	1.57
c_{PA} [kg/s]	2560.6	2524.3	1.44
SSE	0.4334	0.1112	
R-square	0.9807	0.981	
RMSE	0.1291	0.1112	

Conclusions

A damped single degree of freedom model and a damped two degree of freedom model of an elevator car system were used to identify the stiffness and damping characteristics of an elevator car system. With one passenger load on the platform, the passenger stiffness and damping coefficients are determined based on the parameters investigated from the empty platform system. The proposed identification procedure is designed to develop a model to predict the effect of the passenger behaviour on the dynamic response of the elevator car system with good accuracy and repeatability. The model will facilitate a study into the optimal choice of the damping and stiffness characteristics of the elevator car system. The predicted results will be used as benchmarks for further work to develop a software simulation tool for assessing the mechanical behaviour of elevator car systems taking into account the dynamic effects introduced by passenger loads.

References

[1] ISO 2003, Measurement of lift ride quality. ISO 18738: 2003(E), (2003)

[2] I. Herrera and S. Kaczmarczyk: 7th International conference on Modern Practice in stress and Vibration Analysis, Journal of Physics: Conference Series 181 (2009)

[3] I. Herrera: *Elevator Technology*, 17, 152-9, (2008)

[4] I. Herrera: *Elevator Technology*, 17, 160-8, (2008)

[5] C. Silva: *Vibration, Damping, Control and Design*, London CRC Press, (2007)

[6] S. Rao: *Mechanical Vibrations*, Pearson Prentice Hall, (2004)

[7] MATLAB: Curve Fitting Toolbox 2.1, The MathWorks.

Applied Mechanics and Materials Vols. 24-25 (2010) pp 83-88
© (2010) Trans Tech Publications, Switzerland
doi:10.4028/www.scientific.net/AMM.24-25.83

Analysis of Sensors for Vibration and Nip Forces Monitoring of Rubber Coated Rollers

M.C. Voicu[1,2,a], R. Schmidt[1,b], B. Lammen[1,c],

H.H. Hillbrand[1,d] and I. Maniu[2,e]

[1]Faculty of Engineering and Computer Science, University of Applied Science Osnabrück, Albrechtstr. 30, D-49076 Osnabrück, Germany

[2]Faculty of Mechanic, "Politehnica" University of Timisoara, Blv. Mihai Viteazu 1, RO-300222 Timisoara, Romania

[a]c.voicu@fh-osnabrueck.de, [b]reinhard.schmidt@fh-osnabrueck.de,

[c]b.lammen@fh-osnabrueck.de, [d]h-h.hillbrand@fh-osnabrueck.de, [e]inocentiu.maniu@mec.upt.ro

Keywords: piezoelectric sensors, piezoelectric paint, strain gauge flexible rollers, vibration monitoring, printing and coating processes

Abstract

To improve the efficiency of printing or coating processes for paper products the velocity of the web and the roller width can be increased. However, these measures bring about deformations of the rollers, heating effects and streak print defects due to undesirable oscillations. This paper presents new sensor technologies for measuring the axial and circumferential distribution of contact pressure along the nip. This work is required for further research on active vibration damping of rollers. The sensors are applied underneath the elastomer covering of the rollers and must be applied without affecting mechanical features or causing a fall off in the quality of the product. In the paper different new measurement techniques are evaluated and compared to state-of-the-art technologies considering dynamic behaviour, sensitivity, linearity, applicability and accuracy. The sensors are integrated into a test rig simulating the rollers of a printing or coating machine. The results are presented in detail and an outlook is given on further research towards active vibration damping.

Introduction

The work presented in this paper contributes to a research project which aims to enhance the productivity of printing and coating processes at equal or improved quality standards by means of innovative technology. Due to increased roller width and web velocities, the vibrations restrict significantly the quality and efficiency of the printing or coating processes. Approved methods like balancing of the rollers and maximizing the bending stiffness have come to technical limits.

One new approach is online monitoring and optimal adjustment of roller systems. Furthermore active components for vibration damping seem to be promising for optimisation of roller systems. Firstly, active counter oscillations by highly dynamic displacements of the roller bearings by piezoelectric actuators and magnetic bearings are proposed and developed in the project. Fig. 1 shows a roller with piezoelectric actuators in the bearings.

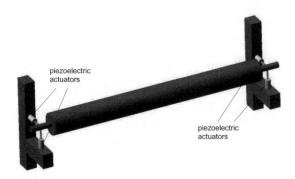

Fig 1: Roller with piezoelectric actuators in the bearings

Measuring the axial and circumferential distribution of contact pressure along the nip is necessary for active vibration damping of rollers. New sensor technologies for measuring the contact pressure distribution both in axial and circumferential direction along the nip are required. The sensors are applied underneath the elastomer covering of the rollers. The measurement results are affected by the pressure transfer through the rubber. Furthermore, the transfer characteristic of the rubber material is influenced by temperature effects in the material. The sensor calibration is a crucial issue and has high relevance for subsequent application in active vibration damping and feedback control.

Moreover, the sensors and the signal conditioning have to be mounted on the rotating rollers. Thus, problems concerning signal transfer and the impact of unbalances caused by the measurement equipment have to be considered.

In the following chapters, different innovative sensor technologies, which are currently under development, are discussed for monitoring and optimal adjustment of roller systems. To test and evaluate the sensor a test rig was set up.

Test rig

The test rig presented in Fig.2 simulates three rollers system of a coating or printing machine and is used to investigate the sensor's efficiency. It contains a pneumatic cylinder (4) which presses the load roller (anilox roll) (1) against the rubber coated roller (plate cylinder) (2) by applying a defined force to (1). Both move to the supporting roller (impression cylinder) (3) until desired contact pressure is achieved. The force sensor (5) is the reference for sensor's calibration.

Fig. 2 Test rig
1-load roller, 2-rubber roller, 3-supporting roller,
4-pneumatic cylinder, 5-force sensor

Sensors

A goal of the research project presented in this paper is to develop a sensor to measure the pressure along the nip during operation. Some of the requirements of the sensor are:

- *Applicability.* The sensor must be thin and applicable to curved surfaces without affecting the quality of printing image in the flexography.
- *Dynamic range.* The sensor should be able to measure the rapid change of the pressure in the nip. The sensor's characteristics must be stable and repeatable with low hysteresis.
- *Sensitivity.* The sensor should be able to detect the contact forces in the nip, by applying the sensor underneath the printing plate made of rubber, plastic, or some other flexible material.
- *Feasibility.* The sensor should be easy to produce, inexpensive and robust.

Different types of sensors have been applied on the test rig:
- underneath the rubber coated roller, as shown in figure 3 are
  a) piezoelectric paint
  b) piezoelectric discs
- on the surface of the load roller
  c) piezoelectric film
  d) strain gauge

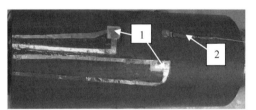

Fig. 3 Applied sensors on the plate cylinder
1 - piezoelectric paint; 2 – piezoelectric disc

The first three sensors contain a piezoelectric material, which creates a measurable charge under force or deformation. Piezoelectric sensors are limited to dynamical measurements because these output signal decay in milliseconds.

a) Piezoelectric paint is a thick-film material used to make dynamic strain sensors to measure vibration [1]. A high quantity of lead zirconate-titanate (PZT) particles 1 μm in diameter was mixed into a water-based paint [2], which can be sprayed or coated on any conductive flat or uneven surface. Successful laboratory tests of the piezoelectric paint have already been realized at the University of Newcastle upon Tyne supervised by Prof. J.M. Hale.

Some problems had to be overcome when applying the water-paint directly on the steel surface. The steel rusted and the paint lost contact. So the piezoelectric paint has been coated by a copper film as shown in fig. 4. The paint creates a dielectric substrate of the piezoelectric sensor, which is actually a plane capacitor. The thickness of the piezoelectric paint is 90 μm, and it is important to achieve a uniform substrate thickness in order to obtain a sensor with a homogeneous sensitivity. The sensor will be poled by applying a high-voltage source onto sensor's wires to orientate the crystal structure of piezoelectric material. Good results are obtained by using a 300 V electrical voltage by a room temperature of 25°C.

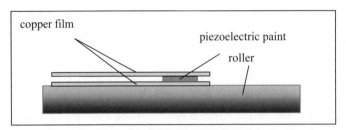

Fig. 4 Piezoelectric paint sensor

b)   Piezoelectric discs are a well known inexpensive technique and form usually actuators by stacking them. The sensors are integrated into the surface of the roller and protected by applying a thin epoxy resin coating [5].

c)   *Piezoelectric film* sensors consist of rectangular piezo-ceramic rods sandwiched between layers of adhesive and electroded polyimide film [6]. They measure distributed solid-state deflection. The film can be a sensor as well as an actuator. It has been applied in circumferential direction on the load roller. The active length is 12 cm, so that only a part of the sensor is pressed in the nip of the rollers.

d)   *Strain gauges* were applied directly on the roller underneath the rubber. Measuring mechanical quantities by strain gauges is a popular and well known technology. It's based on the principle that the resistance of a conductor changes by its stretching or compressing. The presented results come from a sensor applied in circumferential direction.

A comparison of advantages and disadvantages of these sensors are presented in Table 1:

Sensor	*Advantages*	*Disadvantages*
piezoelectric paint	flexible, can be sprayed on any conductive surface, will not affect the quality of printing image	complicate procedure to make a sensor each sensor sensitivity have to be measured is not available on the market no datasheet
piezoelectric disc	inexpensive available on the market	nonlinearity high voltage signals piezoceramics are brittle
piezoelectric film	flexible and very thin available on the market	ceramics can be deteriorated by pressing directly on the sensor
strain gauge	flexible available on the market well known technology	pressing directly on the sensor is no standard application

Table 1 Comparison of sensor's properties

First measurement results with the four different types of sensors are shown in figures 5-7.

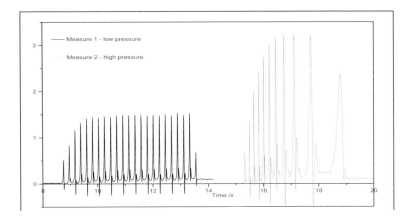

Fig. 5: Piezoelectric paint – the nip pressure is first increasing and then decreasing

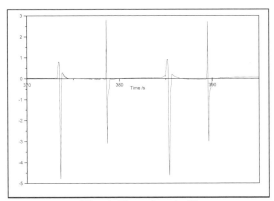

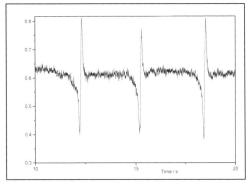

Fig. 6: Piezoelectric film

Fig. 7: Strain Gauge – constant pressure in the nip

These three presented sensors show a good correlation between sensor's signal and the calibrated force sensor.

The most important feature of the sensors for our application is that they have to give a reproducible signal which is depending on the pressure in nip of the rollers. We tested all sensors with different pressure levels. Fig. 5 shows that piezoelectric paint fulfills these requirements in an excellent way. The nip pressure is increasing by bringing the three rotatable rollers in contact and decreasing when the pressure disappears.

The signal of the piezoelectric discs was not repeatable although nip pressure was kept constant. The reason may be that the discs are brittle and cannot be fitted exactly to the roller surface so that there is no defined contact between the rubber coating and the discs. Therefore the results are not presented.

The piezoelectric film sensor also shows good repeatable results and has the advantage that due to its length the nip pressure can be measured for a quite long time so that overlaying vibration (e.g. natural frequencies) can be also measured. Fig. 6 presented measuring results by rotating to the left (negative signal) and by rotating to the right (symmetrical amplitude).

The strain gauge show also repeatable nip pressure dependent results and have the advantage of a simple state-of-the-art technology.

Because it is aimed to apply all presented sensors under the rubber coating and there is no analytical way to determine the sensitivity it is necessary for all sensors to develop a procedure for calibrating the sensors. Durability aspects have to be investigated. There is a lot of experience with strain gages, piezoelectric films and piezoelectric discs in other applications. Therefore no durability problems are to be expected in our application under the rubber coating. The piezoelectric paint, which is very promising for our purpose, is a new technology without experiences concerning durability. Endurance tests in a printing device will be carried out in near future.

Conclusions and outlook

Starting from a set of special requirements for developing an adequate sensor in order to measure the vibration of rubber coated roller some types of sensors are presented in this paper. It could be shown that the piezoelectric paint as well as piezoelectric film and strain gauges show reasonable results and good correspondence with the control measurements with a calibrate force sensor. This is notable, because the sensors were applied under the rubber coating. Solutions have been found for an application of the sensor without affecting the printing image.

The next steps will be the improvement of the new developed sensors and of the calibration procedure for the sensors after implementation in the rubber coated roller. After that these sensors will be integrated into the control loop of the active vibration damping. The actual physical states of the roller measured by the sensors are the feedback information and provide the input signal to a controller which is implemented on a digital signal processor (DSP). The output of the controller drives the actuators, which will be used for realization of active vibration damping of printing or coating roller systems.

References

[1] J. M. Hale, J. R. White, R. Stephenson, F. Liu: *Development of piezoelectric paint thick-film vibration sensors*, Proc. IMechE 2005 Vol. 219 Part C: J. Mechanical Engineering Science

[2] P.N. Raptis, R. Stephenson, J.M. Hale, J.R. White: *Effects of exposure of piezoelectric paint to water and salt solution*, Journal of Materials Science 39 (2004), 6079-6081

[3] U Gabbert, T Nestorovic´, J Wuchatsch: *Methods and possibilities of a virtual design for actively controlled smart systems,* Computers and Structures 86 (2008) 240–250

[4] T Nestorovi´c Trajkov,HK¨oppe and U Gabbert: *Vibration control of a funnel-shaped shell structure with distributed piezoelectric actuators and sensors,* Smart Mater. Struct. 15 (2006)

[5] Information on http://www.smart-material.com

[6] Information on http://www.ceramtec.de

[7] R. Schmidt, H. Wöhrmann: "Grundlagenuntersuchungen zum Vergleich von CFK- und Stahlwalzen in Druckmaschinen", Fachhochschule Osnabrück, 2002, Research Project

Session 3A: Composite and Cellular Materials 1

Applied Mechanics and Materials Vols. 24-25 (2010) pp 91-96
© (2010) Trans Tech Publications, Switzerland
doi:10.4028/www.scientific.net/AMM.24-25.91

Evaluation of edge cracks in cross-ply laminates using image correlation and thermoelastic stress analysis

G.P. Battams[a], J.M. Dulieu-Barton[b] and S.W. Boyd[c]

School of Engineering Sciences, University of Southampton, Southampton, SO17 1BJ, UK

[a]G.Battams@soton.ac.uk, [b]janice@soton.ac.uk, [c]S.W.Boyd@soton.ac.uk

Keywords: Edge cracks, cross ply polymer composite laminate, notched, GFRP, TSA, DIC, thermoelastic stress analysis (TSA), digital image correlation (DIC).

Abstract. The paper describes initial work on using 2D digital image correlation (DIC) and thermoelastic stress analysis (TSA) to obtain data from edge cracks in cross-ply laminates. It is demonstrated that detailed data related to the crack tip stresses can be obtained using TSA. The work reveals some of the limitations experienced when using DIC in applications where high spatial resolution is required. A detailed discussion is provided along with an outline for future work.

Introduction

The increasing use of composites in many crucial engineering applications has placed greater demand on techniques to assess stress concentrations. Damage accumulation in metals is often localised to cracks and can be predicted using various fracture mechanics approaches [1]. In comparison, laminated polymer composite materials display many forms of damage which interact with each another during initiation and propagation. The mechanisms are also dependent on the constituent material and laminate stacking sequence. Matrix cracking is prominent in 90° layers when loaded in tension and occurs well below the failure load of the laminate [2]. Such cracking has a relatively small affect on the stiffness of a structure but does initiate other types of damage such as fibre breakage and delamination. Therefore it is necessary to apply experimental techniques to provide enhanced understanding of the failure mechanisms

Strain measurements using strain gauges or extensometers are useful for obtaining point readings but only provide information at discrete points. Therefore, for complex components full field techniques are more applicable, especially for identification of damage initiation. Two full field techniques that are commonly used are Digital Image Correlation (DIC) and Thermoelastic Stress Analysis (TSA). TSA has been used in relatively few studies on notched or cracked polymer composite specimens [3, 4]. Optical techniques have been used extensively in many materials including metals, ceramics and composites [5]. However, TSA and DIC have not been used in conjunction to obtain data relating to the stress and strain in the component. In the present paper the two techniques will be used to examine the stress/strain distribution produced around an edge notch in a composite specimen. The edge notch provides challenges for both techniques in terms of spatial and temporal resolution. The purpose of the paper is to apply the TSA and DIC techniques to an edge notch in a polymer composite component and to identify the advantages and limitations of using each technique in this application. The motivation is to establish a means of assessing the failure mechanisms at the notch. The failure occurs rapidly and therefore a high speed camera is used with the DIC. It is shown that the TSA provides an excellent validation tool for the DIC.

Experimental Techniques

TSA is a well established technique for providing full-field stress data for orthotropic composite materials [3, 4]. The underlying theory is presented in the review [3]. For an orthotropic material such as a laminated polymer composite the temperature change, ΔT, is related to the change in stress as follows [4]:

$$\Delta T = -\frac{T}{\rho C_P}(\alpha_1 \Delta \sigma_1 + \alpha_2 \Delta \sigma_2) \tag{1}$$

where C_p is the specific heat at constant pressure, T is the absolute temperature on the surface, α denotes the coefficient of thermal expansion and σ is the direct stress. The subscripts 1 and 2 denote the principal material directions.

The infra-red system used in this work is the Silver 480M manufactured by Cedip and has a maximum frame rate of 383 Hz at the full resolution of 320 x 256 pixels. The system is radiometrically calibrated and can deliver values of ΔT directly. In the present work the data will be presented in the form of $\Delta T/T$ to enable data to be compared to other data taken at different room temperatures and also to eliminate the effect of any localised heating close to the damage site. The measured data is T, with ΔT values obtained through performing TSA. This uses the lock-in function, requiring a load reference signal from the test machine used to load the specimens.

Digital Image correlation (DIC) is a technique that utilises standard white light cameras and the contrast of the specimen surface. The correlation was conducted using proprietary software: DaVis by LA Vision. Their correlation algorithm divides the imaged area into a grid of interrogation cells and tracks the movement of each cell in the images collected as a structural test progresses. To enable correlation between the cells, a random surface pattern must be used that has well contrasting pixel grayscale levels; this is achieved using spray paint. The images from the deformed and reference states are correlated, and the displacements obtained. Subsequently the strains are obtained by taking the displacement at the centre of the cell and differentiating over a gauge length defined by the distance between the centres of two adjacent cells. In this paper a single high speed Redlake Motionpro X-3 plus digital camera is used to obtain in-plane 2D deformations and strains. The camera has a maximum frame rate of 2000 Hz at a resolution of 1280 x 1024; higher rates can be achieved if the number of vertical pixels is reduced. The camera stores data internally and has a maximum storage size of 8 GB. The data collection is therefore a compromise between frame-rate, spatial resolution and available recording time.

Specimen manufacture and initial tests

A [90, 0, 90, 0]s sheet of material was produced using Primco UD001/00 unidirectional glass-epoxy pre-impregnated material. The laminate was cured using an autoclave with an applied pressure of 4 bar at 125 °C for one hour. The cured laminate had an approximate thickness of 2 mm. Specimens were cut parallel and perpendicular to the 0° ply to give specimens with 90° and 0° surface layers respectively to specimens of 300 mm long by 25 mm wide. Edge notches were created with a fine piercing saw blade at the centre of the gauge length. The notches have an approximate width of 0.25 mm.

Initial tests were carried out to determine the approximate failure load and extension to failure of the specimens. Unnotched specimens were found to fail at 18.5 kN with 5.53 mm extension, giving a Ultimate Tensile Strength (UTS) of 388 MPa. When a notch of 2 mm was introduced into one side of the specimen, the approximate failure load was found to be 13.0 kN at 3.35 mm extension, giving a critical stress intensity factor of around 34 MPa/mm$^{1/2}$ [1].

Tests were conducted at a ramp rate of 3.33 x 10^{-3} m/s on an Instron 5500 servo-mechanical test machine. During the ramp, white light digital images were collected using the Redlake Motionpro X-3 plus high speed digital camera at a frame rate of 7000/sec. The crack propagated across the width of the specimen in less than a second. Therefore a slower crack growth rate is required if the failure were to be imaged successfully. Decreasing the ramp rate to 0.1 mm/min slowed the crack propagation, but not enough for a fully controlled crack growth. The crack did not grow until the critical stress was achieved and then the crack grew rapidly, increasing in length by 2.5 mm in 0.25 s. Ideally, in the 2D DIC conducted here, the camera would image the deformations at a slower frequency during the initial part of the test to conserve memory and then operate at a higher frequency when the actual failure occurs. However, this camera cannot change frame rate mid test. Furthermore, calculations showed that a wider specimen would be necessary to achieve stable crack growth. It was considered that it is the strain field that occurs prior to crack initiation that is important. Therefore it was decided to concentrate on the imaging of the stress and strain fields of the notched specimens prior to failure rather than during crack growth in the experiments described in this paper.

Thermoelastic stress analysis

A cyclic load of 2 kN with a frequency 10 Hz was applied to both specimen types, i.e. one with a 90° surface ply and the other with a 0° surface, using an Instron servo-hydraulic test machine. During the tests, mean loads of 2, 3, 4, 6 and 10 kN were used to increase the load level in the specimen to initiate failure whilst keeping the background thermoelastic response at the same level. 1080 consecutive infra-red images were taken using a frame rate of 101 Hz with an integration time of 1300 µs. The data were processed so that $\Delta T/T$ was obtained. These are shown in Figure 1 for the 0° surface ply in and Figure 2 for the 90° surface ply.

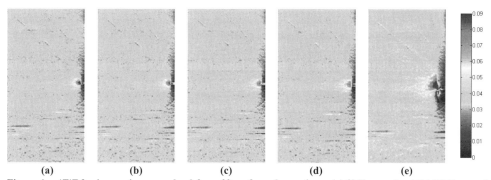

(a)   (b)   (c)   (d)   (e)

Figure 1: $\Delta T/T$ **for increasing mean load for a 0° surface ply specimen (a) 2kN mean load (b) 3 kN mean load (c) 4 kN mean load (d) 6 kN mean load (e) 10 kN mean load (Image area = 1184 mm^2)**

The average $\Delta T/T$ values away from the notch are 0.0375 ±0.0035 K in Figure 1a and 0.0311 ±0.0035 K in Figure 2a. This shows that the response of the surface ply for the same given applied stress is different corresponding to the findings of [4]. As the mean load increases there is little difference in the average $\Delta T/T$ readings for the 0° surface ply i.e. 0.0377 ±0.0034 K for the 3 kN load, 0.0382 ±0.0034 K for 4 kN, 0.0386 ±0.0035 K for 6 kN and 0.0420 ±0.0036 K at the 10 kN load. There is a steady increase in the average value indicating that there is an influence from the damage occurring in the sub surface 90° ply. This is supported by the readings from the 90° surface ply: 0.0314 ±0.0035 K for 3 kN, 0.0339 ±0.0036 K for 4 kN, 0.0322 ±0.0035 for 6 kN and 0.0271 ± 0.0037 K for 10 kN. As damage evolves with increasing mean load, the 0° specimen experiences an increase in thermoelastic response whereas for the 90° specimen the response reduces slightly and when gross damage occurs, reduces significantly. The explanation for this is simple; as the 90° plies fail at lower loads through the creation of transverse cracks, the load carrying capacity is transferred to the 0° plies and hence the response from one decreases and the other increases. The most interesting and revealing difference in the data is the indication of the presence of the notch. For the specimen with the 0° surface ply there is a clear stress concentration at the notch identified by an increasing $\Delta T/T$ value with the increasing mean load. This is very similar to what can be seen in an isotropic material with a crack e.g. [4]. In Figure 1e failure has started and there is a clear indication of longitudinal splitting emanating from the notch tip. For the 90° surface ply, the presence of the notch is indicated only by the reduction in the response above and below the notch where no load is carried. There is a slight increase in response but not at the notch tip. However there are clear indications of transverse cracking between the fibres particularly at the higher mean loads; Figure 2e shows that the transverse cracks have coalesced and the load carrying capability of the ply has diminished significantly. Figures 3 and 4 show plots along a horizontal line through the crack for both specimen types respectively. The peak in response is clearly shown in Figure 3 and is not apparent in Figure 4. There is a difficulty in that the strain in both these specimens is identical and yet the concentration at the crack-tip is not showing in one data set as the response is a result of the surface ply alone.

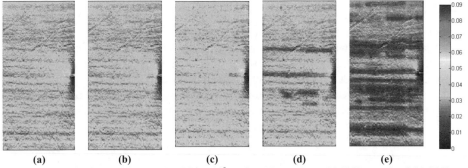

(a)   (b)   (c)   (d)   (e)

Figure 2: $\Delta T/T$ for increasing mean load for a 90° surface ply specimen (a) 2kN mean load (b) 3 kN mean load (c) 4 kN mean load (d) 6 kN mean load (e) 10 kN mean load (Image area = 1115 mm²)

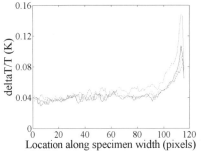

Figure 3: $\Delta T/T$ plot across specimen at notch with increasing mean load for a 0° surface ply specimen

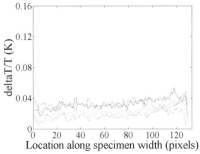

Figure 4: $\Delta T/T$ plot across specimen at notch with increasing mean load for a 90° surface ply specimen

Digital image correlation

0° and 90° surface ply specimens were tested quasi-statically to failure in an Instron 8802 servo-hydraulic test machine. Two ramp rates were used for the test to minimise overall test time and therefore maximise the image frame rate for the given camera storage capacity. Images were captured during the initial ramp of 3.33 x 10^{-5} m/s up to a 3 mm deflection and the final ramp of 3.33 x 10^{-6} m/s to final failure. The camera was set to a frame rate of 30 Hz at a resolution of 1080 x 768 pixels giving a raw image spatial resolution of 0.024 mm. Two 1250 W tungsten lamps were used to illuminate the specimen; load data was simultaneously collected with each image. The speckle pattern was produced by spray paint and the quality of the image was found to be satisfactory based on a histogram of grayscale levels, with a contrast RMS of 176 and an average value of 387 counts. (The manufacturer recommends a minimum of 25 and 80 for RMS and intensity counts respectively.) The images were correlated using the LaVision DaVis DIC software. Initially cell sizes of 64 x 64 pixels were used, as was identified in [7] as providing sufficient strain accuracy with the LaVision software. It should be noted that the resolution of the Motion pro camera is significantly less than those used in [7] for image correlation with low speed data capture. The use of a different camera system prompted a further investigation into the effect of cell size. The LaVision correlation algorithm only allows data to be processed using cell sizes of powers of two; all cell sizes from 4 x 4 to 64 x 64 were examined. All but the 32 x 32 and 64 x 64 contained large amounts of noise. The results for 32 x 32 and 64 x 64 for a 0° and 90° surface ply are shown in Figures 5 and 6 respectively. The data represent the moment before failure as the image before complete specimen failure is used for the correlation. The 90° surface ply specimen failed at a load of 11.3 kN and the 0° surface ply specimen failed at a load of 10.1 kN. It would be extremely difficult to capture this data without a high speed camera. For the 64 x 64 cell size which represented 1.52 mm^2 on the specimen, the strain is uniform but there is no definition around the notch. The 32 x 32 represent a smaller area of 0.76 mm^2. The data has better spatial resolution but more noise. A compromise was found by using 64 x 64 cell size with 75% overlap in the cells; this smoothed the data but enabled much more definition at the notch. The results are shown in Figure 7 for 0° and 90° surface plies respectively.

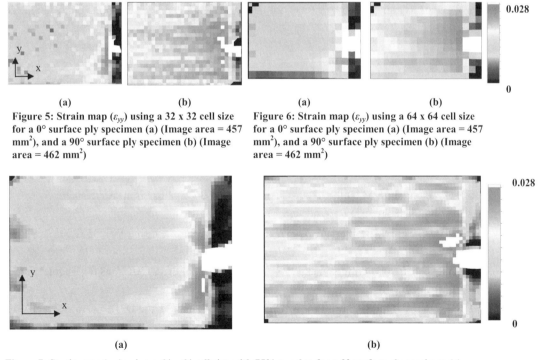

(a)   (b)   (a)   (b)

Figure 5: Strain map (ε_{yy}) using a 32 x 32 cell size for a 0° surface ply specimen (a) (Image area = 457 mm^2), and a 90° surface ply specimen (b) (Image area = 462 mm^2)

Figure 6: Strain map (ε_{yy}) using a 64 x 64 cell size for a 0° surface ply specimen (a) (Image area = 457 mm^2), and a 90° surface ply specimen (b) (Image area = 462 mm^2)

(a)   (b)

Figure 7: Strain map (ε_yy) using a 64 x 64 cell size with 75% overlap for a 0° surface ply specimen (a) (Image area = 457 mm^2) and a 90° surface ply specimen (b) (Image area = 462 mm^2)

Figure 7 shows data related to the longitudinal strain. The TSA data is related to a combination of the principal stresses and coefficients of thermal expansion (see Equation (1)). Therefore a direct comparison of the two is not possible. However it is clear that in Figure 7a there are strain concentrations occurring at the crack and in Figure 7b the data is showing overall ply failure. The resolution of the TSA data is far superior but it does not deliver component direct strains. However, a combination of the two techniques could be used to evaluate the material condition and provide better insight into the material failure.

Conclusions and future work

Some of the limitations and advantages of DIC and TSA have been demonstrated in this paper with several useful properties being confirmed for cross-ply laminates. The decreasing load carrying capacity of a 90° surface layer has been shown through the formation of transverse cracks and inversely the increased stress and consequently a more dramatic stress concentration effect at the notch with a 0° surface layer. Spatial resolution and internal camera storage capacity have been found to be a key limitation for the high speed camera system used. However, it has been clearly demonstrated that the high speed DIC can capture the strain state just prior to material failure. A direct comparison of strain derived from the DIC using high resolution cameras at each load step used in the TSA is the next step in the process. By following the same load steps as in the TSA experiments and calibrating the thermoelastic response with this material, the stress patterns could be compared directly.

Acknowledgements

The work described in this paper is supported by the UK Engineering and Physical Sciences Research Council (Grant number EP/G042403/1) and DSTL under a joint grant managed programme on damage tolerance.

References

[1]   T. L. Anderson: *Fracture Mechanics Fundamentals and Applications,* (Taylor and Francis Group Publishing, 2005).

[2]   J. F. Caron and A. Ehrlacher, Composites Science and Technology Vol. 57 (1997), p. 1261-1270.

[3]   J. Dulieu-Barton and P. Stanley, Journal of Strain Analysis for Engineering Design Vol. 33 (1998), p. 93-104.

[4]   R. K. Fruehmann, J. M. Dulieu-Barton and S. Quinn, Journal of Strain Analysis for Engineering Design Vol. 43 (2008), p. 435-450.

[5]   M. Grédiac, Composites Part A: Applied Science and Manufacturing Vol. 35 (2004), p. 751-761.

[6]   M. Zanganeh, R. A. Tomlinson and J. R. Yates, Journal of Strain Analysis for Engineering Design Vol. 43 (2008), p. 529-537.

[7]   D. Khennouf, J. Dulieu-Barton, A. R. Chambers, F. J. Lennard and D. D. Eastop, Strain Vol. 46 (2010), p. 19-32.

Applied Mechanics and Materials Vols. 24-25 (2010) pp 97-102
© (2010) Trans Tech Publications, Switzerland
doi:10.4028/www.scientific.net/AMM.24-25.97

Mechanical behavior of syntactic foams for deep sea thermally insulated pipeline.

D.Choqueuse[1, a], P.Davies[1,b], D.Perreux[2,c], L.Sohier [3,d], J-Y Cognard [3,e]

[1] IFREMER, Brest Centre, F-29280 Plouzané, France

[2] Ste Mahytec, 210 Avenue de Verdun, 39100, Dole, France

[3] LBMS, ENSIETA –UBO, 2 rue F Verny, 29806 Brest, France

[a]dominique.choqueuse@ifremer.fr, [b]peter.davies@ifremer.fr, [c]dominique.perreux@mahytec.com, [d] laurent.sohier@univ-brest.fr, [e] jean-yves.cognard@ensieta.fr

Keywords: syntactic foam, hydrostatic compression, shear test.

Abstract.

Ultra Deep offshore oil exploitation (down to 3000 meters depth) presents new challenges to offshore engineering and operating companies. Flow assurance and particularly the selection of insulation materials to be applied to pipe lines are of primary importance, and are the focus of much industry interest for deepwater applications. Polymeric and composite materials, particularly syntactic foams, are now widely used for this application, so the understanding of their behavior under extreme conditions is essential. These materials, applied as a thick coating (up to 10-15 cm), are subjected in service to:

- high hydrostatic compression (up to 30 MPa)

- severe thermal gradients (from 4°C at the outer surface to 150°C at the inner wall),

and to high bending and shear stresses during installation. Damageable behavior of syntactic foam under service conditions has been observed previously [1] and may strongly affect the long term reliability of the system (loss of thermal properties).This study is a part of a larger project aiming to model the in-service behavior of these structures. For this purpose it is important to identify the constituent mechanical properties correctly [2, 3]. A series of tests has been developed to address this point, which includes:

- hydrostatic compression

- shear loading using a modified Arcan fixture

This paper will describe the different test methods and present results obtained for different types of syntactic foams.

Introduction.

Syntactic foams are now widely used as passive thermal insulation material for deep sea pipeline and components (Figure 1). They are made by incorporating hollow glass microspheres (150μm > Ø > 10μm, wall thickness around 1.5 μm) in polymers (PU, PP, Epoxy …). Their composition confers good performance in terms of thermal insulation and hydrostatic compression (nominally $0.18 > k_{W/(m.K)} > 0.12$ and max pressure $> 30_{MPa}$). They are applied using different manufacturing process including co extrusion, lateral extrusion, molding….During the lifetime of the structure (installation, in service, …) high mechanical loading can be applied to the material and good knowledge of the mechanical response of such materials is required in order to guarantee long term reliability of the insulated structures

  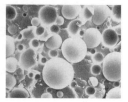

Figure 1 : Thermal insulation structure (doc Socotherm)

Uniaxial compression properties of syntactic foams have already been addressed by different authors [4,5]. In a recent experimental study [6] X ray microtomography has been used to visualize and describe the behavior of these materials during uniaxial loading and confined compression. However all these tests are polluted by edge effects, and a correct description of initial damage appears difficult. In order to model the behavior of the structure during its life time properly it is important to propose mechanical tests giving access to intrinsic properties of the material

Hydrostatic compression

Very few methods are available to determine the hydrostatic compression behavior of materials and "in house" tests are generally used to provide information. The principle is generally to follow the displacement of the piston of a cylinder in which the sample to be tested is placed. The pressure/piston displacement curve can provide a collapse pressure identified by a change of slope of the curve.

In order to improve this characterization Ifremer has developed a specific method based on the direct measurement of the buoyancy of a sample during pressure increase (Figure 2).

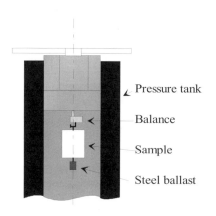

Pressure tank

Balance

Sample

Steel ballast

Figure 2 : Hydrostatic compression test

This test has been developed for materials with density lower than 1000 kg/m^3 but can be adapted to materials with higher specific gravity. It is based on the use of a weight sensor (balance: load range 1N) which can operate under high hyperbaric pressure (up to 100 MPa).

In order to increase the sensitivity of the measurement the procedure consists of following the evolution of the buoyancy of both the sample and the steel ballast, and knowing the specific gravity of the material at atmospheric pressure it is possible to determine the specific gravity of the material versus the pressure. In addition, knowing the evolution of the density of water with pressure [7] it is possible to determine accurately, from the evolution of the buoyancy of the material, the evolution of the volume of the sample versus pressure and then to determine intrinsic parameters such as bulk modulus and collapse pressure.

A material sample, volume around 1 dm^3, is placed below the balance with steel ballast whose weight is calculated in order to give to the ensemble (sample + ballast) a small positive weight in water. All the system is placed in a hyperbaric tank and the pressure is increased by means of an electric pump.

Taking into account the accuracy of the system measurement and of the input data, the accuracy of results obtained for the evolution of buoyancy and bulk modulus can be evaluated : for a 1 dm^3 volume specimen, this is around 1%.

Water uptake of samples during the test can affect the results for bulk modulus. However, the duration of testing, about 1 hour, leads to very little water ingress. Nevertheless systematic weighing of the sample after the test gives an indication of the quantity of water absorbed during the test, to check that it is related to surface absorption only.

Typical curves from the tests are reported below (Figure 3).

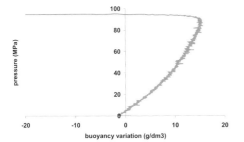 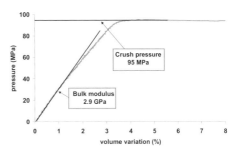

Figure 3: Hydrostatic compression behavior

The curve of variation of volume versus pressure gives access to the intrinsic mechanical behavior of the material. The global behavior of the materials can be characterized by a first linear elastic response, where the bulk modulus of the material is identified (2.9 GPa), and the collapse pressure (95 MPa), called "crush pressure".

A comparison has been made between two materials, Mat A and Mat B, having the same grade of spheres and two different matrix polymers: a rigid matrix (epoxy), Mat A and a soft matrix (PU), Mat B.

It can be noted (Figure 4a) that the initial bulk modulus of the two materials is comparable and that the crush pressure of the material with rigid matrix is significantly higher than for the material with the soft matrix (90 MPa compared to 50 MPa). Up to 80 % of the crush pressure the material with the rigid matrix presents an apparent linear elastic behavior, in strong contrast to the material with the soft matrix which presents a non linear behavior from the beginning of the test. The hydrostatic compression behavior appears to be strongly governed by the nature of matrix.

A study has been made on a material (Mat C: epoxy matrix) at different temperatures (4°C, 50°C and 100 °C) (Figure 4b). This shows a strong influence of the temperature. It can be assumed that the temperature, in terms of micromechanical behavior, mainly affects the behavior of the matrix.

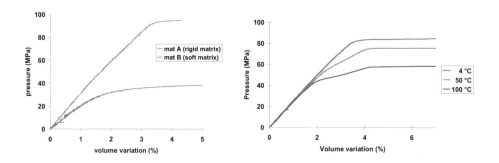

Figure 4: Hydrostatic compression behavior
a – effect of the matrix rigidity b- effect of the temperature

Shear test

To complete the information provided by the hydrostatic compression test it is important to evaluate the shear behavior of the material. For this type of material different solutions have been envisaged: torsion tests on cylinders, Iosipescu test... In a first step, and taking into account the ease of sample preparation (difficulty to machine cylindrical samples) the Iosipescu test has been retained, and tests have been performed on materials A and B (Figure 5). Different material thicknesses (3 and 10 mm) have been tested. These tests provide an apparent shear modulus, but for the rigid matrix the failure appears to be sudden while for soft matrix the strain is too high to be able to reach the failure.

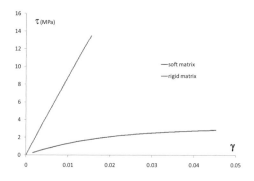

Figure 5: Iosipescu shear test results

In order to address properly non linearity and failure of these materials, which were clearly not obtained with the Iosipescu test, shear tests using the modified Arcan fixture have been evaluated, Figure 6. The Arcan test was initially developed [8] to produce uniform plane stress and a modified version has been used recently, associated with image analysis, to characterize bonded joints [9]. Different geometries of samples have been investigated and finally $0.5*7,6*66mm^3$ specimen bonded on aluminium support have been selected in order to minimize the effect due to the appearance of normal stress.

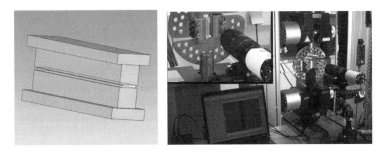

Figure 6: modified Arcan shear test set up

Non linear behavior of the material is shown (Figure 7). It can be noted that the response of the material can be observed right up to cracking initiation, noted around 120% strain for material B. The shear modulus obtained on material A (rigid matrix) is comparable (800 MPa) to the modulus obtained through the Iosipescu test whereas the results obtained on material B are significantly different.

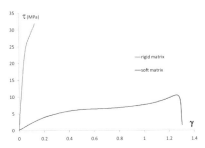

Figure 7: modified Arcan shear test - material behavior comparison

Tests were performed at different loading rates on material B (Figure 8). High sensitivity to loading rate is highlighted, which reveals a viscoelastic/ plastic behavior of the material.

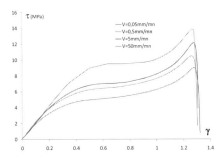

Figure 8: modified Arcan shear test - material B - effect of loading rate

Conclusions

Syntactic foams exhibit complex mechanical behavior (damageable - visco - elasto-plastic behavior) and their characterization needs a careful experimental approach in order to obtain valid data.

Two types of tests are proposed in this paper.

- Hydrostatic compression test. These are based on the measurement of apparent buoyancy during pressure increase, gives pertinent data allowing characterization of the material.
- Shear tests. For these, improvement of the methods is still necessary, but an adapted Arcan fixture appears to be very promising.

This experimental study will be completed by multi axial tests using a specific device allowing mechanical loading to be performed under hydrostatic compression loading [10], in order to cross-correlate the results obtained from the different test methods.

References

[1] F.Grosjean, N.Bouchonneau, D.Choqueuse, V.Sauvant-Moynot "Comprehensive analyses of syntactic foam behaviour in deep water environment", J Mater Sci (2009) 44:1462–1468

[2] D.Choqueuse, et al, "Recent progress in analysis and Testing of insulation and buoyancy materials", Composite Materials and Structure for offshore application CMOO-4, 2004

[3] D.Choqueuse, et al, "Modeling approach for damageable mechanical behaviour of glass/polymer syntactic foams", Syntactic and Composite Foams II, 2007.

[4] N. Gupta, Kishore, E.Woldesenbet, S.Sankaran, et al, "Studies on compressive failure features in syntactic foam material", J Mater Sci (2001) 36:4485–4491

[5] M.Koopman, K.K. Chawla, B.Carlisle, "Microstructural failure modes in three-phase glass syntactic foams", J Mater Sci (2006) 41:4009–4014

[6] J.Adrien, E.Maire, N.Gimenez, V.Sauvant-Moynot, Experimental study of the compression behaviour of syntactic foams by in situ X-ray tomography, Acta Materialia 55 (2007) 1667-1679

[7] UNESCO, 1983, "Algorithms for computation of fundamental properties of seawater", *Unesco technical papers in* marine science

[8] M. Arcan , Z. Hashin, A. Voloshin "A method to produce uniform plane-stress states with application to fiber-reinforced materials. Experimental mechanics 18 (1978): 141-146

[9] P. Davies , L. Sohier b, J.-Y. Cognard, A. Bourmaud, D. Choqueuse, E. Rinnert, R. Créac'hcadec, "Influence of adhesive bond line thickness on joint strength", International Journal of Adhesion & Adhesives 29 (2009) 724–736

[10] Cartié D, Davies P, Peleau M, Partridge I, "the influence of hydrostatic pressure on the interlaminar fracture toughness of carbon/epoxy composites", Composites Part B: Engineering Volume 37, Issues 4-5 , (2006), Pages 292-300

Applied Mechanics and Materials Vols. 24-25 (2010) pp 103-108
© *(2010) Trans Tech Publications, Switzerland*
doi:10.4028/www.scientific.net/AMM.24-25.103

Analysis of the strain and stress fields of cardboard box during compression by 3D Digital Image Correlation

J. Viguié[1,a], P.J.J. Dumont[1,b], P. Vacher[2], L. Orgéas[3], I. Desloges[1], E. Mauret[1]

[1]CNRS / Institut Polytechnique de Grenoble (Grenoble INP), Laboratoire de Génie des Procédés Papetiers (LGP2), 461 rue de la Papeterie, BP 65, 38402 Saint-Martin-d'Hères cedex, France

[2]Université de Savoie, Laboratoire SYMME, Polytech Savoie, Domaine Universitaire, BP 80439, 74944 Annecy-le-Vieux, France

[3]CNRS / Institut Polytechnique de Grenoble (Grenoble INP), Laboratoire Sols-Solides-Structures-Risques (3S-R), BP 53, 38041 Grenoble cedex 9, France

[a]jeremie_viguie@yahoo.fr, [b]pierre.dumont@grenoble-inp.fr

Keywords: G-flute corrugated board, buckling behaviour, digital image correlation, strain and stress fields measurement

Introduction

Corrugated boards with small flutes appear as good alternatives to replace packaging folding boards or plastic materials due their small thickness, possibility of easy recycling and biodegradability. Boxes made up of these materials have to withstand significant compressive loading conditions during transport and storage. In order to evaluate their structural performance, the box compression test is the most currently performed experiment. It consists in compressing an empty container between two parallel plates at constant velocity. Usually it is observed that buckling phenomena are localized in the box panels, which bulge out during compression [1]. At the maximum recorded compression force, the deformation localises around the box corners where creases nucleate and propagate. This maximum force is defined as the quasi-static compression strength of the box. The prediction of such strength is the main topic of interest of past and current research works. For example, the box compression behaviour of boxes was studied by Mc Kee et al. [2] and Urbanik [3], who defined semi-empirical formula to predict the box compression strength, as well as by Beldie et al. [4] and Biancolini et al. [5] by finite element simulations. But comparisons of these models with experimental results remain rather scarce and limited.

2D Digital Image Correlation method was used to study the behaviour of a box panel [6] subjected to compression. However this method does not allow the measurement of the out-of-plane displacement which is significant when the panel buckles. It is now possible to measure the 3D displacement field and the surface strain field of any 3D object by the digital image stereocorrelation technique (DISC or 3D DIC) [7,8]. Hence, the objective of this paper is to revisit the work of [6], by using the 3D DIC technique to better estimate both the 3D displacement field and the in-plane strain field of box panels during the box compression. Furthermore, combined with the elastic orthotropic properties of the outer liner (determined elsewhere), such measurements allows the estimation of the stress field on the surface of the panels.

Materials, specimens and Experimental Procedure

G-flute corrugated board and box manufacturing. The material used in this study is a double-faced corrugated board the flutes of which have a so called G profile. Their typical dimensions are shown in Fig. 1(a). This sandwich structure has a nominal thickness e of 0.77 mm and an average basis weight of 450 g/m². It is composed of a 160 g/m² outer liner made up of chemical pulp (kraftliner), a 110 g/m² corrugated medium essentially made up of recycled pulp and a 140 g/m² inner liner made up of a mix of pulps (testliner). These papers are glued together using a starch-based adhesive. Tested boxes are cubic with dimension $73 \times 73 \times 73$ mm^3. As shown in Fig. 1(b), the box blank includes panels, flaps (panels covering the box), folding lines and a manufacturer joint. The crease depth was chosen equal to 1 mm. Once the box was erected and glued, the inner and outer flaps covered completely the top and bottom box surfaces. Notice that the flutes are vertical in the box panels: this means that the cross manufacturing direction (CD) of the corrugated board is aligned along the compression axis.

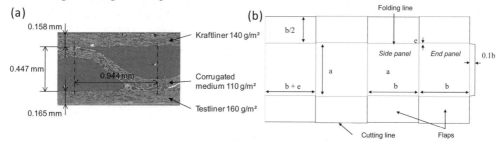

Figure 1. (a) 2D Micrograph showing the cross-section of a G-flute profile corrugated board (obtained from a 3D microtomography volume imaged at the European Synchrotron Radiation Facility (ESRF) on ID19 beamline). Note that the uniform grey zones correspond to voids. (b) Box blank.

Compression tests. Compression tests were performed with an Instron 5569 testing device at a constant compression velocity of 13 mm/min along the $\underline{e}_{y'}$ direction (*cf.* Fig. 2(a)). The compression plates are made up of polished aluminium. A photograph of a box, which is placed between the compression plates before the test, is shown in Fig. 2(a). During these experiments, the compression force F and the axial plate displacement Δa were on-line recorded. This permitted to calculate the following macroscopic box axial stress Σ simply defined as

$$\Sigma = \frac{F}{S_0},\tag{1}$$

where $S_0 = 4e(b-e)$ is defined arbitrarily as the surface of the box cross-section perpendicular to the compression axis $\underline{e}_{z'}$. Similarly, a macroscopic strain ε is defined as

$$\varepsilon = \ln\left(\frac{a+\Delta a}{a}\right).\tag{2}$$

3D Digital Image Stereocorrelation technique. Two of the external panel surfaces were coated with a random pattern made of small black speckles (0.1 to 0.3 mm) obtained spraying a black paint. In order to estimate the 3D displacement field of these two surfaces, the stereocorrelation technique was used [8]. Here, this was performed with the DIC Software 7D [9]. During the compression test, two images of the deforming panels and of their superimposed speckle patterns are recorded simultaneously and sequentially by using, for stereovision, two CCD cameras (3872 $\times$ 2592 pixels), whose relative position and orientation (α angle) are known (cf. Fig. 2(c)) using a calibration procedure [10]. Then, the correspondence between two image points belonging to the pair of simultaneously recorded images was identified by applying an image correlation technique using square windows of 10 $\times$ 10 pixels and subsets of 10 $\times$ 10 pixels centred around the corners of the windows. This stage requires the minimization of a correlation function and involves interpolating the grey levels of the left and right images using bilinear interpolation functions. The error can be estimated to be lower than 0.1 pixel. Then, the 3D coordinates x', y' and z', of a physical point can be computed by triangulation. At each stage of the sequence, the principle of the surface reconstruction is similar. This allows the reconstruction of the whole surface of the studied panels, Fig. 2(d) shows this surface at the maximum compression stress (red point in Fig. 2(b)). The 3D displacement field $\underline{u}(x',y',z') = u'\underline{e}_{x'} + v'\underline{e}_{y'} + w'\underline{e}_{z'}$ of all points of the panels is measured by analysing a sequence of pair of stereo images using the image of the panels taken by the left camera in their initial configuration (before compression) as a reference. As an example, Fig. 2(e) shows the out-of-plane displacement w' of the surface of the two studied panels in the coordinate system (x', y', z') of the left camera. Finally, in this study, the displacement field $\underline{u}(x, y, z) = u\underline{e}_x + v\underline{e}_y + w\underline{e}_z$ was expressed in the panel local coordinate system (x, y, z) (Fig. 2(f)). Thereafter, the components E_{xx}, E_{yy} and $2E_{xy}$ of the Green-Lagrange strain tensor $\underline{\underline{E}}$ can be estimated. The spatial partial derivatives are calculated by using a standard centred finite difference scheme. For the calculation of the strain field, the displacement field was used in its raw form and in a smoothed form (Matlab smooth function). Fig. 2(g-h) were obtained respectively. It is worth noting that the smoothed displacement field allows a less «noisy» strain field to be obtained at the macroscopic scale. The information contained in the smooth displacement field appeared us to be satisfactory for the rest of the analysis.

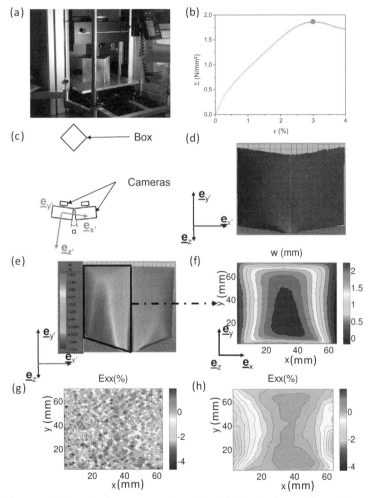

Figure 2. (a) Picture of a box in the compression setup. (b) Typical experimental curve of the box compression stress with respect to the box compression strain. (c) Scheme of the imaging setup (top view). (d) Reconstructed 3D surface (final state). (e) Component w' of the displacement field of the two studied panels in the (x', y', z') coordinate system. (f) Displacement w of the panel with outer flaps (see the arrow) in the panel coordinate system (x, y, z). (g) Component E_{xx} of the displacement field (obtained from the raw displacement field). (h) Same component when the displacement field is smoothed.

Results

Full displacement field of box panels. In this section, the stereocorrelation technique was used to describe the displacement field of a panel with outer flaps at one compression stage corresponding to macroscopic strain $\varepsilon = 3\%$. The corresponding stress-strain curve is given in Fig. 2(b). As shown in this figure, the compression strain $\varepsilon = 3\%$ corresponds to the box compression critical strain (this point is showed by a red disk in the figure). The components u, v and w of the displacement field are displayed in Fig. 3.

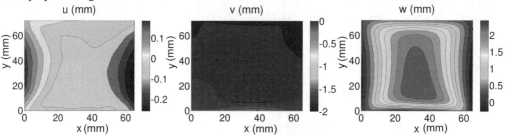

Figure 3. Maps of the components u, v and w of the displacement field of the panel with outer flaps from box with dimensions $73 \times 73 \times 73$ mm^3 at the critical strain ($\varepsilon = 3\%$).

Firstly, it is worth noting that the displacement components are largely heterogeneous. The out-of-plane displacement w and the in-plane displacement v are of the same order and these two components are largely higher than the in-plane displacement u. This indicates clearly that the panel buckles during the compression. The maximum values of the out-of-plane component are obtained for a large zone, centred below the panel centre at the critical strain, as the values of w close to the box edges remained nearly equal to zero or were even slightly negative. Thus, high gradients for this component were found when going through the panel from its edges to the previous central zone, especially along the y-direction. The component v of this displacement field was found to be of the same order as w. But quite astonishingly, it can also be observed that this component is nearly homogeneous over the whole panel surface. This means that the panel exhibits a vertical rigid motion during the box compression. Its origin might be related to both the motion and the compression of the outer flaps joined to the panel, and more presumably to the crush of the junction scores during the compression experiment. Unfortunately, the scores are outside the correlation zones. This point could be improved in future works. The component u of the displacement field was observed to be small compared to v and w, but clearly shows that the vertical edges of the panel do not remain straight and tend to move towards the panel centre as the compression strain increases.

Strain field of box panels. By analysing closely the components of the Green-Lagrange strain tensor, it appeared that the contribution of some spatial gradients of u, v and w could be neglected. This permits to write the strain components E_{xx}, E_{yy} and $2E_{xy}$ adopting the following reduced expressions, which correspond to a von Kármán-type strain field:

$$E_{xx} \approx \frac{\partial u}{\partial x} + \frac{1}{2}\left(\frac{\partial w}{\partial x}\right)^2 ; \ E_{yy} \approx \frac{\partial v}{\partial y} + \frac{1}{2}\left(\frac{\partial w}{\partial y}\right)^2 ; \ 2E_{xy} \approx \frac{\partial u}{\partial y} + \frac{\partial v}{\partial x} + \frac{\partial w}{\partial x}\frac{\partial w}{\partial y}.$$

Figure 4 shows the components E_{xx}, E_{yy} and $2E_{xy}$ of the strain field at the critical compression strain $\varepsilon = 3\%$. Whatever these components, their values are almost equal to zero in the region of the panel centre. In contrast, along the panel edges the strain variations are more pronounced. The variations of E_{xx} are particularly important along the vertical edges of the box, whereas the variations of E_{yy} appear mainly along the horizontal edges of the panel. In both cases, these strain components have negative values, characteristic of a compressive deformation state. The maximum compressive strains are attained at the middle of the edges, and gradually decrease as the corner region is reached. It is also interesting to notice that the compressed zones along the x-axis are wider than along the y-axis, despite this direction corresponds to the box compression axis. The compressive strain E_{yy} is indeed more localized than the E_{xx} one.

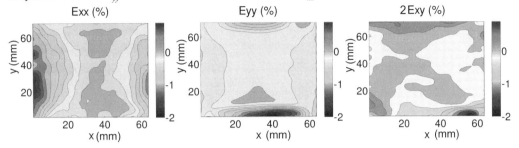

Figure 4. Components of the Green-Lagrange strain tensor at $\varepsilon = 3\%$.

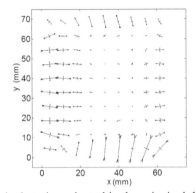

Figure 5. Map of the principal strains oriented in the principal directions of the strain tensor.

Figure 5 shows the magnitude of the principal strains and the orientation of the principal directions. It can be noticed that the principal directions almost correspond to the loading coordinate system in the zones located along the edges, whereas it is not the case in the corner regions. The regions along the vertical edges are compressed in the two directions, but the most significant values are recorded perpendicularly to the compression direction. Along the horizontal edges, only one principal strain is not equal to zero and is higher than those recorded along the vertical edges. This corresponds to a complex strain situation, which might be related to a common influence of the phenomena of compression of the flaps and crush of the horizontal scores, as well as of the neighbouring panels.

Stress field of box panels. The elastic orthotropic properties of the outer liner were determined in a previous work [11]. Combined with the strain field measurements, they allow the estimation of the in-plane stress along the compressed panels, as shown in Figure 6. The evolution of the computed stress field is very similar to that obtained for the strain field. Due to the simplicity of the used elastic model, the stress values are clearly overestimated in the regions close to the edges. But they show that the yield limit is largely exceeded in theses regions. This corresponds to an irreversible damage process of the microstructure of paper. Currently, a Tsai-Wu criterion for the elastic is being built in order to determine where the elastic limit is reached and exceeded in the panel during the box compression.

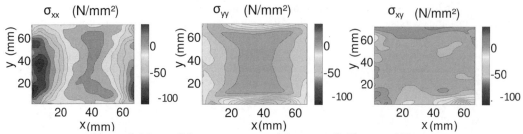

Figure 6. Maps of the components of the stress field at $\varepsilon = 3\%$.

Conclusion

A 3D Digital Image Stereocorrelation technique method was used to analyse the buckling behaviour of box panels during their compression. This technique is highly efficient to provide relevant data on the 3D displacement and strain fields at the surface of the box panels. The analysis of the 3D displacement field provides information on the buckling behaviour of the studied panel. This permitted also to highlight the influence of scores and flaps on this phenomenon. The analysis of the strain field shows that compressive strain states are mainly found to be compressive along the box edges, whereas the central zone is nearly not loaded during the compression. The analysis of the stress field reveals that the regions along the vertical edges are the most stressed zones and emphasizes that the boundary conditions have to be accurately determined in order to predict the buckling behaviour of box panels. This analysis is one part of a large work which investigates the effect of residual stresses, box erection, as well as box compression on the stress field calculated in the outer liner. The effect of box geometries was also investigated. Finally, it provides data that can be used as input for post-buckling analytical or numerical models in order to optimize the box design.

Acknowledgments

The authors would like to thank Cartonneries de Gondardennes (Wardrecques, France) for supplying corrugated boards and the long-term research program "Heterogeneous Fibrous Materials" of the European Synchrotron Radiation Facility (Grenoble, France).

References

[1] J. Viguié, P.J.J. Dumont, I. Desloges, E. Mauret: Packag. Technol. Sci. (2010), on line

[2] R. C. Mc Kee, J. W. Gander, J.R. Wachuta: Paperboard Packag. Vol. 48(8) (1963) p. 149.

[3] T.J. Urbanik: Develop. Validat. Appl. Inelastic Method. Struct. Anal. Des., Vol. 343 (1996) p. 85

[4] L. Beldie, G. Sandberg, L. Sandberg: Packag. Technol. Sci., Vol. 15 (2001), p. 1

[5] M.E. Biancolini, C. Brutti: Technol. Sci., Vol. 16 (2003), p. 47

[6] J.L. Thorpe and D. Choi: Tappi Journal, Vol 75(7) (1992), p. 155

[7] L. Meunier, G. Chagnon, D. Favier, L. Orgéas, P. Vacher: Polym. Test., Vol. 27 (2008), p.765

[8] J.-J. Orteu: Optics and Lasers in Engineering, Vol. 47 (2009), p. 282

[9] P. Vacher, S. Dumoulin, F. Morestin, S. Mguil-Touchal: Instn Mech Engrs Part C ImechE, Vol. 213 (1999), p. 811

[10] T. Couderc, Phd Thesis, Université de Savoie, Annecy, France (2005)

[11] J. Viguié, Phd Thesis, Institut polytechnique de Grenoble (Grenoble INP), Grenoble, France (2010)

Applied Mechanics and Materials Vols. 24-25 (2010) pp 109-114
© (2010) Trans Tech Publications, Switzerland
doi:10.4028/www.scientific.net/AMM.24-25.109

Correlation between full-field measurements and numerical simulation results for multiple delamination composite specimens in bending

C. Devivier[1&2,a], D. Thompson[2,b], F. Pierron[1,c] and M.R. Wisnom[2,d]

[1]Laboratoire de Mécanique et Procédés de Fabrication, Arts et Métiers ParisTech, Centre de Châlons-en-Champagne, Rue St Dominique, BP 508, 51006 Châlons-en-Champagne Cedex, France

[2]Advanced Composites Centre for Innovation and Science, Queen's Building, University of Bristol, University Walk, Bristol, BS8 1TR, United Kingdom

[a]devivier@ensam.fr, [b]daniel.thompson@bristol.ac.uk, [c]fabrice.pierron@chalons.ensam.fr, [d]m.wisnom@bristol.ac.uk

Keywords: Damage, full-field, strains, deflectometry, composite

Abstract. This paper studies the effect of delaminations on strain maps for a simple cantilever beam. The aim is to build an experimental set-up which allows detecting very slight modifications in the strain maps. The case studied is a single delamination on the mid-plane. The measurement method is the deflectometry technique which enables direct slope measurements on a reflective specimen. The comparison with finite element models clearly indicated that the surface strains bear the information of the extent of the delamination. The second step is to use these surface strains to identify a stiffness reduction map for real impact damages.

Introduction

In aeronautical engineering, many surfaces can be assimilated to beam- or plate-like structures like the wings or the blades. In these structures, damage like delamination affects strongly the global behaviour and especially its stiffness hence its remaining life. Therefore, detecting this damage is a crucial point for safety reasons. A large amount of work has been done on this subject. Among the studied techniques, one can find natural frequency changes, mode shapes (changes or derivatives) and modal damping. However, as damage can be slight, the modification of these indicators may not be enough [1]. A summary of the vibration property features applied to damage identification algorithms was given in [2].

As the modifications in the fields caused by the inserted defects are very slight, monitoring of the variation of strains is preferred to that of displacements, so techniques such as curvature mode shape, modal or spectral strain energy and strain frequency response function can be used. All of them use vibrations, but it is also possible to use static techniques such as interferometric methods, digital image correlation (DIC) or deflectometry. Interferometry has a very good resolution but it is very sensitive to any disturbances such as vibrations [3]. DIC requires two differentiations to obtain the strains, therefore the noise will be strongly emphasized. The deflectometry technique requires only one differentiation and is insensitive to vibrations. This paper describes a correlation between numerical and experimental strain fields in bending for a cantilever beam. In the first part the set-up of the measurement post-processing is explained, then the numerical model is described and finally the results are presented and discussed.

Experimental set-up and strain extractions

The deflectometry is based on the specular light reflection principle and its set-up is shown in Fig.1. The specimen is a cantilever beam loaded with 5.5N on the free end as shown in Fig. 2. The distance h is 1.23m, the grid pitch of the reflected grid is 1mm and the samples are 250mm long, 50mm wide and 4mm thick. With these parameters, 8 pixels described exactly one period of the grid. The pictures were taken by a 12Mpixels Nikon D90 camera with a Nikkor 105mm lens. A noise analysis done with this set-up gave a resolution of $2.5 \ 10^{-6}$ rad. The resolution is the standard deviation of the difference between two phase maps at rest, multiplied by the sensitivity. Also because of the clamp, there is no measurement for a 2mm-width band across the whole width.

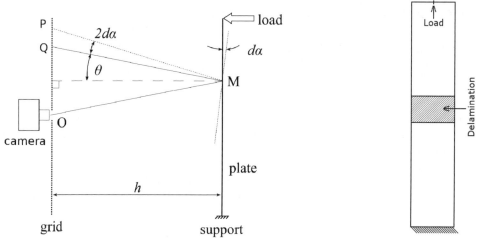

Figure 1: Principle of deflectometry   Figure 2: Model

The method consists of observing the reflection of a grid on the surface of the specimen. As composite components do not have enough specular reflection, a resin (Sicomin Surf Clear) is applied to make a thin reflective coating. At rest, the CCD sensor receives the light coming from P after reflection on M. Once the specimen has been loaded, the slope at the point M has been modified by $d\alpha$ so the sensor records the light coming from Q. Therefore, the camera observes a deformation of the grid. This deformation will be evaluated using spatial phase stepping at rest and under loading for each direction. The phase at rest will be subtracted from the phase maps in the loaded state to obtain the phase modification introduced only by the loading. As these two phase maps are wrapped, an unwrapping process is needed. For that part, a routine developed by Bioucas-Dias and Valadao [4] has been used.

An example of a fringe analysis is shown in Fig. 3. Fig. 3 (a) shows the intensity field of the unloaded beam on the left and the field for the loaded case on the right. From these two states, the phases (transverse and longitudinal) are extracted using a spatial phase stepping algorithm, Fig.3 (b). For each direction, the phase at rest is subtracted from the phase in the loaded state, Fig. 3 (c). As the obtained phase maps are wrapped, the phases are then unwrapped Fig. 3 (d). The results are divided by the sensitivity to scale them, as in [5]. To obtain the curvatures, the slopes are differentiated using a routine developed by [6] with a radius of 15pixels. As it fits locally the slopes with a second order polynomial function, it eliminates the high frequency noise from the grid and smoothes the results. The curvatures are multiplied by half the thickness to obtain the strains, based on thin plate theory.

The samples have a quasi-isotropic lay-up, $([0\ 45\ -45\ 90]_4)_s$, with 32 plies and made with the IM7/8552 prepreg. The ply thickness is 0.125mm. The delaminations inserted are made from a single layer of PTFE film with a thickness of 30μm. They are 3 types of samples: undamaged, ones with a 30mm delamination across the whole width and with a 50mm delamination across the whole width. For each type of specimen, 2 samples have been manufactured. The delaminations have been positioned in the centre of the studied area, the striped zone in Fig. 2.

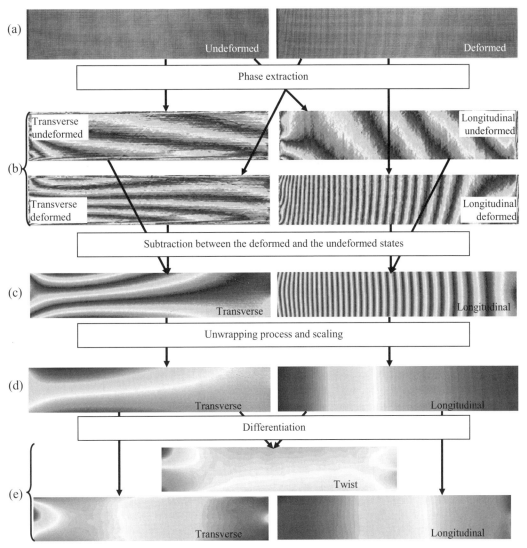

Figure 3: Fringe analysis (a) Intensity pattern (b) phase field (c) corresponding wrapped phase map (d) unwrapped phase maps (e) curvatures maps

Numerical model

The finite element model uses 8 nodes linear elastic bricks, named C3D8R in Abaqus. The element is 1mm long, 1mm wide and the thickness depends on the layer as shown in Fig. 4. Each thin element is 0.125mm thick, and each thick one is 1.75mm which is 14 plies. This way of modelling has been chosen to speed up the computation as the thick elements only transfer the shear stress. The material properties are shown in Table 1.

The longitudinal and through-the-thickness translations have been blocked at the cantilever end. At the free end, a 5.5N load perpendicular to the surface is applied on a single point on the central node. To model the delamination, coincident nodes are unconstrained.

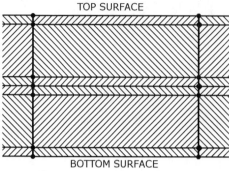

TOP SURFACE

BOTTOM SURFACE

Figure 4: Side view of the meshing.
Each striped zone represents an element.

E_1	61.6 GPa
E_2	61.6 GPa
E_3	11.3 GPa
G_{12}	23.4 GPa
G_{13}	3.92 GPa
G_{23}	3.92 GPa
v_{12}	0.319
v_{13}	0.319
v_{23}	0.319

Table 1: Material properties.
Subscript 1 stands for longitudinal, 2 for transverse and 3 for through the thickness

Single delamination in mid-plane

The following graphs represent the strain fields in microstrains. Each group of graphs (e.g. Fig. 5.a) has the same colour scale which is indicated only once. The zone where there is no information in the measurements has been deleted in the FE results.

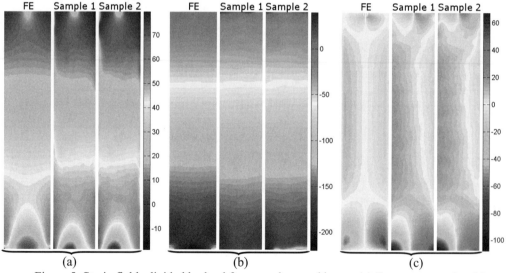

(a)   (b)   (c)

Figure 5: Strain fields divided by load for an undamaged beam, (a) Transverse strains (b) Longitudinal strains (c) Twist strains

For the undamaged beam, there is very good correlation. The small discrepancy at the root on the longitudinal strains is explained in part by the fact that the applied load was not exactly 5.5N and also that the material used did not have exactly the properties used in the FE model.

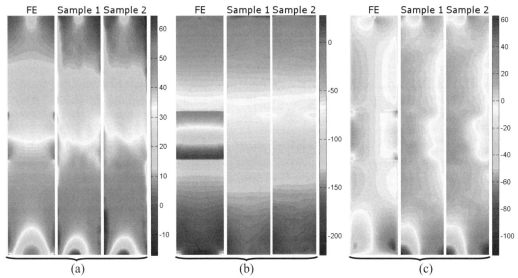

Figure 6: Strain fields divided by load for a beam with a 50mm delamination, (a) Transverse strains (b) Longitudinal strains (c) Twist strains

From the FE results, the delamination is the most visible in the longitudinal strain maps but the experimental results do not show it. In the twist strains, the pattern predicted by the FE model has been seen. As the delamination has been realised by a single layer of PTFE film, it is possible that the resin stuck to the film.

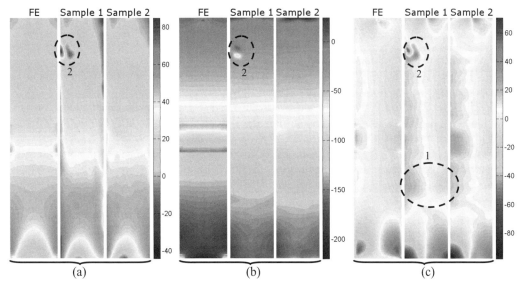

Figure 7: Strain fields divided by load for a beam with a 30mm delamination, (a) Transverse strains (b) Longitudinal strains (c) Twist strains

For "sample 1" with the 30mm delamination, it appears from the Fig. 7 (c) that the delamination is not centred (circled zone number 1). This was confirmed by examining the specimen again. On the same sample, there was a defect in the coating (circled zones number 2) therefore the missing information was interpolated by the derivative routine.

These results showed a modification in the strain fields due to the delamination. The modifications are clearly visible for the twist strains and less clear for the longitudinal ones.

Conclusions

This paper showed a good correlation between the numerical results and the measurements for 2 out of 3 strain components. For these two strain components, a delamination can be clearly identified and located.

This study needs to be broadened to include other cases such as multiple delaminations or real barely visible impact damage. The results from these test measurements can be used with a virtual field routine to extract the orthotropic bending stiffnesses of the damaged and undamaged zones. These strain maps could potentially be used to identify interply properties such as the location, extent and depth of delamination damage. Each of these delamination parameters is potentially uniquely identifiable in the features of the surface strain map.

Acknowledgements

Daniel Thompson would like to acknowledge Rolls-Royce Plc for their support for his research.

References

[1] S.W. Doebling, C.R. Farrar, M.B. Prime, *A summary review of vibration-based damage identification methods*, The Shock and Vibration Digest, vol.30, pp.91–105, 1998

[2] L.S. Lee, V.M. Karbhari, C. Sikorsky, *Investigation of integrity and effectiveness of RC bridge deck rehabilitation with CFRP composites*, Report No. SSRP-2004/08, Department of Structural Engineering, University of California, San Diego, 2004

[3] Y. Surrel, *Deflectometry: a simple and efficient noninterferometric method for slope measurement*, Xth SEM International congress on experimental mechanics, 2004

[4] J.M. Bioucas-Dias, G. Valadao, *Phase Unwrapping via Graph Cuts*, IEEE Transactions on image processing, vol.16, no.3, pp.698-709, March 2007

[5] Kim J.-H., Pierron F., Wisnom M., Syed-Muhamad K., Identification of the local stiffness reduction of a damaged composite plate using the virtual fields method, **Composites Part A: Applied Science and Manufacturing**, vol. 38, n° 9, pp. 2065-2075, 2007.

[6] Avril S., Feissel P., Pierron F., Villon P., Comparison of two approaches for controlling the uncertainty in data differentiation: application to full-field measurements in solid mechanics, **Measurement Science and Technology**, vol. 21, 015703 (11 pp), 2010.

Applied Mechanics and Materials Vols. 24-25 (2010) pp 115-120
© *(2010) Trans Tech Publications, Switzerland*
doi:10.4028/www.scientific.net/AMM.24-25.115

Towards a Planar Cruciform Specimen for Biaxial Characterisation of Polymer Matrix Composites

M. R. L. Gower[1,a] and R. M. Shaw[1,b]

[1]Materials Division, National Physical Laboratory, Hampton Road, Teddington, Middlesex, TW11 0LW, United Kingdom.

[a]michael.gower@npl.co.uk, [b]richard.shaw@npl.co.uk

Keywords: polymer matrix composites, biaxial, cruciform, digital image correlation

Abstract. This paper details work undertaken towards the development of a standard test method for the biaxial response of planar cruciform specimens manufactured from carbon fibre-reinforced plastic (CFRP) laminates and subject to tension-tension loading. Achieving true biaxial failure in a cruciform specimen without the need for the inclusion of a stress raiser, such as a hole, in the gauge-section, is a subject attracting much research globally and is by no means a trivial exercise. Coupon designs were modelled using finite element analysis (FEA) in order to predict the stress and strain distributions in the central region of the specimen. An Instron biaxial strong-floor test machine was used to test the specimens. Strain gauges were used to measure the strain in the specimen arms and to assess the degree of bending. Digital image correlation (DIC) was used to measure the full-field strain distribution in the central gauge-section of the specimen and this was compared to values measured using strain gauges. The strain readings obtained from strain gauges, DIC and FEA predictions were in good agreement and showed that the strain distribution was uniform in the central gauge-section, but that strain concentrations existed around the tapered thickness zone. These regions of strain concentration resulted in interlaminar failure and delamination of the laminate propagating into the specimen arms.

Introduction

Most composite components experience multi-axial loading during their service life and are often fabricated from multi-directional laminates. Existing composite test standards for generating mechanical design data are predominantly uniaxial in scope, and although several laboratories are proposing standard methods for biaxial characterisation, there is currently no standard protocol in place. A world-wide exercise to compare different failure theories for multidirectional polymer composites has already been undertaken [1] and it is particularly important that the experimental data used for these intercomparisons are robust. There are several methods suitable for creating multi-axial loading, including the use of axial forces and pressure (internal/external) using tube specimens, plate or cruciform type biaxial configurations, and full rig systems applying combinations of axial, bending and/or twisting loads. Whilst considerable work has been conducted using tubular specimens, recent moves by industry towards both larger and lighter composite structures requires the planar performance of these materials to be better understood. Currently, the biaxially loaded cruciform specimen has been identified as of most interest with much effort targeted toward achieving a valid failure mode in an un-notched specimen. A potential means of achieving a valid failure without the need for notching is through the use of specimens with a locally thinned gauge-section formed by either milling or ply drop-off. Milling material from specimens is not only laborious and expensive, but it is also difficult to machine the gauge-section to the desired ply level. The alternative technique, whereby plies are dropped-off during the lay-up process has the advantage that the surface of the gauge-section is guaranteed to be of the desired ply orientation. However, this production technique is also laborious and it is difficult to accurately position plies around the centre of the specimen.

Specimen Design

In general, planar specimens suitable for biaxial characterisation should satisfy a number of requirements, namely; (i) the shape of the test specimen should be cruciform with a central gauge-section, as shown in Fig. 1, (ii) fibre reinforcement should be present in both the loading directions, (iii) the shape and dimensions of the test specimen should be such that when loaded in tension in the primary and secondary loading directions simultaneously, a uniform biaxial strain field is produced within a centrally located gauge-section of minimum 20 mm diameter, and (iv) final failure should occur within the gauge-section and is deemed to have occurred when the load carrying capacity of the specimen drops by 20%.

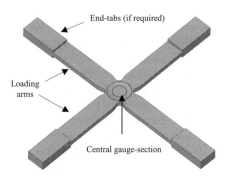

Figure 1. Generic form of biaxial cruciform specimens

With these stipulations in mind, coupons were designed with specimen arms all of the same length and a circular thickness waisted central gauge-section. The lay-up of the laminates for all specimens was $[+45°/0°/-45°/90°]_{2s}$. The size and geometry of the test specimens was limited by the maximum size of square panel (300 x 300 mm) that could be manufactured in the NPL autoclave. The size of the gauge-section and arm width was limited by the load capacity (50 kN) of the multi-axial test facility.

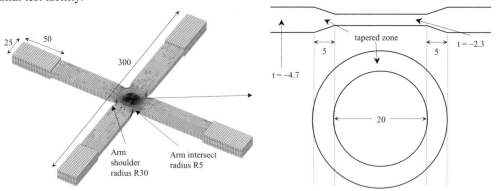

Figure 2. Schematic of 25 mm specimen design and detail of gauge-section

Initially, a coupon with 50 mm wide arms and a circular gauge-section 40 mm in diameter was designed in order to give a relatively large area of uniform strain in the centre of the specimen in which biaxial strain measurements could be made. However, in order to reduce the load required to fail the biaxial specimens, a smaller gauge-section design was adopted (Fig. 2). This design has 25 mm wide arms and a circular gauge-section 20 mm in diameter. The specimen width/length was 300 mm. The design of the radius at the intersection of the arms was changed to an intruding radius of 5 mm with an arm shoulder radius of 30 mm. The protruding radius was chosen to reduce the stress transfer between specimen arms via the ±45° plies and promote failure in the gauge-section rather than at the radius of the arms. This approach had shown potential in work reported by Smits et al [2]. The thickness of the central gauge-section was reduced from 4.7 to 2.3 mm via a tapered zone 5 mm in length.

Specimen Preparation. The material used was SE84 LV unidirectional carbon fibre-reinforced epoxy pre-preg supplied by Gurit Holdings AG. Test panels, 300 x 300 x ~4.7 mm, were autoclave manufactured at NPL following material supplier's guidance. The waisted gauge-section was created by milling away material in 0.1 mm increments from both sides of the specimen until the correct depth profile had been achieved. All specimens were extracted from cured laminates using water jet profiling and end-tabbed using 2 mm thick Tufnol 10G/40® woven glass fibre-reinforced epoxy material.

Finite Element Analysis (FEA). A 3D elastic finite element analysis was undertaken for the biaxial coupon design using LUSAS Composite software. End-tabs and a 0.25 mm thick adhesive bond-line were included in the model. Elastic ply data for the SE84 LV material system (measured at NPL) were used as input into the FE model. A 20 MPa through-thickness stress was applied to the end-tab regions to replicate the clamping pressure applied in the experimental tests via hydraulic grips. The FEA analysis was undertaken to predict strain distributions for comparison to values measured using strain gauges and digital image correlation (DIC).

Biaxial Testing and Results

Multi-axial Test Facility. The multi-axial test facility (Fig. 3(a)) consists of a tee-slotted cast iron strong floor, upon which are mounted four hydraulic actuators each with a dynamic load rating of ±50 kN and a stroke of ±50 mm. Fig. 3(b) shows the set-up used for biaxial testing.

(a)   (b)

Figure 3. (a) NPL multi-axial test facility and (b) biaxial test set-up with hydraulic grips

Strain Measurements. A combination of strain gauges and DIC were used for strain measurements. In the central gauge-section, one side of the specimen was used for DIC measurements whilst strain gauges were used on the opposite face. Specimens analysed with DIC were sprayed with white, grey and black paint to achieve a unique surface finish with a variety of grey levels. A LaVision® DIC system was used with a 5 megapixel video camera to map 2D strain distributions. Images were recorded throughout the tests and all strain results were calculated relative to the first image recorded at zero load. DIC was used to measure the strain over the central gauge section of specimens.

A series of 2 mm gauges were bonded to all 4 arms in the positions shown in Fig. 4. These gauges were used to monitor the degree of in-plane and through-thickness bending. The configuration of the gauges, bending analysis and limits for an acceptable level bending were adopted from the uni-axial tensile standard ISO 527-5 [3]. It is noted that the combined (in-plane and through-thickness) limit level of bending of <3% is considered too strict for a test where load is applied in orthogonal directions. However, the analysis is of value in that it provides a means for assessing the alignment of the test set-up and specimen. In addition to the arm gauges, a single triaxial gauge was used for measuring strain in this region of the specimen.

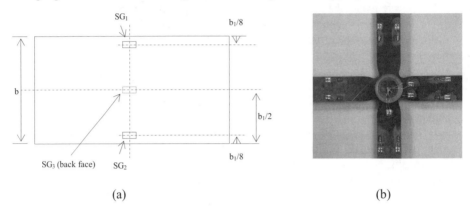

(a)   (b)

Figure 4. Strain gauge positions: (a) arm positions and (b) strain gauged 25 mm specimen

Elastic Measurements. Hydraulic grips were fitted to the biaxial set-up so as to improve the alignment of the biaxial set-up and uniformity of gripping. The load and therefore strain ratio in the two orthogonal directions was set as 1:1. The load profile for both axes was a monotonic ramp up to 10 kN. The tests were undertaken in position control using a displacement rate on each axis of 1 mm/min until the 10 kN load had been reached. The specimen was then held under load for 10 seconds before unloading. Images for DIC analysis were recorded throughout each test at a frequency of 2 Hz.

Initial elastic trials showed a difference in the measured strains in the central gauge section in the two orthogonal directions. This was an unexpected result as the strains should be the same in both directions. On closer inspection of the specimen it was observed that the top surface of the locally thinned gauge-section was not that of a +45° ply, but of a 90° ply instead. Additional machining was undertaken to remove the remainder of the 90° ply on both sides of the specimen and to reveal the +45° ply surface. Subsequent loading trials of the re-machined specimen resulted in strain measurements in much better agreement in both directions indicating that the specimen machining was improved compared to the initial state. A 1:1 loading ratio test was undertaken and the degree of bending was checked. The bending results at the maximum load of 10 kN (in both axes) are shown in Table 1.

Table 1. Assessment of bending in 25 mm specimen at 10 kN

Bending component	Bending (%)			
	Arm 1	Arm 2	Arm 3	Arm 4
In-plane, B_b	2.75	3.94	5.81	9.12
Through-thickness, B_h	0.46	0.95	3.02	0.45
Total, $B_b + B_h$	3.21	4.89	8.84	9.56

With the degree of bending deemed to be as good as could be achieved with the experimental set-up used, the strains in the central gauge-section were measured using strain gauges and DIC on the same specimen. These strain measurements were then compared to the FEA model. The comparisons of measured and predicted strains were performed at 2.5, 5, 7.5 and 10 kN. The results are shown in Table 2 and generally the strains are in good agreement. The FEA predicts the strains in the x- and y- directions should be the same, however there were differences in the strains in the two directions as measured by the strain gauges and DIC, with the strain in the x- direction lower than that in the y- direction. This was reasoned to be due to difficulties in accurately machining the gauge-section to the exact depth across the gauge-section area.

Table 2. Comparison of ε_{xx} and ε_{yy} strains in gauge-section

Load (kN)	ε_{xx} (microstrain)			ε_{yy} (microstrain)		
	Gauge	DIC	FEA	Gauge	DIC	FEA
2.5	403	447	426	442	484	427
5.0	804	872	850	886	943	854
7.5	1206	1298	1280	1330	1402	1280
10.0	1608	1723	1703	1774	1860	1709

Fig. 5 and 6 show comparisons of the FEA predicted and DIC strain maps in the x- and y-direction at 10 kN, respectively. It was not possible to use the same contour levels for the DIC and FEA strain maps, and therefore regions of equivalent strain level do not exactly colour match. The predicted and measured strain maps were in good agreement and show a fairly uniform strain field in the gauge-section. Also, regions of high strain were observed corresponding to the tapered section and at the radius of the specimen arms.

(a)   (b)   (a)   (b)

Figure 5. ε_{xx} (%) by: (a) FEA and (b) DIC   Figure 6. ε_{yy} (%) by: (a) FEA and (b) DIC

Having completed the elastic measurements a specimen was loaded to failure using a loading ratio of 1:1 under position control. A displacement rate of 1 mm/min was used in both axes. The specimen failed at a load of 28.9 kN as a result of interlaminar failure occurring in the tapered thickness section causing delamination failure to propagate into the specimen arms. Also, failure was observed at the radius of one of the arms. The failure locations are in good agreement with the strain concentrations predicted using FEA and measured using DIC. The specimen did not fail in the specimen gauge-section, and thus the mode of failure was not considered to be representative of a true biaxial failure.

Conclusions

DIC has been successfully used to measure the full-field strain distribution and showed that the strain was fairly uniform in the gauge-section of the specimens. DIC also detected the presence of high interlaminar strains around the tapered thickness zone, which was in good agreement with FEA predictions. These strain concentrations resulted in interlaminar failure and delamination of the laminate propagating into the specimen arms. This failure mode has also recently been observed in specimens of similar geometry by Makris et al [4] and Ramault et al [5]. To overcome this failure mode, Ramault et al [5] have attempted to use a specimen that does not feature a thickness reduction in the gauge-section. Instead, the specimen is clad with an adhesively bonded end-tabbing material everywhere except the gauge-section. However, it was shown that specimen failure then occurred in the adhesive bond-line between the tabbing material and the base laminate and again, true biaxial failure did not result. It is clear from this study that more work is required before a standard biaxial characterisation method for planar specimens can be recommended.

Acknowledgements

The authors acknowledge the financial support provided by United Kingdom Department for Business, Innovation and Skills (National Measurement Office), as part of the Characterisation Programme. The authors would like to express their gratitude to Gurit Holdings AG, Qinetiq, Vrije Universiteit Brussel (VUB) and NPL colleagues Dr Bill Broughton and Dr Graham Sims.

References

[1] M. J. Hinton, P. D. Soden and A. S. Kaddour, 'Failure Criteria in Fibre-Reinforced-Polymer Composites: The World-Wide Failure Exercise, (Oxford UK, Elsevier Ltd., 2004).

[2] A. Smits, D. Van Hemelrijck, T. Philippidis and A. Cardon, 'Design of a cruciform specimen for biaxial testing of fibre reinforced composite laminates', Composites Science and Technology V66, pp 964-975, 2006.

[3] BS EN ISO 527-5 Plastics - Determination of tensile properties - Part 5: Test conditions for unidirectional fibre-reinforced plastic composites

[4] A. Makris, C. Ramault, D. Van Hemelrijck, E. Lamkanfi and W. Van Paepegem, 'Damage Evolution on Composite Cruciform Specimens Under Quasi-Static Biaxial Loading', ICCM-17, 27-31 July 2009, Edinburgh, UK.

[5] C. Ramault, A. Makris, D. Van Hemelrijck, E. Lamkanfi and W. Van Paepegem, 'Effect of Tab Design on the Strain Distribution of a Biaxially Loaded Cruciform Composite Specimen', ICCM-17, 27-31 July 2009, Edinburgh, UK.

Session 3B: Novel Sensor Technology

Applied Mechanics and Materials Vols. 24-25 (2010) pp 123-128
© (2010) Trans Tech Publications, Switzerland
doi:10.4028/www.scientific.net/AMM.24-25.123

Measurement of Mechanical Strain using Chromatic Monitoring of Photoelasticity

Garza C.[1,a], Deakin A.G.[2,b], Jones G. R.[2,c], Spencer J. W.[2,d] and Hon K.K.B.[1,e]

[1] Department of Engineering, The University of Liverpool, Brownlow Hill, Liverpool, L69 3GH, UK

[2] CIMS, Centre for Intelligent Monitoring Systems Research Group, Department of Electrical Engineering and Electronics, Brownlow Hill, Liverpool, L69 3GJ, UK

[a] claudia.garza@liverpool.ac.uk, [b] anthonyd@liverpool.ac.uk, [c] grjones@liverpool.ac.uk, [d] joe@liverpool.ac.uk, [e] honkk@liverpool.ac.uk

Keywords: Chromatic monitoring, photo-elasticity, Strain Measurement

Abstract. The present contribution describes a chromatic processing approach for quantifying the two dimensional, polychromatic interference patterns produced by a strained photo-elastic element and recorded with a CCD camera. The outputs from the three R, G, B channels of the camera covering a selected area of the interference pattern are processed to yield three chromatic parameters which are H (dominant signal wavelength), L (nominal signal strength), S (effective wavelength spread of signal). It is shown that the value of each of the three parameters varies with strain in a quasi cyclical manner, all being out of phase with each other. Consequently the strain measurement range and sensitivity can both be optimized by the use of the appropriate chromatic parameter within different strain ranges.

Introduction. There are several experimental methods for measuring strain such as foil strain gauges [1] split Hopkinson pressure bar (SHPB) or Kolsky bar technique [2], velocity interferometers [3] etc. There are also several optical methods such as the caustics method [4] Moiré techniques (interferometry or photography), speckle interferometry, photo-elasticity [5], etc. The latter may be deployed in one of several forms, e.g. Tardy method, use of compensators, digital photoelasticity [6] or chromatic evaluation of optical fibre based photo-elastic elements [7]. More recently Madhu and Ramesh [8] have considered colour adaptation in three fringe photoelasticity. The present paper is concerned with a further evaluation of the chromatic addressing [9] of photo-elastic elements subjected to mechanical strain. A detailed investigation is made of the strain levels over which each of three parameters (H, L, S) produced by chromatic processing [9] can be used to advantage for optimizing the measurement range and the sensitivity. For this purpose use has been made of two dimensional, polychromatic, photo-elastic interference patterns recorded with a CCD camera.

Measurement System. The experimental system used for the present tests was based upon monitoring the mechanical strain in a cantilever beam when loaded to various extents (Fig. 1). A polychromatic light source (Tungsten halogen lamp) was used to illuminate a 3mm thick photo-elastic element (PS-1A [10]) bonded to the strained beam with PC-1 Vishay Epoxy [10] resin adhesive. Images of the polychromatic fringes produced by the strained photo-elastic element were obtained using a 7 MegaPixels CCD camera in conjunction with an optical polarizing filter. The outputs from the CCD camera were transferred to a personal computer for chromatic processing. Simultaneous to capturing the images of the photo-elastic fringes, the corresponding strain values to which the beam was exposed were recorded with a strain gauge bonded with M-BOND 200 Vishay Adhesive [10] to the beam which was situated under the beam and below the photo-elastic sensing element (Fig. 1).

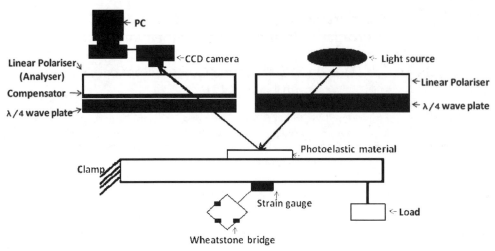

Figure 1.- Measurement System and Cantilever Beam

Chromatic Processing. A polychromatic signal produced by a photo-elastic element (e.g. Fig (2) interference signal) may be captured with a CCD camera via a matrix of three pixel detectors (R, G, B) at each spatial location and which have overlapping wavelength responses (Fig. 2)[9]. The outputs from the detectors (r, g, b) can be transformed into three signal defining, chromatic parameters which are the dominating wavelength (H), nominal signal strength (L) and effective wavelength spread of the signal (1-S) [9]. Values of each of these chromatic parameters (H, L, S) may then be calibrated against strain values measured simultaneously with a strain gauge (Fig. 1). The algorithms used for the r, g, b transformation to H, L, S are given in Appendix I.

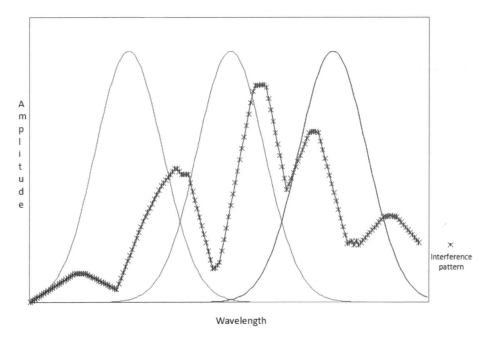

Figure 2.- Responses of 3 Non-Orthogonal Filters (R, G, B) as a function of wavelength superimposed upon a polychromatic signal (×)

Experimental procedure and calibration. The experimental procedure involved loading the cantilever to various extents and capturing the corresponding images of the polychromatic fringes produced by the strained photo-elastic element. The output values r, g ,b obtained from each image pixel area were transferred to the computer and transformed into corresponding values of the chromatic parameters H, L, S for each pixel area using the algorithms given in Appendix I. Values of H, L, S were space averaged across a chosen sector of the image (314 pixel extent [4mm^2] over 20 * 20 mm actual area), using specially developed software (Droptri. [11]). This sector was located directly above the foil strain gauge which was situated under the beam and below the photo-elastic sensing element (Fig. 1). Examples of H, L and S values determined in the above manner during a single test used for calibration are shown for different strain levels applied to the beam on Fig. 3 (a), (b), (c) respectively. These results exhibit quasi cyclic features with the maximum value plateau progressively shifted to higher strain values in order H, L, S. In particular:

- H is insensitive to strain in the range 0 – 4.5 E-05 micro-strain but thereafter reduces monotonically ($20° < H < 60°$) as the strain increases further.
- L is insensitive to strain in the range 5 E-05 – 1.0 E-04 micro-strain but corresponds to two micro-strain values outside this range ($0.2 < L < 0.8$)
- S is insensitive to strain in the range of 0 – 7.5 E-05 micro-strain but thereafter increases monotonically ($0.2 < S < 0.42$) as the strain increases further.

By using a look up table for the H, L, S parameters and strain measured with the strain gauge, strain calibrated values Hm, Lm, Sm for each of the three chromatic parameters can be obtained as indicated on the Flow Diagram of Fig. 4. The regions of high sensitivity for each parameter indicated on Fig. 3 can then be selected to provide a preferred strain value. As a result, the measurement range and sensitivity can both be optimized.

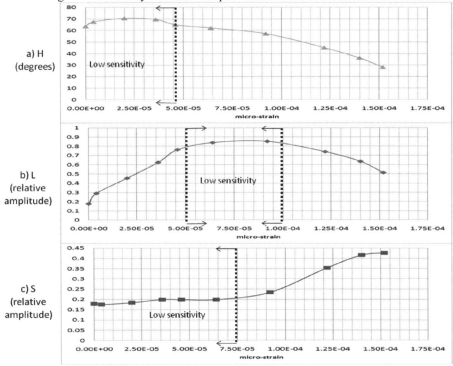

Figure 3.- Variation of Chromatic Parameters H, L, S with Measured Mechanical Strain
(a) H: Strain (b) L: Strain (c) S: Strain

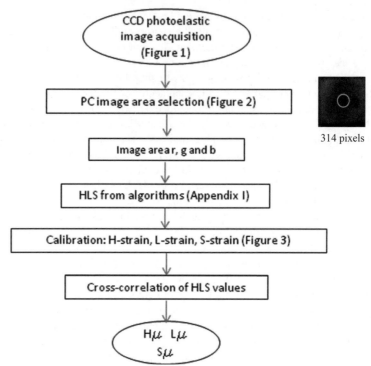

314 pixels

Figure 4.- Data Processing Flow Diagram for Obtaining Micro-strain from an Image of the
Photo-elastic Interference Pattern

Experimental Results. Three additional tests were performed on the cantilever beam loaded to different extents. Values of H, L, S were determined from the resulting camera records. These were then converted into strain values (Hm, Lm, Sm) using the results from the calibration test (Fig. 3). The calculated strain values (Hm, Lm, Sm) for a variety of imposed strains from the three different tests are shown (vertical axis) as a function of the imposed strain measured with the strain gauge (horizontal axis) on Fig. 5. Chromatic parameters results from each of the three tests are represented by different symbols as defined in the figure caption.

The results show an overall linear correlation between the chromatically derived (Hm, Lm, Sm) and the imposed strain values with a coefficient of determination $R^2=0.955$ [12]. The scatter for each chromatic parameter is more pronounced in the strain range within which it exhibits low resolution (Fig. 3 (a), (b), (c)). i.e. at lower strain level for Hm (<4.5 E-05) and Sm (<7.5 E-05), but within the mid strain range for Lm (6 E-05 – 1.1 E-04) which shows highest scatter. Overall, the best behaved of the three chromatic parameters appears to be Hm with a coefficient of determination $R^2= 0.988$ compared with $R^2=0.974$, 0.906 for Lm and Sm respectively. Thus, a preferred option would be to rely upon the use of Lm at low strain values, Hm for the mid range and a combination of Hm, Lm, Sm for the high range (Fig. 5).

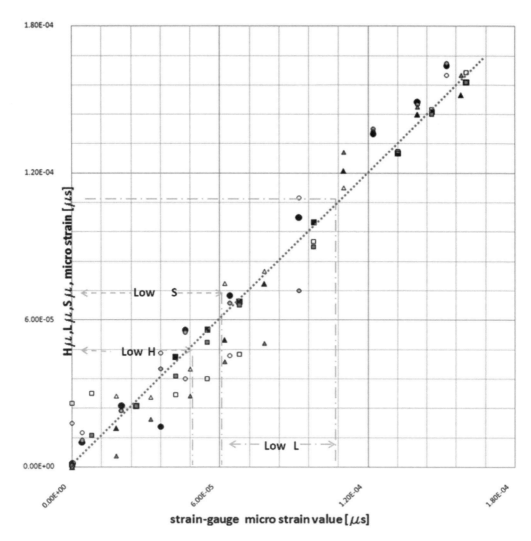

Figure 5.- Strain from chromatic parameters H, L, S as a function of directly measured strain for three separate tests. ●, ◕, ○ Test 1: Hm, Lm, Sm; ■, ▨, ▢ Test 2: Hm, Lm, Sm; ▲ ▲ △Test 3: Hm, Lm, Sm

Conclusions. Results have been presented for the variation of three chromatic parameters (H, L, S) with micro-strain, the chromatic parameters being derived from CCD camera images of the interference pattern produced by a photo-elastic element. It has been shown that the variation of these parameters with strain have quasi-cyclical features, each being phase shifted with respect to the others. This has the potential for providing an extended measurement range at optimum sensitivity by automatically selecting the appropriate chromatic parameter most suitable for particular ranges of strain.

Inspection of the raw outputs (r, g, b) from the R, G, B camera pixels indicates that although these outputs also show a wavelength dependent cyclical behavior, the phase differences are considerably less than those observed with the chromatic parameters H, L, S, so making the latter the preferred choice. Other chromatic parameters (e.g. $x=r/(r+g+b)$, $y=g/(r+g+b)$ [7]) also show some cyclical behavior but cannot be so conveniently processed as H, L, S to provide range extension and optimum sensitivity. Ongoing work shows that the calculation procedure involving H, L, S described here can be significantly simplified using secondary chromatic processing [9] to produce a linear relationship between chromatic properties and micro-strain.

References

[1] Perry, C. C., 1984, *The resistance strain gauge revisited*, Experimental Mechanics, 24(4), pp. 286 - 299.

[2] Frew D.J., Forrestal M.J., Chen, W., 2002, *Pulse Shaping Techniques for Testing Brittle Materials with a Split Hopkinson Pressure Bar*, Experimental Mechanics, 42, pp. 93-106.

[3] Huang H., Asay J.R., 2005, *Compressive strength measurements in aluminium for shock compression over the stress range of 4–22 GPa*, J. Appl. Phys. 98 (3).

[4] Wallhead I. R., Edwards L., 1997, *A practical guide to the measurement of the elastic stress intensity factor in engineering materials by the method of caustics*, The Journal of Strain Analysis for Engineering Design, 32 (4), pp. 253-266.

[5] Zandman F., Render S., Dally J., 1977, *Photoelastic Coatings*, Society for Experimental Mechanics press.

[6] Ramesh K., 2000, *Digital Photoelasticity*, Springer.

[7] Murphy M. M., Jones G. R., 1992, *Polychromatic birefringence sensing for optical fibre monitoring of surface strain*, Eurosensors V Conference, 2, pp. 691–695.

[8] Madhu, K.R., Ramesh, K., 2007, *Colour Adaptation in Three Fringe Photoelasticity*, Experimental Mechanics 47(2) pp. 271–276.

[9] Jones G.R., Deakin A.G., Spencer J.W., 2008, *Chromatic Monitoring of Complex Conditions*, CRC press.

[10] Information found at http://vishay.com

[11] Deakin A.G. , CIMS, Centre for Intelligent Monitoring Systems, The University of Liverpool.

[12] Allen M. P., 1997, Understanding Regression Analysis, Springer US.

APPENDIX I   R,G, B – H, L, S TRANSFORMS [9]

$$L = \frac{R+G+B}{3}$$

$$S = \frac{\text{Maximum (R, G, B)} - \text{Minimum (R, G, B)}}{\text{Maximum (R, G, B)} + \text{Minimum (R, G, B)}}$$

$$H = 240 - 120 * \frac{g}{g+b} \qquad\qquad r = 0$$

$$H = 360 - 120 * \frac{b}{r+b} \qquad\qquad g = 0$$

$$H = 120 - 120 * \frac{r}{r+g} \qquad\qquad b = 0$$

where: $r = R - \min(R, G, B)$, $g = G - \min(R, G, B)$, $b = B - \min(R, G, B)$

Applied Mechanics and Materials Vols. 24-25 (2010) pp 129-134
© *(2010) Trans Tech Publications, Switzerland*
doi:10.4028/www.scientific.net/AMM.24-25.129

Development of a Methodology to Assess Mechanical Impulse Effects Resulting from Lightning Attachment to Lightweight Aircraft Structures

C. A. Featherston[1, a], M. Eaton[1,b], S. L. Evans[1,c], K. M. Holford[1,d], R. Pullin[1,e] and M. Cole[2,f]

[1]Cardiff School of Engineering, Cardiff University, Queens Buildings, The Parade, Cardiff, CF24 3AA, UK

[2] EADS Innovation Works, The Quadrant, Celtic Springs, Coedkernew,Newport, NP10 8FZ, UK

[a]FeatherstonCA@cardiff.ac.uk, [b]EatonM@cardiff.ac.uk, [c] EvansSL6@cardiff.ac.uk, [d]Holford@cardiff.ac.uk, [e]PullinR@cardiff.ac.uk, [f]Matthew.Cole@eads.com

Keywords: Lightning, lightweight structures, digital image correlation, finite element analysis.

Abstract. The effect of lightning attachment to structures and vehicles is a cause of major concern to a number of different industries, in particular the aerospace industry, where the consequences of such an event can be catastrophic. In 1963, a Boeing 707 was brought down in Maryland killing 81 people on board, triggering the improvement of lightning protection standards. However, commercial jets are still struck on average once every 10,000 hours of flight time and between 1963 and 1989 forty lightning related accidents were recorded within the U.S.A alone. The rapid increase in the use of composite materials in aircraft design and the consequent increase in complexity when determining the effects of a lightning strike, has led to new challenges in aircraft protection and the requirement for improved understanding and standardisation.

The attachment of lightning to a structure causes damage through three mechanisms. Primarily a supersonic acoustic shock wave, caused by rapid heating of the arc channel during initial attachment, resulting in a large and rapid overpressure. Secondly a magnetic force generated by the fields developed in the high current areas around the lightning attachment point. Finally a mechanism specifically related to composite materials, where the rapid vaporisation of an expanded copper foil layer (designed to quickly transmit current across a structure, thus reducing its focus) trapped between the composite material and the protective paint layers causes an additional overpressure, which is exacerbated by additional paint layers acting to contain the explosion and direct it inwards.

The work described in this paper looks to develop a technique to measure these forces in order to better understand and assess their effects. A novel methodology has been developed to allow the estimation of peak overpressure forces produced by the acoustic shock wave resulting from the attachment. The methodology utilises Digital Image Correlation and ultra-high speed photography to acquire full-field displacement measurements of panel deflections at frame rates of up to 1,000,000 frames per second. The experimental results are used in the optimisation of a finite element model, outputting the parameters of an acoustic shock that produces representative displacements, velocities and accelerations. The method is currently being validated by performing a series of tests on aluminium panels subject to instrumented impact testing. The next stage will be to use the technique developed to aid a program of investigation into the effects of artificial lightning strike events on aluminium and composite panels.

Introduction

Lightning parameters A lightning strike is essentially a high amplitude direct-current pulse. Its waveform is well-defined and is divided into four parts known as components A to D as shown in Fig. 1. Component A is a high-current pulse. It is a direct current transient typically in the order of 200 kA lasting for 50 µs and having a current rise rate of 3 $\times 10^{10}$ A/s. Component B is a transition phase generally in the order of 2kA. Component C is a continuing current of approximately 200-800A and lasting up to 0.75 s. The last component, D, is a restrike surge which is typically half the amplitude of component A in a given strike and has the same duration as component A. On average 3 or 4 restrikes will occur in one lightning event but where the maximum observed is 26 restrikes in one event.

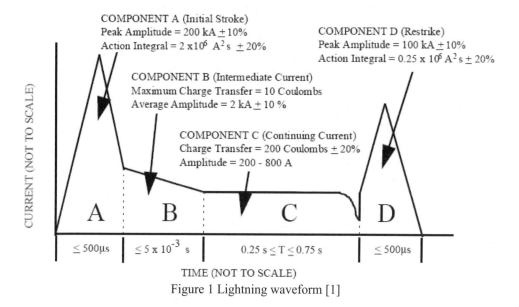

Figure 1 Lightning waveform [1]

Indirect effects The multiphysics nature of the lightning strikes causes a range of different effects including heating, a pressure wave resulting from the acoustic shock and magnetic forces.

On initial attachment the high current return stroke generates a large acoustic shock wave caused by the rapid heating of the lightning arc which propagates cylindrically at an initially supersonic speed. The resulting axial pressure at the root of the arc can cause damage in the form of permanent deformation or in the case of composites matrix cracking. Initial estimates of this overpressure and its propagation for a 30kA stroke have been made by Hill [2] who calculated a maximum overpressure of 30 MPa at a radial distance of 10mm from the point of attachment, reducing to 20 MPa at 20mm and 10 MPa at 40mm. However in order to fully understand the effects of this pressure wave for different levels of lightning strike, this data must be derived for a range different currents.

Further forces are generated due to the magnetic fields developed around the point of attachment of the arc. Since following attachment current flows away from this point in many directions each of which can be considered as a separate current path and when two or more electric currents flow in the same direction through two or more parallel wires (or paths), attractive forces act on the wires (or the different parts of the structure), forces are generated which can deflect the structure. Although it is generally believed that this effect is much smaller than the effect of the acoustic shock wave it must still be quantified in order to understand the total effects of the lightning strike.

Finally the localised heating occurring in the vicinity of the attachment point can also cause problems particularly in the presence of paint finishes and in composites the Expanded Copper Foil layer due to the generation of explosive forces as explained previously. This also needs to be quantified.

Only by gaining an insight into the effects of these distinct forces and their interdependencies can the effects of different levels of lightning strike be adequately predicted. The work presented in this paper represents a pilot study to determine the feasibility of using an explicit finite element model describing the response of a specimen during lightning attachment in order to determine the mechanical effects of such a strike. This model, once optimised using experimental data on full field deformation and hence velocity and acceleration profiles during impact obtained using high speed DIC will provide the loading parameters required.

Methodology

Due to the level of complexity involved in such a multi-phenomenal event in which the energy input due to the various effects has the ability to cause a combination of elastic and potentially plastic deformation and damage and which is truly dynamic and therefore incorporates inertial effects it was concluded that direct measurement of the forces involved was not feasible. For this reason a methodology was developed in which indirect effects such as full-field displacement and hence the velocity and acceleration of the plate would be measured using DIC. Until recently measurements of this kind would not have been possible due to limitations in available camera hardware, i.e. maximum frame rates achievable at a suitable resolution. However with the advent of Ultra-High speed cameras, such as the Shimadzu HPV-2, suitable images can now be captured at up to 1M fps on a single CCD chip. High speed DIC has previously been utilised to monitor dynamic fracture events in polymers [4] and to aid the calculation of material parameters at high strain rates using the virtual fields method [5]. The full-field data would then be used to optimize the parameters within a dynamic FEA model created using a combination of the pre and post processor Oasys Primer and D3 Plot and the analysis code LS-DYNA. The technique would be validated based on a simple test involving impacting a metallic plate. This methodology is illustrated in Fig. 2.

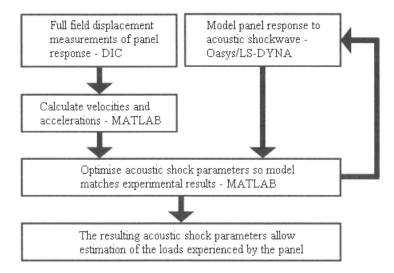

Figure 2 Methodology

Experimental Work

Set-up As stated, the methodology is being validated based on a high speed impact test on a rectangular plate. The plate was manufactured from 0.6mm thick aircraft grade duraluminium BS1470 6082-T6 with an unsupported length of 287mm and a width of 254mm (based on the dimensions of a typical aerospace panel (Jegley [3])). It was built-in along all four edges by bolting through the plate to a pair of frames 25mm x 25mm in cross section. Full field displacement data was captured using digital image correlation. This was performed using VIC 3D (Correlated Solutions Inc, USA) system, capturing data at a rate of 1000fps using a pair of AOS Technologies MOTIONeer high speed cameras synchronised using Correlated Solution's VIC Snap software to gather data for post processing using VIC 3D software. The test was performed using a Instron Dynatup 9250HV accelerated drop test machine, with a series of impacts based on dropping a mass of 5.7kg from a variety of heights from 0.05 to 0.2m. The test set-up is illustrated in Fig. 3.

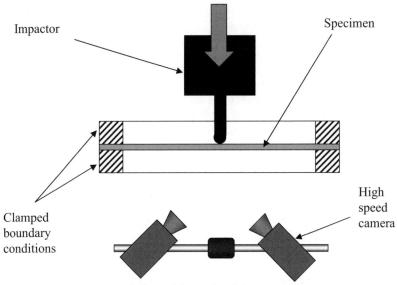

Figure 3 Schematic of drop test set-up

Test results A series of profiles of displacement against time at the point of impact taken from the results of the digital image correlation are presented in Fig. 4, for drop heights of 0.05, 0.1 and 0.2m. Velocity profiles again corresponding to behaviour at the point of impact are presented in Fig.5. Finally full field displacement contours taken from the tests performed with a drop height of 0.2m corresponding to different times during the impact event are provided in Fig. 6.

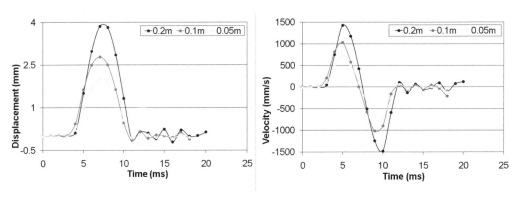

Figure 4 Displacement at the point of impact   Figure 5 Velocity at the point of contact

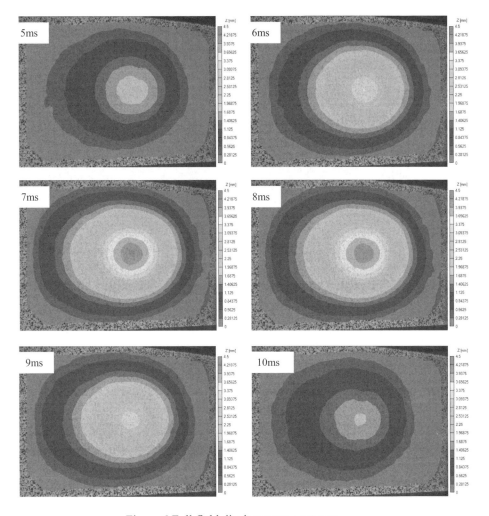

Figure 6 Full field displacement contours

Modelling

LS-Dyna is an explicit solver developed for the analysis of high speed short duration events in which the effects of inertia are important. It is capable of analysing time steps of less than 1μs. Initially developed to aid crash testing in the automotive industry Dyna is capable of dealing with linear and non-linear transient dynamics, non-linear material behavior, contact surfaces and crucially for this work solid-fluid interactions. Utilising these features LS-Dyna has been used to model blast effects and explosive forming of sheet metals and hence is ideally suited to the proposed application.

As discussed above initial validation of the proposed methodology for measuring the effects of lightning strike is being conducted using an aircraft grade aluminium panel subject to a drop weight impact. Utilising symmetry, impact at the centre of the panel has been represented based on a one quarter model. To represent the impactor, a rigid body with the geometry of a sphere the same diameter as the impact tup and the equivalent mass of the impact carriage is placed a small distance above the panel and given an initial velocity which can be varied to represent different drop heights. An image of the model is presented in Fig. 7.

Figure 7 Finite Element model of panel impact.

Optimisation

The aim of the optimisation process in this context is to minimise the difference between the experimental measurements and the FEA results by adjusting parameters related to loading. This is achieved through use of a minimisation function in MatLab which measures the difference between the experimental and FEA data sets, be those displacements, velocities or accelerations, and minimises this difference by iteratively adjusting the loading parameters. The resulting parameters provide an estimation of the load experienced by the panel. In this case this is the initial velocity of the impact sphere, however once validated the same approach can be extrapolated to determine for example parameters describing the acoustic shock.

Conclusions

A methodology has been developed to enable the forces experienced by a structure during lightning strike to be determined using indirect effects which can be measured. A pilot study is being performed to validate this methodology based on a metallic plate subject to impact from a drop test rig. A series of tests based on different impact energies have been performed. Full field displacement and velocity data have been successfully obtained using high speed digital image correlation. A model of the experimental set-up has been developed using the explicit finite element analysis code LS-Dyna. Loading parameters within the model are currently being optimized to obtain a fit with the experimental data. Comparison of the loading parameters with measured data will validate the suitability of the technique for the measurement of load.

References

[1]   Department of Defence: *Electromagnetic Environmental Effects*, Military Handbook 464, (1997).

[2]   R.D. Hill: *Channel Heating in Return Stroke Lightning*, Geophys., Vol. 76, (1971).

[3]   D.C. Jegely: *Behavior of compression-loaded composite panels with stringer terminations and impact damage*. AIAA/ASME/ASCE/AHS 39th Structures, Structural Dynamics, and Materials Conference, Long Beach, California, (1998).

[4]   M. S. Kirugulige, H. V. Tippur and T. S. Denney: *Measurement of transient deformations using digital image correlation method and high-speed photography: application to dynamic fracture.* Appl. Opt., Vol. 46 (22), (2007).

[5]   S. Avril, F. Pierron, M. A. Sutton and J. Yan : *Identification of elasto-visco-plastic parameters and characterisation of Lüders behaviour using digital image correlation and the virtual fields method.* Mechanics of Materials, Vol. 40, (2008).

Applied Mechanics and Materials Vols. 24-25 (2010) pp 135-140
© (2010) Trans Tech Publications, Switzerland
doi:10.4028/www.scientific.net/AMM.24-25.135

A Laser Speckle Method for Measuring Displacement Field. Application to Resistance Heating Tensile Test on Steel

C. Pradille[a], M. Bellet[b] and Y. Chastel[c]

MINES-ParisTech, CEMEF, UMR CNRS 7635, Sophia Antipolis, France

[a]christophe.pradille@mines-paristech.fr, [b]michel.bellet@mines-paristech.fr,

[c]yvan.chastel@mines-paristech.fr

Key words: full field measurement, laser speckle, cross-spectrum, cross-correlation

Abstract. Strain field measurement with non intrusive techniques is needed in order to characterize the behaviour of steels at high temperatures subjected to small displacements. In this work we present a technique based on laser-produced speckles coupled with a cross-correlation cross-spectrum method. This method proves more accurate than cross-correlation for small displacements. The laser wave length used (532 nm) allows to perform strain measurements, even with heat radiation.

Introduction

Understanding of solidification crack phenomena is a challenge. On one hand, industrial concerns are considerable in terms of productivity and product quality. On the other hand, physical mechanisms of hot tearing are far from being well understood, which makes it a scientific challenge. For this highly multiphysics problem, one of the main difficulties lies in the rheological characterization of metallic alloys, at high temperature and in the semi-solid state. Actually, there are only few data available for constitutive laws of industrial steel for solidification conditions (ie high temperature and very low strain and strain rate). Rheological characterization under tensile conditions has been carried out using a direct resistance heating machine named Taboo. The Taboo set-up allows cyclic loadings at very low strain rate (10^{-3}-10^{-4} s^{-1}) between 900°C and the solidus temperature. A problem with resistance heating techniques is the significant thermal gradient along the sample, which causes heterogeneous strains. Due to the spatial heterogeneity of the strain field, the displacement field needs to be known on the entire sample. The speckle correlation photography has several advantages like full field and non destructive measurement. However, this method is classically limited in temperature (around 900°C). The reason for this is that the speckle is usually produced with painting [1]. Furthermore, at high temperatures these methods based on optical measurements are also very sensitive to the visible spectrum, in particular in the red wave length. All these limitations lead us to develop a new method based on laser-produced speckles. Since, as for steel, radiation is absent in the green wave length around 1500°C, a green laser is used. In this work we also present a technique of image analysis based on a cross-spectrum method. This method proves more accurate than cross-correlation technique for small displacement and more efficient for quickly analysing two pictures within only a few milliseconds. In this paper, we first describe the experimental set-up; then we present the measurement technique. Thanks to this method we obtain full field measurements.

Experimental Set-up

Taboo Tensile Test Machine. The "Taboo" machine (Fig. 1) is a thermo-mechanical simulator which has been developed at Cemef to investigate the behaviour of steels at very high temperature up to the solidus temperature and possibly in the semi-solid state. Before applying the loading cycle, the sample is directly heated by Joule effect using a continuous current. The temperature monitoring of the sample is achieved as follows. A thermocouple is welded in the middle of the specimen. During the heating and loading phases of the test, the current intensity is dynamically controlled in order to prescribe a desired temperature. An alternative monitoring procedure consists in using a thermal camera to measure the temperature field on the specimen surface. During the heating stage, the mobile grip displacement is monitored so that no external force is applied on the specimen. Once the desired temperature is reached, the mechanical test starts and the tensile force is measured by a load cell. The measurement of force vs displacement curves requires a high temperature extensometry technique. For that purpose, a non-contact method has been developed and is presented in the next section.

Fig. 1: "Taboo" tensile machine (left), laser speckle equipment (centre) and a typical laser speckle pattern (right).

Speckles Set-up. A laser is projected on the area in which the displacement field needs to be measured. The laser beam is enlarged by a microscope lens that allows illuminating a large area on the central part of the tensile sample (typically 10 x 6 mm, the specimen being 90 mm long and 6 mm wide). Speckle images arising from local interference phenomena due to surface roughness are captured by a CCD camera and are simultaneously transferred to a computer. The size of analysed pattern areas is 1920 x 1080 (about 2 Mpixels). Each pixel can store a grey scale 8-bit value ranging from 0 to 255. During the experiment, a series of patterns can be either collected for post treatment later on or correlated and analysed for the control loop. The correlation between initial and current speckle patterns provides the displacement, as explained in the next section.

Image Correlation Techniques

The surface is considered to change relatively slowly during the experiment. Consequently, over a small time step, strains are low and the observed images are simply translated. In the following paragraph a technique of image analysis based on a coupled cross-correlation / cross-spectrum technique is presented. This technique is used to obtain the displacement field from two images. This coupled method allows characterizing the displacement within a few milliseconds with a sub-pixel resolution for a large range of displacement (from less than 1 pixel to 20 or more pixels).

The cross-spectrum and the cross-correlation methods have different advantages. Cross-correlation method is less accurate for small displacements and less rapid than cross-spectrum method but it is more accurate for large displacement.

The Cross-spectrum Method [2]. Consider a plane surface covered by a speckle pattern where the image intensity at time t_0 is defined by $f(x,y)$. The Fourier transform $F(u,v)$ is defined as:

$$TF\{f(x,y)\} = F(u,v) = \iint f(x,y)e^{-2i\pi(ux+vy)}dxdy. \tag{1}$$

Consider now a second image where the intensity at time $t_1 = t_0 + \Delta t$ is defined as $g(x,y)$. If function $g(x,y)$ is written as a translation of $f(x,y): g(x,y) = f(x-x_0, y-y_0)$, and defining the cross-spectrum $I(u,v)$ as the multiplication of $F(u,v)$ by the complex conjugate of $G(u,v)$ (denoted $G*(u,v)$), we get:

$$I(u,v) = F(u,v)G*(u,v) = F(u,v)F*(u,v)e^{2i\pi(ux_0+vy_0)} = |F(u,v)|^2 e^{2i\pi(ux_0+vy_0)}. \qquad (2)$$

Eq. 2 shows that the displacement can be expressed according to two components corresponding to two different directions. The phase can consequently be decomposed into two expressions. The displacement is obtained from a linear fit of the expression of the phase (Fig. 2a). It seems quite simple to obtain a measure of the displacement between two different pictures with a cross-spectrum method. But, at high spatial frequencies, the noise and/or a bad sampling can lead to a zero value of the phase instead of showing the expected random variation (between $-\pi$ and $+\pi$). Moreover, fitting the phase slopes is not precise when the displacement is larger than 1 pixel. In fact, in this case the phase varies between $-\pi$ and $+\pi$. A displacement can not be accurately determined when a poor sampling is added to a slope of the phase, as illustrated in Fig. 2b.

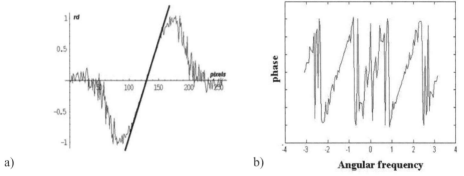

a)   b)   **Angular frequency**

Fig. 2: a) Example of displacement calculations. The phase falls down to 0 for high spatial frequencies; b) Example of the phase for a displacement larger than 1 pixel.

The Cross-correlation Method. In most papers dealing with field measurement (e.g. [3, 4]), the cross-correlation technique is reported to be used. In practice, the Wienner-Kinchine theorem states that the cross-correlation can be written as the inverse Fourier transform of Eq. 2 and these two functions contain equivalent information. However, it is more convenient to use the cross-spectrum method to measure very small displacements (smaller than pixel). Defining the Fourier transform of $I(u,v)$ as $I(\xi,\eta)$, we write:

$$I(\xi,\eta) = \iint I(u,v)e^{2i\pi(\xi u+\eta v)} dudv = \iint |F(u,v)|^2 e^{2i\pi(u(\xi-x_0)+v(\eta-y_0))} = \overline{I}(\xi-x_0,\eta-y_0). \qquad (3)$$

A Coupled Cross-correlation - Cross-spectrum Technique. As mentioned above and shown in Fig. 2, for a large displacement (over 10 pixels), the use of the cross-spectrum method becomes inappropriate. Moreover, a shear mode might be difficult to analyse. Indeed, a minor displacement in the non-longitudinal direction creates a noise for the phase in the principal direction. In this case, it is suggested to use a coupled cross-correlation - cross-spectrum technique (CC-CS). The principle of the method is presented in Fig. 3. As shown is this figure, a cross-correlation is first used to determine a displacement of the order of one pixel. One of the two images is shifted about one pixel and a cross-spectrum analysis is then used to determine the sub-pixel scale displacement.

Finally, displacement between two images is calculated by summing the displacements obtained by the two methods. In the technique presented here, a single image is divided into a grid of sub-pictures. Their size varies between 8x8 and 256x256 pixels. The algorithm is then applied to each couple of sub-pictures to determine the local displacement field between them. Therefore, the size and number of sub-pictures give the maximum spatial resolution of the displacement field. Sub-elements can be perfectly placed side-by-side or overlapped.

In the next sections, a numerical validation and an application of real tensile tests are presented, to demonstrate the reliability of the proposed method.

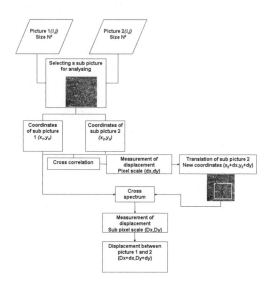

Fig. 3: Schematic algorithm of the coupled cross-correlation cross-spectrum (CC-CS) technique.

Numerical Validation

Two tests are defined hereafter. They consist in applying a virtually prescribed displacement field to the speckle image (800 x 800 pixels) as shown in Fig. 1, using the software GIMP.

Numerical Rigid Body Rotation. The image is numerically rotated by 2° around the image centre. Fig. 4a shows the displacement vectors which were calculated with the hybrid CC-CS algorithm shown in Fig. 3. The displacement is small in the centre of specimen and increases linearly towards the border of the analysed box, as expected.

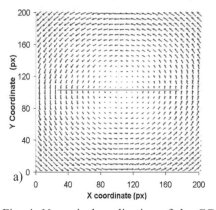
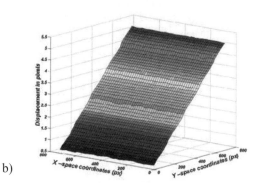

Fig. 4: Numerical application of the CC-CS technique: a) Displacement vectors for a prescribed rigid body rotation (central region only), b) Horizontal displacement field after tensile test.

Numerical Tensile Test. The same initial picture is elongated along one direction, by 5 pixels. The displacement after coupled CC-CS analysis is presented in Fig. 4b. A linear variation of the displacement field from the left side to the right side boundary can be observed. It can be noted that the displacement at the fixed boundary is found to be very close to zero and that the displacement field is continuous throughout the image. This demonstrates the sub-pixel accuracy of the proposed hybrid method.

Application to a Real Tensile Test

A notched thin sample (25 mm wide, 120 mm long, 1 mm thick) is subjected to a tensile load. One side of the specimen is illuminated by the laser, while the other side is covered by a speckle-like pattern created using black and white spray paints. All images are recorded on both sides of the sample using two cameras. Fig. 6a presents both speckle patterns. Different comparisons are applied on captured images, first for different correlation algorithms, and second for different types of speckles.

Comparison of Results Obtained on the Painted Surface. The presented algorithm and a commercial algorithm are both used to determine displacement fields on the painted surface of the sample. Fig. 5a and b show a comparison of the fields determined by two different algorithms, and Fig. 5c presents more precisely a displacement curve along the Y transverse axis. In both cases the results are similar.

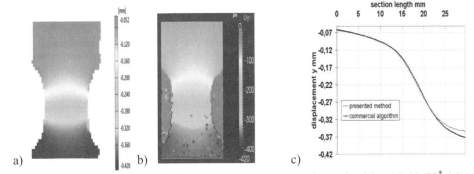

Fig. 5: Distribution of vertical displacement component, as determined by ARAMIS[*] (a) and the hybrid algorithm (b) on the painted face. On the right (c), comparison between displacement curves.

Comparison of Displacement Obtained with the Two Methods. The analysed area is not the same on the two faces of the sample; only a quarter of the sample is analysed. Fig 6b shows the comparison of displacement curves obtained with the laser speckle method and the painted speckle.

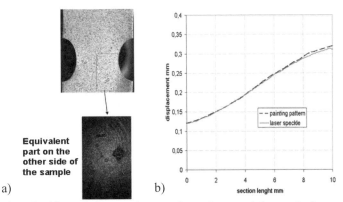

Fig. 6: a) Black and white painting on the sample surface, and the equivalent area on the other face over which the laser is projected. b) Comparison of displacements obtained with the two techniques on the two faces of the sample.

Application to Hot Tensile Test

Following previous validations, the hybrid CC-CS technique is now applied to tests performed at high temperature (1200°C). Fig 7 shows the analysis of a laser speckle pattern in order to determine a displacement field. As shown in Fig 7b, accurate correlation is possible except in areas exposed to severe oxidation (red circle area). As expected, the method remains reliable for high temperature except in the region where the surface aspect changes notably between two shots. Therefore, a careful protection against oxidation like argon gas along the surface is necessary. Another concern at very high temperature remains thermal gradients in gas boundary layers, or thermal plume detachment that could affect the accuracy of the speckle method. Using argon protection may help avoid such problem, but this needs to be confirmed.

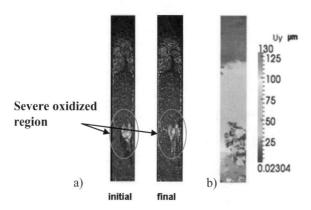

a)   b)

initial   final

Fig. 7: a) Example of speckle pattern on a hot sample (1200°C). b) Measured vertical displacement field.

Summary

This paper presents a non-contact method using laser speckles to evaluate the displacement field during a mechanical test. As shown, the method was validated on virtual and real tensile tests. Moreover an original algorithm based on a coupled correlation cross-spectrum method proves efficient for sub-pixel scale measurement and faster than algorithms commonly used. Given the calculation times, dynamic control is possible with such a speckle method. A first approach also shows the possibility to apply this method at high temperature. Besides, a finite element method is currently being developed to simulate a combined electrical thermal and mechanical test. And an inverse analysis procedure will be developed to determine high temperature behaviour.

Acknowledgement

This study has been conducted in the framework of the cooperative project "Cracracks" supported by the French Agence Nationale de la Recherche (ANR).
* ARAMIS is a registered trademark of GOM GmbH (http://service.gom.com)

References

[1] R. Knockaert, PhD thesis, MINES-ParisTech, (2001)

[2] C. Grec, C. Aime, M. Faurobert, G. Ricort, F. Paletou, A&A, Vol. 463, p. 1125-1136, (2006)

[3] L. Chevalier, S. Calloch, F. Hild, Y. Marco, Eur J mech A/solids, Vol. 20, p.169-187,

[4] H.J. Huan, A. Waas, Optical Eng., Vol. 46, p. 051005, (2007)

Applied Mechanics and Materials Vols. 24-25 (2010) pp 141-146
© (2010) Trans Tech Publications, Switzerland
doi:10.4028/www.scientific.net/AMM.24-25.141

Characterisation of full-field deformation behaviour using digital imaging techniques

Y. H. Tai, M. Zanganeh, D. Asquith and J. R. Yates

Department of Mechanical Engineering, University of Sheffield,

Mappin Street, Sheffield, S1 3JD, UK.

y.tai@sheffield.ac.uk

Keywords: full-field measurements, diametral contraction, DIC, edge detection

Abstract. Optical techniques for displacement measurements have become more common in recent years. The current preferred technique is digital image correlation (DIC) which works very well but has limitations for measuring diametral contractions in cylindrical specimens using a standard 3D system. To overcome the limitations of using either a diametral clip gauge or standard 3D DIC, a method has been developed for measuring diametral contractions simultaneously in two directions using a standard 3D DIC system in conjunction with an edge detection algorithm. Results have shown the method to work well.

Introduction

Optical techniques for full-field measurements have been gaining popularity in the recent years especially with advances in digital imaging and computational capabilities. One method which has seen significant increase in use and capabilities is digital image correlation (DIC). It works very well for both 2D and 3D full-field measurements given unlimited resources. However, for a standard 3D DIC system with a pair of cameras, the measurement area for a cylindrical specimen is limited to approximately one third of the circumference. This is due to the maximum overlap area between any given pair of images.

Accurate measurement of the diametral contraction of cylindrical tensile specimens with various notch geometries are used in the development of local damage models of fracture. Due to the anisotropic material properties encountered in many structural metals, the diametral contraction should be measured in at least in the x and z directions (please refer to Figure 3 for illustration). A 3D DIC system can provide these measurements but the tests have to be repeated using different specimens.

This paper describes a technique developed based on edge detection used in conjunction with a standard Vic3D DIC system to obtain diametral contraction data in two orthogonal directions simultaneously during a tensile test. Tensile tests were done on cylindrical specimens of an aluminium alloy with both plain and notched profiles.

Experimental work

Material

The material used for this study was 2050-T851 aluminium lithium alloy in 15mm thick plate form. Chemical composition of the plate is shown in Table 1 below.

Table 1: Chemical composition of 2050-T851 alloy used (wt%)

Ag	Zr	Mn	Cu	Si	Mg	Li	Zn	Al
0.36	0.09	0.38	3.51	0.03	0.40	0.91	0.02	Balance

Specimen geometry

Three different geometries of tensile specimens were used to allow for different states of triaxiality required for tuning the damage models. Specimen geometry G1 was a standard plain smooth tensile specimen. Specimen geometry G2 and G3 were notched with two different radii. Figure 1 illustrates the design and dimensions of the three specimens used which were machined from the longitudinal (L), transverse (T) and 45° directions. Table 2 summarises the specifications and number of specimens used in this work. Further information on the design of the specimens can be found in literature [1]

Table 2: List of specimens and specifications

Specimen type	No of Specimens	Gauge Length (mm)	D (mm)	R_o (mm)	ρ_o (mm)	R_o/ρ_o
G1 – L	2	40	10	-	-	-
G2 – L	2	40	10	4	6	0.7
G3 – L	2	40	10	3	2	1.5
G1 – T	2	40	10	-	-	-
G2 – T	2	40	10	4	6	0.7
G3 – T	2	40	10	3	2	1.5
G1 – 45	2	40	10	-	-	-
G2 – 45	2	40	10	4	6	0.7
G3 – 45	2	40	10	3	2	1.5

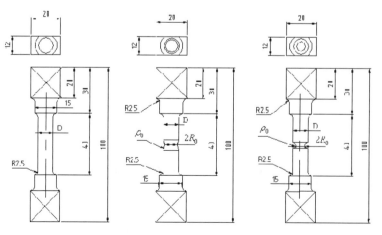

Figure 1: Specimen geometry G1, G2 and G3 (left to right)

Experimental set-up

The primary aim of the experiments was to obtain diametral contraction measurements in both x and z directions as illustrated in

Figure 3. In previous work, this was done using a single axis diametral clip gauge which meant the tests had to be performed twice [1]. With the advances in digital imaging and measurement techniques, it was worth investigating the feasibility of using digital image techniques for measuring diametral contractions and preferably simultaneously in both directions.

For contraction in the z-direction, 3D DIC was used which in this case was a standard Vic3D system where two cameras were used to give a 3D perspective of the specimen surface. Figure 2 shows the images of a prepared specimen obtained by camera 1 and camera 2. Comparison of the two images shows that the overlap between two images is limited and hence the effective area available for deformation measurement is constrained to the area of overlap as shown in Figure 2. Unless the system is expanded with more cameras which incur significant cost, a standard 3D DIC system for diametral contraction measurements is effectively a substitute for a clip gauge albeit negating the need for repositioning when the position of the final necking is not known *a priori*.

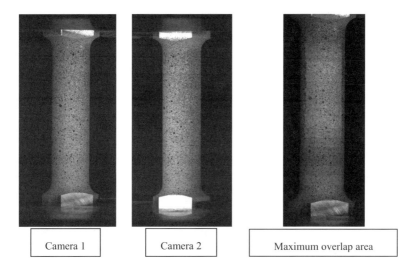

| Camera 1 | Camera 2 | Maximum overlap area |

Figure 2: Illustration of images obtained with stereo set-up and resultant overlap area

To overcome the limitations of using a standard 3D system, a third digital camera (3) was positioned between the two cameras used for DIC. All three cameras had the same sensor types with a resolution of 5 MP.

Figure 3 illustrates the position of all three cameras with respect to specimen. The position of camera 3 allowed for diametral contraction in the x-direction to be measured by analyzing the digital images obtained using an edge finding routine which is described in the following section. Synchronisation of the images was done by matching the recorded load signals. The main advantage of whole set-up is that diametral contractions can be measured simultaneously on the same specimen in both x and z directions. The tests were carried out using an Instron 8501 servo-hydraulic machine fitted with hydraulic grips under displacement control at a constant rate of 0.01mm/s for all the tests.

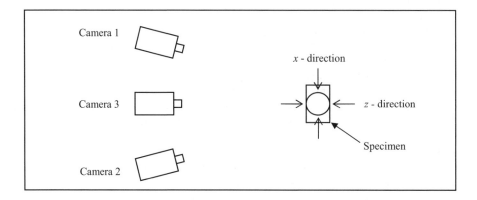

Figure 3: Plan view schematic of camera set-up w.r.t specimen

Edge finding routine

As shown in Figure 4, the edges of the specimen and consequently the load-diametral contraction of the specimens were measured using the Digital Image Processor 1.3 (DIP) [2]. The edge finding algorithm used in DIP is based on the extraction of the specimen from its background. The image obtained from camera 3 is an intensity image. To eliminate the background this intensity image needs to be converted to a binary image. This was carried out using a global threshold by minimizing the intraclass variance of the black and white pixels.

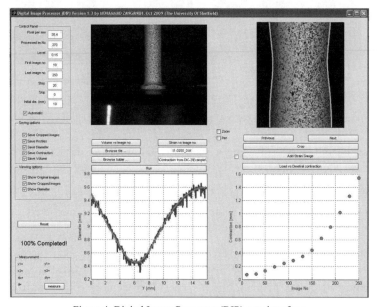

Figure 4: Digital Image Processor (DIP) user interface

Since the digital image correlation technique is being used simultaneously, a pattern of speckles is necessary on the surface of the specimen. Such speckles introduce some noise into the image. It was not possible to use back illumination for the 3rd camera used for DIP to reduce the noise levels because it would interfere with the cameras used for DIC. The noise was eliminated by using a higher value for global threshold or a flood-fill operation based on morphological reconstruction.

This would yield a matrix containing only zeros and ones where the zero elements are the background and the non-zero elements represent the specimen. However, the detected edges using above algorithm are still noisy, mainly because of required speckles on the specimen and hence the need for smoothing. To ensure that the outliners in the noisy data do not distort the behaviour of the smoothed curve, a locally weighted regression along with a robust smoothing procedure was used in DIP. A detailed explanation of this algorithm can be found in [2].

Apart from edge detection, DIP is capable of determining the Load vs Diametral contraction curve, true stress-strain curve, cross section area and volume of the specimen in the field of view. Furthermore, DIP can be used as a video extensometer in any direction using a two dimensional image correlation algorithm available in DIP V1.3.

Results and discussion

Images obtained from three cameras were processed using 3D DIC and DIP. Images from the first two cameras were correlated using Vic3D software. Since the eccentricity in loading is negligible and there is no buckling in the specimen (the loading condition is predominantly tension) the out-of-plane displacement (z - direction as shown in Figure 6) is representative of the radial contraction of the specimen. On the other hand, the x - direction diametral contraction can be measured using DIP as discussed in previous section. Experiments were performed in different material rolling directions with different triaxiality levels. Figure 5 to Figure 8 show the Load versus Diametral contraction determined using both techniques.

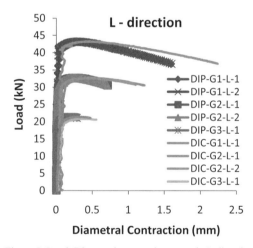
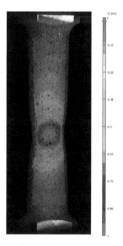

Figure 5: Load: Diametral contraction curve in L direction   Figure 6: Out-of-plane deformation measured using DIC

Although, compared to DIP results, DIC results are slightly scattered in elastic region of loading, results obtained for different repetitions of the experiments are very consistent using either DIC or DIP. In plastic and post necking regions both techniques are able to represent the material behaviour very smoothly. As it is evident from Figure 7 and Figure 8 the load-diametral contraction curves determined for T and 45° directions, using both techniques, matched well in most of triaxiality levels examined. However, there are some discrepancies in results obtained for L direction (Figure 5). This is mainly due to anisotropic contraction behaviour of this type of material. One the advantages of the technique explained is this paper is capability of measuring anisotropic behaviour.

The results also show that the edge detection technique is comparable to DIC for diametral contraction measurements. The maximum errors encountered in the technique were of the order of 3 pixels which in this particular case corresponded to 50 microns.

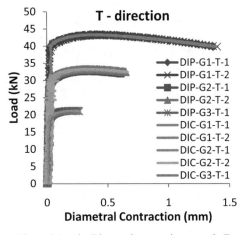

Figure 7: Load – Diametral contraction curve in T direction

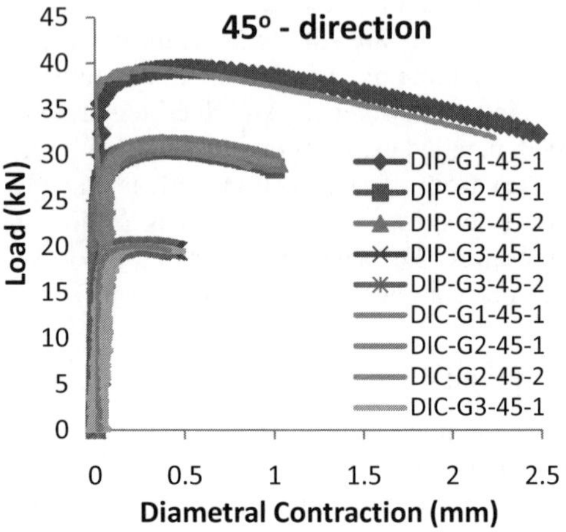

Figure 8: Load – Diametral contraction curve in 45° direction

Conclusion

The combination of measuring diametral contraction by the edge detection technique and DIC has identified the extent of the deformation anisotropy in the 2050-T851 aluminium alloy used. Work is currently underway to better integrate the two techniques and generate 3D profiles of the surface of the specimen during deformation.

References

[1]   Ayvar-Soberanis S, *3D CAFE modelling of ductile fracture in gas pipeline steel*, (2008). Dept of Mechanical Engineering, University of Sheffield.

[2]   Zanganeh M, Tai Y H, and Yates J R, *An optical method of measuring anisotropic deformation and necking in material testing,* (In press).

Applied Mechanics and Materials Vols. 24-25 (2010) pp 147-152
© (2010) Trans Tech Publications, Switzerland
doi:10.4028/www.scientific.net/AMM.24-25.147

Digital holographic interferometry
by using long wave infrared radiation (CO$_2$ laser)

Igor Alexeenko[1,a], Jean-François Vandenrijt[2,b], Marc Georges[2,c],

Giancarlo Pedrini[1,d], Thizy Cédric[2,e], Wolfgang Osten[1,f], Birgit Vollheim[3,g]

[1] Institut für Technische Optik, Pfaffenwaldring 9, Stuttgart, 70569, Gemany

[2] Centre Spatial de Liège, Avenue du Pré Aily, B-4031 Angleur (Liège), Belgium

[3] InfraTec GmbH, Gostritzer Str. 61-63, Dresden, 01217, Germany

[a]itoaleks@ito.uni-stuttgart.de, [b]jfvandenrijt@ulg.ac.be, [c]mgeorges@ulg.ac.be,

[d]pedrini@ito.uni-stuttgart.de, [e]Cedric.Thizy@ulg.ac.be,

[f]osten@ito.uni-stuttgart.de, [g]B.Vollheim@InfraTec.de

Keywords: holography, digital holographic interferometry, non-destructive testing.

Abstract. We show how digital holographic interferometry in the Long Wave InfraRed spectral range (LWIR) can be used for the investigation of mechanical structures. The 10.6 μm radiation is produced by a CO$_2$ Laser. Experimental results showing that the method can be used to locate defects in a panel are presented and advantages and disadvantages of this approach are discussed.

Introduction

Digital holographic interferometry is commonly used for non-destructive testing of mechanical structures and dynamical stress analysis (e.g., vibrations, non stationary dynamical processes). The technique is based on the comparison of wave fronts recorded at different instants of time [1] where the recordings correspond to the deformation states of the investigated object. The result of the wave fronts comparison is the calculated phase difference containing the information about the object deformation.

In the early years of holography, the holograms were recorded by using photographic plates or films only, but the CCD, CMOS and other detectors (having sensitivity in the visible, ultraviolet and infrared spectral range) developed during the past 20 years, allowed the electronic digital recording and processing and thus a dramatic speed-up of the measurements and information retrieval.

Long wave radiation in the infrared spectral range (LWIR) may be used in order to decrease the sensitivity of the holographic method. Some investigations about this subject have been already published [2-6]. We developed digital holographic interferometry at 10.6 μm for diffuse reflecting objects, the sensitivity is 20 times lower compared with methods using visible light and thus allows the recording of large deformations as they usually occur in aeronautical structures.

Infrared digital detectors can be used for the registration of holograms. Two types of such detectors are commercially available. One type uses an uncooled microbolometer array, the other is based on IR photodiode technology and is cooled down with a cryogenic system, this last detector has higher sensitivity and speed compared with the bolometers but it is more expensive.

The digital holographic interferometry can be used only if a speckle field is scattered by the object surface and this appears when the roughness reaches the magnitude of the used wavelength [7]. Since aeronautical structures have a roughness of approximately 10 micrometers, they can be easily investigated by using LWIR digital holographic interferometry [8].

Industrial applications usually require the measurement of large scale objects. Consequently CO_2 lasers are well suited for this purpose since they may produce radiation having power up to kW and thus allow the investigation of surfaces having size of some m².

Set-up

Figure 1 shows the set-up to record the image plane holograms [9] by using a CO_2 laser with wavelength 10.6 μm and coherence length 1-1.5 m and an uncooled microbolometer infrared camera (VarioCam-hr) manufactured by InfraTec. The Focal Plane Array (FPA) detector has 640x480 pixels (physical dimension of each pixel is 25x25 μm²), its dynamical range is 16 bit and the frame rate is 50 Hz. The imaging system is composed by a Germanium objective with 50 mm focal length. The basic parts of the interferometer have been isolated inside a beam delivery system in order to avoid that unwanted reflections from the surrounding environment hit the detector which is very sensitive and could be destroyed irreversibly by the CO_2 laser radiation. Attenuators were used to reduce the output intensity for the illumination and reference beams. Finally the reference beam was delivered by using a flexible hollow silica fiber with core diameter 300 μm. The use of a fiber together with the beam delivery system makes the laboratory unit compact, flexible and reliable. The investigated object (part of an aircraft) was a carbon composite 25x25 cm plate with a delamination defect. The loading system was an infrared lamp located at the back side of the object that increases the temperature of the sample and produces its deformation.

If we denote with $R(x,y)$ and $U(x,y)$ the reference and the object waves, then the intensity recorded on the sensor is:

$$I(x,y) = |R_H(x,y)|^2 + |U_H(x,y)|^2 + R_H^*(x,y)U_H(x,y) + U_H^*(x,y)R_H(x,y),\tag{1}$$

where the subscript H indicates the sensor plane and * denotes a complex conjugation. The third term contains the information about amplitude and phase of the object wavefront and the determination of the phase change allow us to calculate the deformation of the object as a function of the time. If we consider two states, the phase difference $\Delta\varphi$ depends on the sensitivity vector **s** and the object deformation **d**:

$$\Delta\varphi = \frac{2\pi}{\lambda}\mathbf{d}\cdot\mathbf{s},\tag{2}$$

$\mathbf{s} = \mathbf{k_1}-\mathbf{k_2}$ is determined by the geometry of the set-up where $\mathbf{k_1}$ and $\mathbf{k_2}$ are the unit vectors of illumination and observation respectively. Wave length and sensitivity vector determine the sensitivity of holographic set-up.

Two arrangements have been developed, in the first one a tilted reference wave (off-axis set-up) has been used, the second is an in-line configuration where a phase shifting of the reference is applied. When we use the off-axis configuration we are restricted by the spatial resolution of the detector and for recording the hologram the maximal spatial frequency should satisfy the condition $f_{max} < 1/(2\Delta)$ where Δ is the camera pixel size. If θ_{max} is the maximal angle between reference and object beam we get:

$$f_{max} = \frac{2}{\lambda}\sin\left(\frac{\theta_{max}}{2}\right) \approx \frac{\theta_{max}}{\lambda} \le \frac{1}{2\Delta}\tag{3}$$

The spatial frequency of the object beam has to be restricted by an aperture in order to satisfy this condition, then by taking the Fourier Transform of the intensity (Eq. 1) it is possible to select the third term. Its inverse Fourier transformation restores the complex amplitude of the object wave, thus the phase can be calculated as:

$$\varphi = arctan\left(\frac{Im(U_H(x,y))}{Re(U_H(x,y))}\right)\tag{4}$$

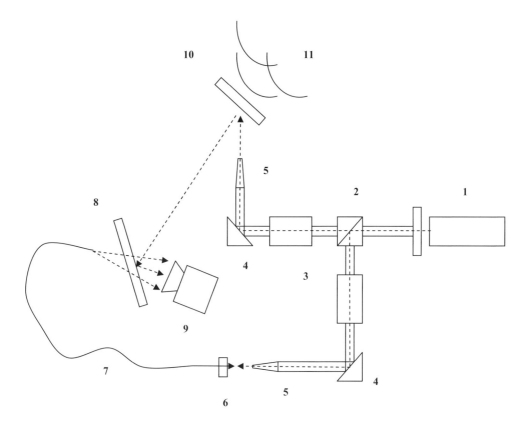

Fig. 1: Experimental set-up: 1. CO_2 Laser, 2. Beam splitter, 3. Intensity attenuators, 4. 45° Mirrors, 5. Output nozzles, 6. Piezo with fiber, 7. Hollow fiber, 8. Beam combiner, 9. IR camera,10. Object, 11. Thermal loading (2,3,4,5 are included in a closed compact beam delivery system)

For the in-line configuration, the calculation of the phase is done by using the four steps algorithm:

$$I_i = I_R + I_U + 2\sqrt{I_R I_U}\, cos(\phi + i\pi/2),\qquad(5)$$

where i=0, 1, 2, 3, I_0, I_1, I_2, I_3 are intensities recorded with phase shifted reference: 0°, 90°, 180° and 270° respectively. I_R and I_U are the intensities of reference and object beams. The phase is given by [10]:

$$\varphi = arctan\left(\frac{I_3 - I_1}{I_0 - I_2}\right),\qquad(6)$$

For the four steps algorithm the phase shifting of the reference beam was carried out by shifting a fiber mounted on a piezo stage. During the acquisition of the interferograms the trigger regime of the camera has been used. The system was full automated by using LabView software and NI data acquisition boards (DAQ). The phase difference calculation was implemented as post processing module in the main program.

Experimental results of digital holographic interferometry in LWIR

The object was illuminated by CO_2 laser radiation (power 300 mW). The interference between the speckled radiation field reflected by the object and the reference was registered by the FPA. We measured the roughness (RMS) of the object with the white light interferometer (Zygo New View 6300) and found that the roughness parameter R_Q was about 12 μm. This is approximately the magnitude of the wavelength and produces fully developed speckle which are ideal for the successful investigation of object deformations.

At first we used the Fourier transform method to calculate the phase difference due to the thermal loading on the object. The advantage of this method is that in one single hologram we have the information necessary for calculating the phase of the wavefront, it is thus well suited for the investigation of dynamical and high speed processes.

In order to restrict the spatial frequency of the object beam we reduced the camera aperture to 5 mm, since the default minimum aperture size of the IR camera was 7 mm and thus not enough to decrease the wavefront spatial frequency and satisfy Eq. 3. The reference beam was transmitted through the objective and focused into the aperture plane. The small aperture reduced the resolution of the recorded images and the quality of the phase difference map. It was not possible to get good results by using these methods, as shown in Fig. 2 the obtained phase map has poor quality and the defect recognition is not easy.

a)   b)

Fig. 2. Fourier transform method: a) FFT of the recorded intensity, b) Phase difference due to the thermal deformation, loading duration: 15 sec, investigated area: 15x15 cm^2

The four steps algorithm gives better results. Figure 3 represents the phase difference calculation due to the displacement under the thermal loading. High density circular fringes on the phase map and surface distortion on 3D representation can be observed in the delaminated zone. The spatial resolution of the image system is about 100 μm and the smallest detectable deformation is approximately $\lambda/10 = 1$ μm.

The experimental results (Fig. 3) show deformation magnitudes of about 15 μm. It is very difficult to measure such large deformations by using interferometers based on visible light. Figure 4 shows the mathematically modeled results for the same deformation but obtained by using different wave lengths. The calculations were done by simulating the effect produced by the wave length change on the phase difference. The twenty fold sensitivity (Fig. 4.c) obtained by using visible light (500 nm) leads to unresolved fringe distributions and is not well suited to detect deformations of about 15 μm. Large deformations (e.g. 5-100 μm) could be measured and properly represented by using interferometers based on visible radiation but in this case much more holograms should be recorded during the object displacement (e. g. by using a high speed camera) in order to get a reduced deformation between two consecutive holograms; further addition of the phases is necessary. However this approach is time consuming and increases the errors due to the addition of speckled phase map sequences.

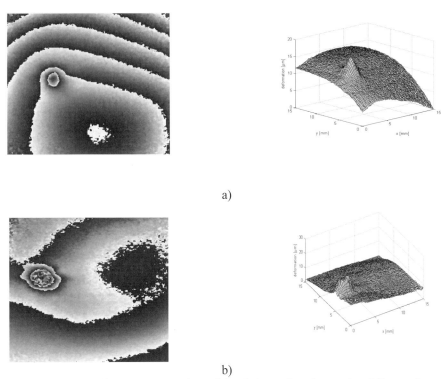

a)

b)

Fig. 3. Phase difference and 3D representation of the deformation due to the different thermal loading durations: a) 15 sec, b) 60 sec. The investigated area was 15x15 cm^2.

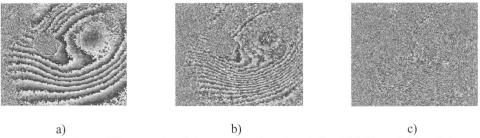

a)   b)   c)

Fig. 4. Modeled phase difference for different wave lengths. a) fivefold, b) tenfold and c) twenty fold sensitivity (investigated area: 15x15 cm^2)

Conclusion

We have developed optical NDT methods where LWIR radiation is used. These methods may be used successfully in measurements of large deformations. The algorithm for recording and processing is fully automated and allows a real time representation of the results.

We have shown that the Fourier transform method is not well suited to be used in our set-up due in particular to the relatively low spatial resolution of the IR detectors. The in-line configuration with the four step algorithm gives better results but may be used only for the investigation of relatively slow processes, because four interferograms need to be recorded in order to calculate the phase, which has to remain approximately constant during the image capturing. In order to speed up the method: short integration times and high frame frequency of the camera, are needed.

The disadvantage of the LWIR holographic interferometry is the high price of the detectors and some optical components, which is slightly counterbalanced by the low price of CO_2 lasers. However the detectors are becoming cheaper together with increased performances and resolution.

The advantage of our technique is the possibility of measuring larger deformations compared to equivalent methods using visible light, furthermore the use of LWIR allows measurements in more perturbed environmental conditions. However our first experiments are performed in laboratory conditions.

Acknowledgments

These works are funded by the FP7 European project FANTOM (ACP7-GA-2008-213457).

References

[1] C. M. Vest: *Holographic Interferometry* (Wiley & Sons, New York 1979).

[2] M. M. Rioux, M. Blanchard, M. Cornier, R. Beaulieu, D. Bélanger, in: Plastic recording media for holography at 10.6 μm, Appl. Opt. Vol. 16 (1977), pp. 1876-1879.

[3] E. Allaria, S. Brugioni, S. De Nicola, P. Ferraro, S. Grilli, R. Meucci, in: Digital Holography at 10.6 μm, Opt. Commun. 215 (2003), pp. 257-262.

[4] N. George, K. Khare, W. Chi, in: Infrared holography using a microbolometer array, Appl. Opt. Vol. 47, N° 4 (2008), pp. A7-A12.

[5] J.-F. Vandenrijt, M. Georges, in: Infrared Electronic Speckle Pattern Interferometry at 10 μm, Proc. SPIE Vol 6616 on Optical Measurement Systems for Industrial Inspection V, paper 6616-72, Munich, June (2007).

[6] S. De Nicola, P. Ferraro, S. Grilli, L. Miccio, R. Meucci, P. K. Buah-Bassuah, F. T. Arecchi in: Infrared digital reflective-holographic 3D shape measurements, Optic Communication, 281 (2008), pp. 1145-1449.

[7] I. Yamaguchi, in: Fundamentals and applications of speckle, Proc. SPIE, Vol. 4933 (2003), pp.1-8

[8] J.-F. Vandenrijt, C. Thizy, I. Alexeenko, I. Jorge, I. López, I.S. de Ocáriz, G. Pedrini, W. Osten, M.Georges, in: Electronic Speckle Pattern Interferometry at Long Infrared Wavelengths. Scattering Requirements, Fringe 2009 – 6th International Workshop on Advanced Optical Metrology, Stuttgart, September (2009).

[9] G. Pedrini and H. Tiziani, in: *Digital Speckle Pattern Interferometry and Related Techniques*, ed. P. K. Rastogi / Wiley & Sons, New York (2001).

[10] J. M. Huntley, in: *Digital Speckle Pattern Interferometry and Related Techniques*, ed. P. K. Rastogi / Wiley & Sons, New York (2001).

Session 4: Civil Engineering

Applied Mechanics and Materials Vols. 24-25 (2010) pp 155-160
© *(2010) Trans Tech Publications, Switzerland*
doi:10.4028/www.scientific.net/AMM.24-25.155

Size of the fracture process zone in high-strength concrete at a wide range of loading rates

R. C. Yu, X. X. Zhang[1] , G. Ruiz, M. Tarifa and M. Cámara

Universidad de Castilla-La Mancha, 13071 Ciudad Real, Spain

[1]zhangxiaoxinhrb@gmail.com

Keywords: Fracture process zone, loading rate, high-strength concrete

Abstract. Compared with the extensive research on properties of the fracture process zone (FPZ) under quasi-static loading conditions, much less information is available on its dynamic characterization, especially for high-strength concrete (HSC). This paper presents the very recent results of an experimental program aimed at disclosing the loading rate effect on the size and velocity of the (FPZ) in HSC. Eighteen three-point bending specimens were conducted under a wide range of loading rates from from 10^{-4} mm/s to 10^3 mm/s using either a servo-hydraulic machine or a self-designed drop-weight impact device. Four strain gauges mounted along the ligament of the specimen were used to measure the FPZ size. Surprisingly, the FPZ size remains almost constant (around 20 mm) when the loading rate varies seven orders of magnitude.

Introduction

For cementitious materials, the inelastic zone around a crack tip is denoted as the Fracture Process Zone. The extention and location of the FPZ is often dominated by complicated mechanisms, such as micro-cracking, crack bridge and friction. Due to the close relationship between the FPZ size and the characteristic length of a material, the FPZ evolution under different loading conditions has been the object of countless research efforts for decades [1-4].

In the present work, we chose the strain-gauge technology to measure the FPZ size in HSC at a wide range of loading rates, from 10^{-4} mm/s to 10^3 mm/s. Two testing devices, a hydraulic servo-controlled testing machine and a self-designed drop-weight impact device were adopted. Furthermore, the detailed information from the strain history records will undoubtedly facilitate the validation of numerical models aimed at disclosing rate dependency.

Experimiental procedure

Material characterization. A single HSC was used throughout the experiments, made with porphyry aggregates of 12 mm maximum size and ASTM type IV cement, I42.5L/SR. Micro silica-fume slurry and super plasticizer (Glenium ACE 325, B255) were added to the concrete composition. The mixing proportions by weight were 1:0.336:3.52:1.62:0.3:0.043 (cement: water: coarse aggregate: sand: micro-silica fume slurry: super plasticizer).

Compressive tests were conducted according to ASTM C39 and C469 on 75 mm ×150 mm (diameter × height) cylinders. Brazilian tests were also carried out using cylinders of the same dimensions and following the procedures recommended by ASTM C496. Eight cylinders were cast, four for compressive tests and four for splitting tests. The mechanical properties as determined from various characterization and control tests are shown in Table 1.

Three-point-bend fracture tests. As aforementioned, in order to study the loading-rate effect in HSC, three-point bending tests on notched beams were conducted over a wide range of loading rates, from 10^{-4} mm/s to 10^3 mm/s. Two testing apparatus were employed, one was a hydraulic servo-controlled testing machine, the other was a self-designed drop-weight impact instrument. The beam dimensions were 100 mm×100 mm (B×D) in cross section (beam width and depth), and 420 mm in total length L. The initial notch-depth ratio a_0/D (a_0: initial notch length) was approximately

0.5, and the span S was fixed at 300 mm during the tests, see Figure 1. Each specimen was removed from the moist room one day before the test and restored to the chamber after bonding the strain gauges. The specimen surface was polished and all four strain gauges (SG01-SG04, Model: LY 11 6/120A, 6 mm in length and 2.8 mm in width) were bonded to that surface, with a distance of 10 mm between each neighbouring gauge. Since a running crack in concrete is often deflected by aggregates along its path, the four strain gauges were bonded 10 mm apart from the centerline of the beam, see Figure 1. Those strain gauges provided not only the strain history at the bonded positions, but also the time at which the crack tip of the FPZ passed each strain gauge.

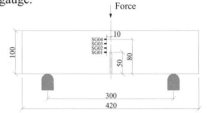

	f_c [MPa]	f_t [MPa]	G_F [N/m]	E [GPa]	ρ [kg/m^3]
Mean	102.7	5.4	141	31	2368
Std. Dev.	2	0.8	9	2	1

(Note: f_c: compressive strength; f_t: indirect tensile strength;

G_F: fracture energy; E: elastic modulus; ρ: density)

Table 1: Mechanical and fracture properties of the HSC tested.

Figure 1:Specimen with bonded strain gauges (units in mm).

Tests under loading rates from 10^{-4} mm/s to 10^1 mm/s. The tests were performed employing the hydraulic servo-controlled testing machine under position control. Three loading rates, from quasi-static level (5.50×10^{-4} mm/s) to rate dependent levels (0.55 mm/s and 17.4 mm/s), were applied. Three specimens were tested at each loading rate.

Tests under loading rates from 10^2 mm/s to 10^3 mm/s. All tests were conducted using the instrumented, drop-weight impact apparatus, which was designed and constructed in the Laboratory of Materials and Structures at the University of Castilla-La Mancha. More details are given in reference [5].The apparatus was employed to drop from three heights 40, 160 and 360 mm. The corresponding impact speeds were 8.81×10^2 mm/s, 1.76×10^3 mm/s and 2.64×10^3 mm/s, respectively. Three specimens were tested at each impact speed. A detailed description of the instrument is given in reference [6].

Crack-velocity measurement. When the fracture initiates, an unloading stress wave is generated and travels to the strain gauge, the sudden decrease of strain as a function of time indicates the crack initiation, see Figure 2 for a typical strain history record from one of the four strain gauges.

The crack velocity naturally refers to the speed in which this initiated cohesive crack tip, i,e. the FPZ front, will propagate. The time interval t_f is the crack initiation time. Additionally shown in Figure 2 are $t_{\varepsilon max}$ and $t_{\varepsilon 0}$, which indicate the time at peak strain and the time at which the strain is relaxed to zero, respectively. We define the time interval between $t_{\varepsilon max}$ and $t_{\varepsilon 0}$ as the strain relaxation time t_r.

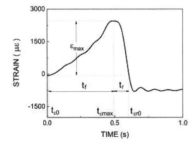

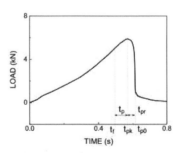

Figure 2: A typical strain versus time curve (shown in the record of SG01), taking the example of the loading rate at 0.55 mm/s.

Figure 3: The typical load history for low loading rates, taking example of 0.55 mm/s.

Since the stress wave speed is much greater than the crack propagation velocity [7], the time taken by the unloading stress wave to propagate from the crack line to SG0n (the offset distance from the center line is 10 mm) need not be taken into account. Thus an average crack-velocity between two neighboring strain gauges can be obtained through dividing the distance in between–10 mm– by the time interval across the two corresponding peak signals recorded.

Furthermore the peak load is also an important parameter, which reflects the loading capacity of a given structural element, in our case, a three-point-bend beam, consequently all the information related to the peak load is also essential. In Figure 3, we give all the peak-load related information in one typical load history curve for low loading rate. The terms t_p and t_{pr} are defined as the pre- and post-peak crack propagation time. The elapsed time between t_{max} at SG04 and t_{p0} is used to obtain the crack velocity along the last 20 mm where no strain gauge was bonded. In addition, knowing the crack length at peak load a_p, the pre- and post-peak crack propagation velocity v_1 and v_2 are also calculated as a_p/t_p and $(D-a_0-a_p)/t_{pr}$ respectively and given in the next section.

FPZ measurement. Hillerborg et al. [8] first proposed a fictitious crack model for fracture of concrete as shown in Fig. 4. In this model the newly formed crack surfaces and the corresponding fracture process zone are simply simulated by a cohesive zone located in the front of the initial crack tip. As a result, the energy dissipation can be represented by a material-specific cohesive law within the FPZ.

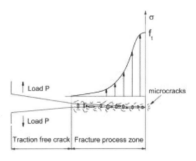

Figure 4. Sketch of concrete crack and FPZ

Here, we explore the advantage of the strain-gauge technology, having in mind that the attainment of peak strain signals, the passing of a cohesive tip, and strain values relaxed to zero represent a traction-free crack tip.

Results and discussion

The measured load histories are depicted in Figure 5. It needs to be pointed out that under high loading rates, the load refers to the impact force, i.e., the inertial force is also included.

Information related to the peak load, such as the dynamic increase factor (DIF), the time intervals t_{pk}, t_p and t_{pr} are reported in Table 2. The measured velocities v_{sg}, the pre- and post-peak crack propagation velocities v_1 and v_2 are all listed in Table 3.

Loading rate effect on peak loads. From Figure 5, note that the peak load increases proportionally with the loading rate, such rate effect is minor at low loading rates while it is pronounced at high loading rates. We define the dynamic increase factor (DIF) as the ratio of peak load and its corresponding quasi-static value (5.50×10^{-4} mm/s in this case). The DIF for peak loads are 1.4 and 25.0, for the loading rates of 17.4 mm/s and 2.64×10^3 mm/s, respectively. In other words, the DIF at high loading rates is approximately one order higher than that at low loading rates.

It also needs to be pointed out that in Figure 5 (bottom row), we have scaled the load-axis by a factor proportional to its loading rate. Note that the peak load increases slightly faster than its loading rate. This is mainly due to the significant increase of inertia forces, see [9].

It is noteworthy that, at low loading rates, when the load peak is achieved, the crack length (a_p, notch length not included) increased from 10 mm and 4 mm (5.5×10^{-4} and 5.5×10^{1} mm/s) to 37 mm (17.4 mm/s); while at high loading rates, the crack length varied from between 5 to 14 mm for all three cases, see Table 2. In particular, for the loading rate of 17.4 mm/s, when the peak load is achieved at t_{pk} of 21 ms, SG02 is deformation free at $t_{g\cdot02}$ of 19.9 ms, this shows the first 10-mm stretch from the notch tip is already traction free.

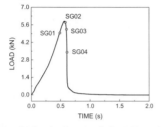 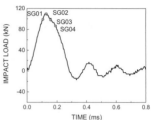

Figure 5: Load history for low loading rate (left): 0.55 mm/s, and high loading rate (right): 2.64×10^3 mm/s, where SG0n marks the time at which the strain peak is obtained for strain gauge SG0n (n=1, 2, 3,4). Note that for each graph, the load-axis is proportionally scaled to its loading rate.

Loading rate	Peak load	DIF	t_f	t_{pk}	t_p (t_{pk}-t_f)	t_{p0}	t_{pr} (t_{p0}-t_{pk})	a_p
[mm/s]	[kN]		[s]	[s]	[s]	[s]	[s]	[mm]
5.5×10^{-4}	4.4	1.0	432	494	62	512	18	10
[mm/s]	[kN]	-	[ms]	[ms]	[ms]	[ms]	[ms]	[mm]
5.5×10^{-1}	5.9	1.3	490	567	77	614	47	4
1.74×10^{1}	6.3	1.4	15.8	21	5.2	23.8	2.8	37
[mm/s]	[kN]	-	[μs]	[μs]	[μs]	[μs]	[μs]	[mm]
8.81×10^{2}	30.3	6.9	168	200	32	428.5	228.5	11
1.76×10^{3}	63.4	14.4	128	172	44	331.0	159	14
2.64×10^{3}	109.9	25.0	108	120	12	284	164	5

Table 2: Peak load and information related to peak load.

Loading rate effect on crack velocity. The crack velocities are listed in Table 3. In the low loading rate range, on the one hand, for each loading rate, the crack advances with increasing speed; on the other hand, as the loading rate increases, the crack velocity increases proportionally. For instance, at 5.5×10^{-4} mm/s, the crack velocity increased by a factor of 38 from 0.19 mm/s for v_{sg1} to 7.3 mm/s for v_{sg3}; while at the loading rate of 2640 mm/s, the crack speed varied from 417 m/s to 387 m/s. When the loading rate increased by a factor of 1000 (from 5.5×10^{-4} mm/s to 0.55 mm/s), the first-stage crack velocity v_{sg1} increased by 4100, while the late-stage velocities v_{sg3} and v_{sg4} only increased by a factor of 1369 and 1476 respectively. This indicates that, when the loading condition changes from quasi static to low loading rates, the loading rate effect on the early-stage crack velocity is almost three times stronger than its effect on the late-stage crack propagation; however, within the low loading rate range, when the loading rate increased by 34, from 0.55 mm/s to 17.4 mm/s, the increase factor from v_{sg1} to v_{sg3} remained practically the same (from 14.4 to 17.3). Within the high loading rate range, on the contrary, the crack advances with decreasing speed, and as loading rate increases, the crack propagation speed tends to be uniform, this is clearly seen from the pre and post-peak crack velocities. The maximum crack velocity reached approximately 20.6% of the Rayleigh wave speed.

Comparing the numerically-predicted two-stage crack propagation in [9], the experimentally observed pre- and post-peak velocities in Table 3 suggest that, at low loading rates, pre-peak crack propagation is stable in a sense that, continuous loading is necessary for continuous crack advancing, whereas post-peak one is unstable, since less external load leads to faster crack propagation. On the contrary, at high loading rates, impact loads result fast crack propagation from the very beginning, less external load at post-peak is accompanied by a slower crack extension.

Loading rate (mm/s)	V_{sg1} SG01-SG02 (m/s)	V_{sg2} SG02-SG03 (m/s)	V_{sg3} SG03-SG04 (m/s)	V_{sg4}* (m/s)	v_{max}/v_R %	Pre-peak V_1 (m/s)	Post-peak V_2 (m/s)
5.5×10^{-4}	1.9×10^{-4}	2.7×10^{-4}	7.3×10^{-3}	2.1×10^{-3}	-	2.3×10^{-4}	1.2×10^{-3}
5.5×10^{-1}	0.78	0.73	1.05	3.1	-	0.58	0.73
1.74×10^{1}	11.2	12.6	16	4.2	-	6.8	4.2
8.81×10^{2}	292	250	208	138	14.4	344	171
1.76×10^{3}	357	278	357	187	17.6	327	224
2.64×10^{3}	417	417	387	200	20.6	417	275

* V_{SG4}, crack velocity along the last 20 mm distance

Table 3: Average crack velocity evolution.

Loading rate effect on the size of FPZ. Figure 6 shows the method to determine the growth and development of the FPZ, taking the example of the loading rate at 2640 mm/s.

The upper half of the figure gives four strain histories recorded in the four strain gauges, with the time at peak strain $t_{\epsilon max}$ and the time when the strain relaxed to zero $t_{\epsilon r0}$ marked with filled squares and circles respectively. The time at peak load t_{pk} is also shown to distinguish the pre and post-peak crack propagations. The lower half of Figure 6 shows the FPZ evolution with time during loading. Again, the crack velocity between two neighbouring strain gauges is the average , the variation of the velocity was not taken into account along this distance.

Table 4: FPZ size

Loading rate (mm/s)	FPZ size (mm)
5.5×10^{-4}	14-20-17
5.5×10^{-1}	25-47
1.74×10^{1}	17-21-14
8.81×10^{2}	23-21-16
1.76×10^{3}	16-19-15
2.64×10^{3}	18-21-16

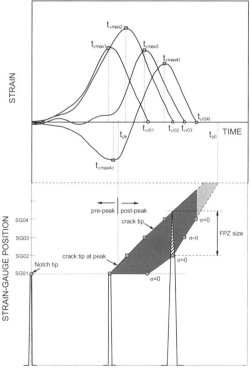

Figure 6: Methodology to estimate the development and growth of FPZ. Filled square symbols represent time at peak strain, whereas filled circles stand for time when the strain relaxed to zero. The upper half shows the strain histories recorded in the four strain gauges; the lower part illustrates the initiation and propagation of the main crack, where the shaded zone is the evolution of the FPZ during loading. The dashed-line-surrounded shadow indicates unconfirmed information due to lack of further measurements. Shown is the case of the loading rate at 2640 mm/s.

The upper limit of the shaded zone shows the evolution of the cohesive crack tip, while the lower one represents the traction-free crack tip. For instance, in order to know the FPZ ended at SG02, i.e., when $\sigma = 0$ is reached at $t_{\varepsilon r02}$, one needs to know the current location of the cohesive crack tip. From the upper part of Figure 6, we find the intersection point between the line $t = t_{\varepsilon r02}$ and the upper limit of the dark shaded zone, the distance between this intersection point and SG02 is the sought FPZ size. Note that, the FPZ was not completely developed either within the first nor the last 20 mm due to boundary effects. Since four strain gauges were employed to measure the strain history, at most three FPZ sizes can be directly obtained, more values can be obtained through interpolation as in Figure 6. We nevertheless list only those directly obtained FPZ sizes in Tab. 4 separated by a dash "-" sign. If we exclude the possible boundary effects of the notch and final ligament of each specimen, the central FPZ size in Tab. 4 should be considered as the material FPZ size. Surprisingly, the FPZ size remained almost the same when the loading rate varied seven orders of magnitude. This is clearly different from NSC, in which the FPZ size actually decreased with loading rate, see Du et al. [10,11] and Wittman [6].

Conclusions

Using strain-gauge technology, employing a servo-hydraulic machine and a drop weight impact device, we have measured crack propagation velocities and the size of the FPZ for a HSC loaded over a wide range of loading rates, from 10^{-4} mm/s to 10^{3} mm/s. The following conclusions can be drawn. (a) The peak load is sensitive to the loading rate. Under low loading rates, the rate effect on the peak load is minor, while it is pronounced under high loading rates. (b) The measured time to peak load t_{pk}, a measure of the initial CMOD rate, varied from 0.12 ms to 494 s. (c) Unlike normal strength concrete, the FPZ size varied only slightly for loading rates of seven orders of magnitude. (d) Under low loading rates, the main crack advances with increasing velocity, the late-stage velocity is one-order higher than the early-stage one; the rate effect on the crack velocity is remarkable. At high loading rates, the main crack propagates with a decreasing crack velocity of several hundred m/s, the rate effect on crack velocity is minor. In addition the crack propagation velocity in the high loading-rate range reached 20% of the material's Rayleigh wave speed.

References
[1] Cedolin, L, Dei Poli, S., and Iori, I. Cement and Concrete Research, 13, pp. 557-567, (1983).
[2] Castro-Montero, A., Shah, S.P., and Miller, R.A., Journal of Engineering Mechanics, 116, pp. 2463-2484 (1990).
[3] Swartz, S.E., and Go, C.G., Experimental Mechanics, 24, pp. 129-134 (1984).
[4] Hu, X., and Wittmann, F.H., Journal of Materials in Civil Engineering, 2, pp. 15-23 (1990).
[5] Zhang, X.X., Ruiz, G., and Yu, R.C.: submitted to *Strain, doi:10.1111/j.1475-1305.2008.00574.x.* (2008).
[6] Wittmann, F.H., Sadhana - Academy Proceedings in Engineering Sciences, 27, pp. 413-423 (2002).
[7] Mindess, S., Canada Journal of Physics, 73, pp. 310-314 (1995).
[8] Hillerborg, A., Modéer, M., and Petersson, P.E., Cement and Concrete Research, 6, pp. 773-781, (1976).
[9] Zhang, X.X., Ruiz, G., and Yu, R. C., Journal of Materials in Civil Engineering, 20, pp. 544-551 (2008).
[10] Yon, J.H., Hawkins, N.M., and Kobayashi, A.S., ACI Materials Journal, 89, pp. 146-153 (1992).
[11] Du, J., Yon, J.H., Hawkins, N.M., Arakawa, K., and Kobayashi, A.S., ACI Materials Journal, 89, pp. 252-258 (1992).

Applied Mechanics and Materials Vols. 24-25 (2010) pp 161-166
© (2010) Trans Tech Publications, Switzerland
doi:10.4028/www.scientific.net/AMM.24-25.161

Practical in-situ applications of DIC for large structures

N. J. McCormick [1, a], and J. D. Lord [1, b]

¹National Physical Laboratory, Hampton Road, Teddington, Middlesex, TW11 0LW, UK

ᵃnick.mccormick@npl.co.uk, ᵇjerry.lord@npl.co.uk

Keywords: digital image correlation, DIC, in-situ, crack measurement, civil engineering

Abstract. The paper describes recent use of Digital Image Correlation (DIC) for in-situ measurements of deformation and cracking of large civil engineering structures like bridges and power plant. Recent work at NPL has demonstrated the potential of DIC as a novel NDT tool for measuring deformation and cracking in reinforced concrete structures. This has particular application where the area of interest is in a region where inspection is difficult or costly and where direct access may have safety implications. In this case accurate measurements from pairs of images can be very cost effective.

Work is on-going in establishing the effects on the measurements caused by environmental effects and the requirement for repositioning of the camera during image capture. Techniques for mitigating these effects on accuracy will be reported. This paper will present data from some initial studies and discuss some of the factors that influence the accuracy of the technique when used outside the laboratory.

Introduction

Large infrastructure like power plant, bridges and tunnels need robust monitoring techniques to maximise lifetime and to economically schedule maintenance. Changing staff demographics means fewer skilled personnel will be available for inspection in the future and so automatic inspection techniques are required.

Full field optical monitoring techniques like digital image correlation (DIC) have potential for low cost examination of large structures where access may be difficult or costly. Full field techniques can also highlight where continuous point monitoring sensors may be best placed and the effect of repair on a structure.

DIC has had limited use outside the laboratory, but recent developments in cameras and the ready availability of cheap processing means it has considerable promise for practical in-situ measurements of large structures.

Digital Image Correlation

Digital image correlation is a technique that uses computer software to compare two images, such as digital photographs at different stages of a mechanical test. The technique was originally developed in the early 1980's [1,2] although more recent developments reviewed in [3] have enabled displacement accuracies to approach 0.01 pixels. This combined with the relative low cost of high resolution digital cameras means measurements can be made with resolutions approaching 1:100,000 of the field of view. The increase in resolution of camera images has been matched by increased low-cost computing power and this has enabled more cost effective, rapid measurements to be made.

The technique works by matching small subset images taken from two images being compared, that may be 11x11 pixels, for example. Providing there is sufficient contrast and hence uniqueness within each subset image block then a successful match can be made. By first carrying out a quick integer level comparison using a correlation function, an approximate map of the deformations required to match these subset images across the whole image field can be made. This can then be refined by using interpolation techniques to allow sub-pixel displacements combined with strain, shear and rotations to be added, thus increasing the deformation resolution of DIC.

NPL has developed analysis algorithms to allow accurate measurements to be made even from 60MPixel images and with displacement resolutions of up to 0.01 pixels. Using a distributed parallel computing system, the NPL Grid, can speed up calculations by up to 300x over a single PC.

Practical In-situ Examples

Until recently the achievable image resolution and difficulties of making suitable high quality in-situ measurements limited DIC measurements to laboratory measurements. These were mainly for materials characterisation or measurements on structures where there were large or rapidly varying strain distributions. There are examples of DIC being used for component testing of civil engineering structures [4] but relatively few examples where measurements have been made out of the lab or in-situ [5, 6]. The following examples have been chosen to give an idea of the range of structures for which DIC may be used.

Large Scale Structures: Bridges. Non-contact displacement measurements were made by mounting a camera close to a highway bridge during load testing. The measurements were made during the night and both flash lighting and ambient lighting from streetlights was used. The differences in deformation were measured between images captured with no-load and loading produced by four 32 tonne lorries being parked on the bridge structure in different locations.

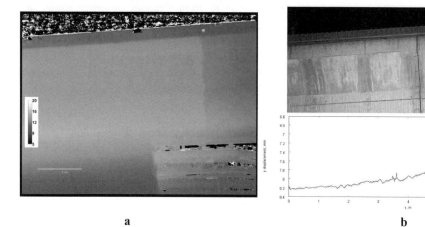

a   b

Figure 1. Displacement (in pixels) in the vertical (y) direction of a highway bridge during a static load test, Fig. 1a. Displacement (in mm) in the vertical (y) direction, Fig. 1b. The difference in vertical displacement across the bridge joint can clearly be seen in the profile measured along the red line.

The images were captured using a medium format PhaseOne camera with a 39 MPixel digital camera back using a 80mm lens at a working distance of about 15 metres. Results of the measurements can be seen in Fig. 1a and 1b these were consistent with laser tracker measurements taken at the same time.

Large Scale Structures: The London Eye. In this example two pictures were taken of the London Eye a few seconds apart, Fig. 2. By using DIC the magnitude of movement of the objects in the image can be calculated from the x and y displacements. As can be seen, Fig. 2b the edge of the London Eye has moved equally. The large amounts of noise in the background is due to the limited contrast available in the cloudy sky, and as expected the buildings behind and to the side of the London Eye show no movement.

The rim of the London Eye has moved about 0.5 m between the two images which were taken two seconds apart which implies a speed of 0.25 m/s, this compares favourably with the published speed of 0.26 m/s.

 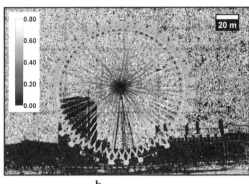

a   b

Figure 2. The London Eye, image taken from the Embankment using a 6 MPixel Nikon D40, Fig. 2a. Fig. 2b an image showing the magnitude of movement of different parts of the London skyline shown in Fig. 2a. The colour scale is in metres.

Small Scale Structures: Vibrating Pipes. In this example a pair of images were taken of vibrating pipes used for cooling water, Fig. 3. These are typical of insulated pipes that might be found in many process plant. Fig 3b shows the displacement in the y direction, i.e. downwards in the image, the two pipes that were vibrating can be clearly seen, the images were taken using a 60MPixel PhaseOne Camera at a distance of about 5 metres, the amplitude of the motion is less than 1 mm.

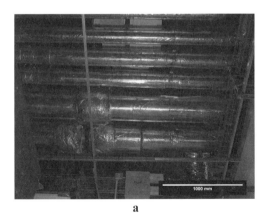 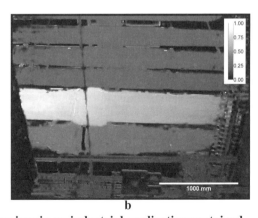

a   b

Figure 3. An image of insulated cooling water pipes in an industrial application contained within a roof space, Fig. 3a. Fig. 3b shows the deformation in the y direction (downwards in the picture) and the scale is in mm. The vibrating pipes can clearly be identified and the consequent motion of the adjacent pipes compared to the black (zero movement) background can also be seen.

Crack Opening Measurement in Reinforced Concrete. DIC can be used to detect cracking and measure crack opening in reinforced concrete. The surface finish of typical concrete structures that have naturally weathered provide a sufficiently random surface that enables accurate full field deformation measurement using DIC.

Fig. 4a shows a typical crack in concrete, as can be seen the width varies across the length of the crack making a crack width difficult to define. However often the change in crack width is more important for assessing structures and in this case capturing two images and applying DIC techniques allows high accuracy measurements of change in crack opening to be made.

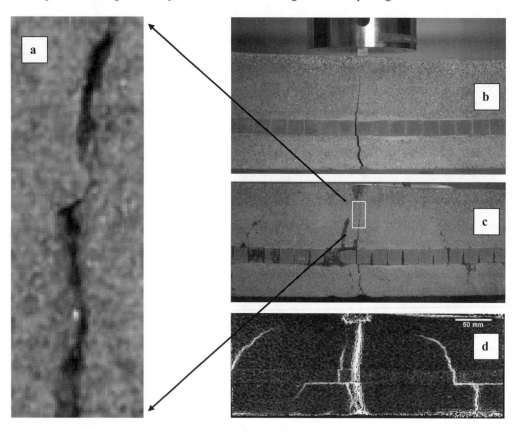

Figure 4. A cracked concrete specimen, Fig. 4a shows a close up of the crack in Fig. 4c. Fig. 4b shows the concrete specimen being cracked. Fig 4c shows the cracks identified using dye penetrant, Fig. 4d shows a DIC image plotting "strain" or local deformation, light areas are where there is high "strain".

Fig. 4 shows a cracked reinforced concrete specimen that has had controlled cracks introduced using three point bending, Fig.4b. To highlight the position of the cracks red dye penetrant has been used, Fig 4c. Local deformation or "strain" is derived from DIC measurements calculated from images of the concrete specimen before loading and after cracks have formed. In Fig. 4d. the light areas clearly show the position of large crack opening and smaller cracks that couldn't be identified using the dye penetrant method but can be clearly seen in this 14 MPixel image.

Techniques For Compensating For In-situ Effects

There are many additional factors that need to be considered when making DIC measurements outside the laboratory and in-situ. Weathering of common civil engineering materials, like concrete, provides surfaces with sufficient texture and randomness to allow full field measurements to be made, and the NPL DIC code has been designed to allow for variations in lighting between images. A strategy for coping with variable surfaces like wet or dry surfaces has also been developed and will be implemented in the next version of software. Imaging at an angle can be compensated successfully for using geometric models and measurements of camera position and subject position using standard surveying techniques.

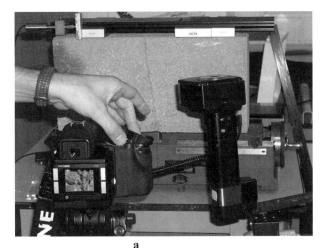

a   b

Figure 5. Fig. 5a shows the experimental arrangement used to explore repositioning of the camera between digital photographs. Fig. 5b shows the kinematic mount used to mount the camera and to help reposition the camera accurately.

One of the biggest issues with in-situ measurements is the problem introduced by repositioning a camera, as might be required when using a fixed camera position with photographs only taken every few months. To investigate solutions to this experiments were carried out making measurements of a controllable crack, see Fig. 5a, by repositioning the camera each time an image was captured using a kinematic mount similar to Fig. 5b. This type of solution will aid control of the camera position in the x, y and z directions and will control orientation around these coordinate axes also. If the lens used is a fixed focal length then the camera auto-focus will result in the same field of view.

Fig. 6 shows measurements of crack opening plotted against a LVDT displacement sensor. The choice of the axes units was chosen to show the scale of the measured crack opening in pixels to reinforce the idea that whilst different optical setups can be used to change the field of view in terms of linear distances, it is the pixel displacements that are the fundamental units of resolution. In this case the crack opening measurement resolution is much less than 1 pixel and can be as high as 0.1 to 0.01 pixels in optimum conditions, so for a horizontal field of view that can be up to 9000 pixels this implies a resolution greater than 1 part in 90000 of the horizontal field of view. As the measurement uses images these allow easy calibration of the system. Measurements of crack opening made using the highest resolution camera systems indicated that resolutions of better than 0.5 micrometres could be achieved in optimal circumstances.

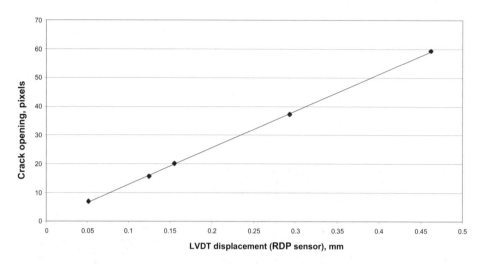

Figure 6. Crack opening measured using DIC versus crack opening measured using a LVDT displacement sensor fixed to the concrete specimen shown in Fig. 5a

Conclusions

This paper has demonstrated that DIC can be used for in-situ measurement of large civil engineering structures measuring full-field deformations, displacements and crack opening. Recent industrially funded feasibility studies carried out for transport infrastructure owners and nuclear waste storage specialists have identified the metrological challenges that need to be overcome to enable more widespread adoption of DIC for large structure measurements in an out-of-laboratory environment.

Acknowledgements

The authors would like to acknowledge the support from URS Corporation, Carillion plc, Sellafield Ltd. and the UK Government National Measurement Office.

References

[1] M. A. Sutton and W. J. Wolters: Mater. Image Vision Computing Vol. 1 (1983), p. 133

[2] T. C. Chu, W. F. Ranson, M. A. Sutton and W. H. Peters: Experimental Mechanics Vol. 25 (1985), p. 232

[3] P. Bing, X. Hui-min, X. Bo-gin, and D. Fu-long: Measurement Science Technology Vol. 17 (2006), p. 1615

[4] C. De Roover, J. Vantomme, and J. Wastiels: Experimental Techniques Vol. 25 (2002), p. 37

[5] S. Yoneyama, A. Kitegawa, S. Iwata, K. Tani, and H. Kikuta: Experimental Techniques (2007), p. 35

[6] M. Kuntz, M. Jolin, J. Bastien, F. Perez, and F. Hild.: Canadian Journal of Civil Engineering Vol. 33 (2006), p. 1418

Applied Mechanics and Materials Vols. 24-25 (2010) pp 167-171
© *(2010) Trans Tech Publications, Switzerland*
doi:10.4028/www.scientific.net/AMM.24-25.167

From the Inside, Out

Use of Optical Measuring Techniques for Wind Turbine Development

A. Stanley[1a,], M. Klein[2b]

[1]GOM UK

[2]GOM mbH

[a]a.stanley@gom.com, [b]m.klein@gom.com

Keywords: Optical measurement, 3D scanning, Photogrammetry, DIC

Abstract. Current trends in the wind turbine industry demand that manufacturers and related research facilities increase development efficiency whilst minimizing production costs. A key component developed is the blade, which needs to efficiently capture the wind energy whilst being able to survive severe weather conditions. More recently larger blades have been used, where main advantages include an increase in wind capture efficiency, a reduction in the installation and reduction in maintenance costs per MW. This paper outlines the main principles behind optical measurement techniques employed, influencing factors for using this technology, and presents the importance of the results achieved thus far and describes ongoing development activities.

Introduction

In order to better understand the local material and component response and the global behaviour of wind turbines, GOM's Optical Measuring Techniques were employed. Components were measured for the determination of raw material parameters, component response under loading, as well as in 3D quality control applications.

Component Deformation Measurement

With larger blades being implemented, use of lightweight material affords reduced inertial forces on the drive train. However, consideration needs to be taken of non linear failure modes resulting from increased flexibility [1].

The ARAMIS 3D deformation measuring system uses digital image correlation. Using a black and white speckled deterministic pattern, grey value distributions are calculated for a large amount of small subsets in each camera image and provide sub-pixel accurate positions about corresponding points between all images (Fig.'s 1 and 2).

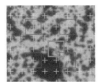

Figure 1: Undeformed patterned specimen with regular subsets

Figure 2: Undeformed patterned specimen with, now with irregular subsets

The mathematical model of the sensor setup, the digital image correlation method and a triangulation calculation are combined to derive highly accurate 3D coordinates. Subtracting the surface information in all loading stages in 3D space provides precise X, Y and Z displacement values.

Strains are calculated considering the component's geometry and plasticity theory. As the image acquisition is time-based, even 3D velocities and strain rates are automatically achieved.

Out of Plane Flexure on light weight Internal Components
Composite material is essential for modern wind turbine blades due mainly to its high strength-to-weight ratio. Understanding deformation of the load carrying spars that run the length of the blade is a key step in understanding blade behaviour. Past research has shown how these structures are subject to out of plane flexure due to the Brazier effect [1], and the optical measurement technique has been used in conjunction with coupon spar specimens under flexure to analyse how this out of plane deformation affects the anisotropic blade structure [2].

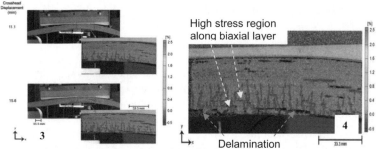

Fig.: 3) bend test on composite turbine blade material; 4) Display of biaxial strain and delamination in the vicinity of matrix cracks.

The measurement results showed numerous delaminations, mainly at the highest bending strains (Fig. 3). Closer examination shows that at high strains along the biaxial layers, with cracks present, the biaxial layers themselves still hold the majority of the bending load (Fig. 4) [2].

Large scale tests and FE verification:
The measurement principle remains constant for all test types, only here, a 34 meter glass fibre core of a wind turbine blade was bended to failure, where a measurement area of 3x3m was set up (Fig. 5).

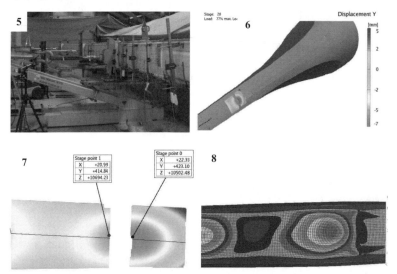

Fig. 5. Turbine blade test set up; Fig. 6. Coordinate system for Aramis plot on CAD dataset; Fig. 7.Aramis results for bending effects of blade surface; Fig. 8; Result from FE simulation.

The results obtained give integral information regarding bending effects on the surface of the blade (Fig. 7). However, it goes one step further than that in being used for the integrated verification of simulations (Fig. 8).

Furthermore, it is possible to import CAD datasets, which in this case have been used to set a coordinate system for the measurements results (Fig. 6).

Wind turbine off-shore mounting
Using the GOM TRITOP coordinate measurement technique, a measurement object is recorded with a high-resolution digital camera.

For the setup of offshore wind turbines the components are built on land then are tugged to their destination ashore and where the foundations are anchored to the sea bed (Fig. 9). For mounting the lowest tower segment, 120 steel bolts are recessed in the concrete foundation, positioned in two circles with a diameter of approx. 4 meters (Fig. 10).

Fig. 9: Offshore anchored concrete foundation

Fig. 10: Position of mounting bolts in the concrete foundation and preparations for the measurement

The mounting bolts in the concrete foundation have to be verified to the required accuracy of 1/10 millimetre. The knowledge about the 3D positions of the bolts enables the verification of these against the nominal positions directly after the completion of the foundations ashore as well as after the anchoring on high seas (Fig. 11).

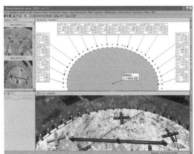

Fig. 11: Inspection of a single and all 120 mounting bolts positions.

Falsely positioned or tilted mounting bolts can be identified quickly and clearly with the required accuracy eliminating problems before starting the mounting process [3].

Wind operation analysis
The GOM PONTOS approach is one of point target tracking. It uses a calibrated pair of cameras to triangulate the X, Y and Z coordinates of the centres of circular dot targets with an accuracy of approximately 1/50 pixel with a nominal accuracy of 30 microns per meter of field of view (Fig. 12). The coordinates of each target at a reference condition are subtracted from the current coordinates in each test image to compute the 3D displacements.

The optical system setup for measuring deflections on a wind turbine is seen in Figure 13 [4, 5] - targets are recorded at a speed of 100 Hz.

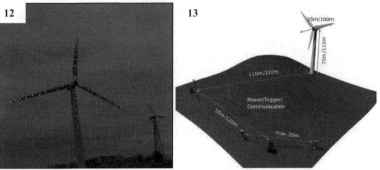

Fig. 12. Experimental layout of camera system; Fig. 13, Measuring targets on wind turbine.

The initial measurement data shows all motion relative to the camera system, where a coordinate system transformation is applied to set the global coordinate system relative to the turbine.

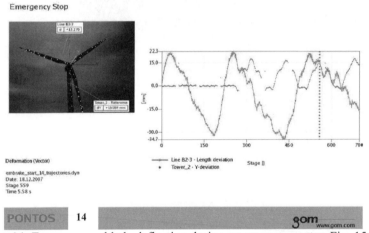

Fig. 14: Tower versus blade deflection during an emergency stop Fig. 15: Vector plots that can be animated during review of captured data show 3D resultant displacements for three similar points on each blade. The phases are very orderly, with some amplitude variations (next page).

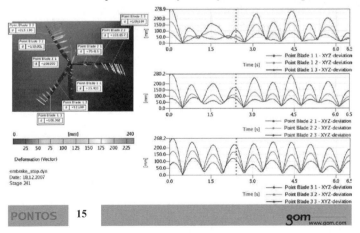

Summary:
The utilisation of optical measuring systems contributes strongly in the process of development and manufacturing of wind turbines. Material parameters and components' response are determined and used for the development of new material models and the verification of FE models. 3D digitizing and coordinate measurements are part of the quality control, calculating the full field deviation to CAD models and determining quickly areas where changes need to be done. Further assembly applications are served and full-scale deformation and vibration analysis can be carried out [6].

References:
[1] Jensen, F: *Ultimate strength of a large wind turbine blade.* Technical University of Denmark (2008);
[2] Puri, A et al: *Analysis of Wind Turbine Material using Digital Image Correlation.* Imperial College London (2009).
[3] Radke, M: *Mobile Optical Coordinate Measuring Technology Used in Offshore Wind Turbines Setup.* I.T.A., (2009)
[4] Schmidt Paulsen, U., Erne, O., Klein, M: *Modal Analysis on a 500 kW Wind Turbine with Stereo Camera Technique.* I.O.M.A.C., (2009)
[5] Schmidt, T., Tyson, J: *Some Common and Not So Common Applications of 3D Image Correlation.* Proceedings of the GOM International User Meeting, Braunschweig, Germany, (2007).
[6] Erne, O., Friebe, H., and Galanulis, K., *Is it possible to replace conventional displacement and acceleration sensor technology? Solution methods using optical 3D measuring technology.* GOM white paper, (2007).

Applied Mechanics and Materials Vols. 24-25 (2010) pp 173-178
© (2010) Trans Tech Publications, Switzerland
doi:10.4028/www.scientific.net/AMM.24-25.173

Details of temperature compensation for strain measurements on NPL bridge - demonstrator for SHM

E. N. Barton[1, a], B. Zhang[1,b]

[1]National Physical Laboratory,

Hampton Rd, Teddington, Middlesex, UK TW11 0LW

[a]elena.barton@npl.co.uk, [b]bufa.zhang@npl.co.uk,

Keywords: bridge monitoring, temperature compensation, fibre Bragg gratings, solar radiation sensors, structural health monitoring

Abstract. This paper is part of ongoing structural health monitoring in the National Physical Laboratory. The work is aimed at improving reliability of sensors and increasing the uptake of monitoring systems by improving the understanding of the fundamental interactions between the sensors and their environments. It is important to optimize and validate the selection of the sensors for each application in order to ensure that the results are meaningful. An example of unexpected aspects of a sensor performance exposed to temperature and solar radiation is demonstrated and the importance of temperature compensation is highlighted. More work is required fully to explain this phenomenon.

Introduction

A 15 tonne, 5 metre high, 20 metre long reinforced concrete footbridge is one of the largest specimen ever created at NPLand is ideal as a demonstrator for this project. Built in the mid 1960s and used for more than 40 years, the bridge has been moved and more than 12 different techniques have been selected for long-term structural health monitoring. In order to capture as much data as possible during the controlled damage and repair tests on the bridge, and to showcase several different available technologies, the footbridge has recently been covered with a variety of sensors, both traditional and novel, and measured externally using a wide range of techniques. Examples of techniques are listed below in alphabetical order:

· 3D Laser Scan provided during the tests by Dr N Eden from John Moore University,
· Acoustic Emission to detect the formation of cracks, supplied and installed by Physical Acoustics and Dr R Pullin from Cardiff University,
· Digital Image Correlation widely used in NPL by Dr N McCormick,
· Digital Leveling of foundations performed before and after each test by Smithers Purslow Property Services,
· Electrolevel Beam Sensors measuring tilt, provided and installed by Soil Instruments,
· Fibre Bragg Gratings written by City University and patches installed by Smart Fibres
· Resistance Strain Gauges in the form of standard Vishay gauges, installed by NPL experts
· Time-Domain Reflectometry Distributed Crack Sensor provided by Strainstall
· Vibrating Wire Strain Gauges measuring strain on both sides of the deck, installed by Soil Instruments,
· Video Gauge Technique to measure displacements during the tests using Imetrum cameras
· Wireless Accelerometer Sensors used as 2 dimensional tilt sensors supplied by Sencieve
· Wireless Magnetic Induction Sensors using SIM cards from Smithers Purslow Property Services
· Numerous environmental sensors including a weather station and surface wetness sensors from Delta-T Devices, and solar radiation sensors from Campbell Scientific.

Strain Measurements.

Various strain sensors have become widely available for structural health monitoring, especially in civil infrastructures. Most of these devices cannot sense strain and temperature independently. Therefore, their temperature response and mechanisms for its compensation are of critical concern for externally installed sensors.

Temperature Variation due to Sensor Location.
During our monitoring program, significant daily temperature variations of about 15-20 °C on average were observed between day and night and about 5 °C between the sunny and shady sides due to location of the bridge on NPL site, close to the parkland.

Fig. 1. Bridge location showing an open front side and parkland.

Having relatively slender dimensions, with the deck being about 950 mm wide and 350 mm thick, the thermal load was very significant, as shown experimentally and supported by modeling [1], [2]. For example, a vertical displacement at the end of the cantilever due to average daily temperature variation of 2.7mm was approximately equivalent to an applied vertical load of 4.5kN at the free end of the cantilever [3].

Among hundreds of other sensors the FBG patches were installed on the bridge in the locations shown on Fig. 2. Each patch contained two gratings: one to measure strain and the other to provide temperature measurement for temperature compensation. It is clear from Fig. 3 that the sensors 3, 4, 5, 6 and 7, on the top of the deck, are fully exposed to the sun and the sensors 1, 2, 8, 9 and 10, in shady areas, are protected from the direct sunlight most of the time.

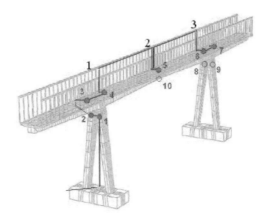

Fig. 2 Position of FBG patches on the bridge shown in red and joined by a line. Solar radiation sensors locations on the top of the deck are shown as black numbers on top of the railings (1, 2 and 3).

Fig.s 3 and 4 show typical examples of temperature and strain data before temperature compensation over a one day period. (Sensor 4 is not included here as it was broken soon after installation.) It is clear that all sensors exposed to the direct sun have additional features described here as a fine structure, whereas the ones in the shade do not.

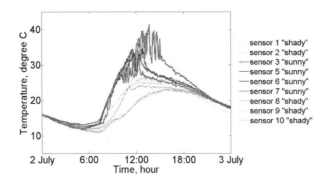

Fig. 3   Temperature for all FBG sensors.

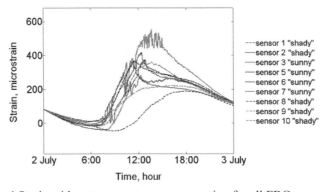

Fig. 4 Strain without temperature compensation for all FBG sensors

Three solar sensors were positioned along the deck at both ends and in the middle as shown in Fig.2. The FBG temperature data was compared with the solar radiation sensor output. The additional features in the FBG temperature response correspond well with solar radiation measured in similar locations. An example of solar radiation sensor 3 and FBG 7 data is shown in Fig.5.

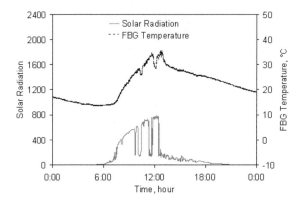

Fig. 5. Typical examples of FBG temperature (line on the top) and solar radiation sensors (line on the bottom).

Moreover, the FBG sensor 3 is located on the sunniest end of the bridge and the time period of solar activity is therefore the longest, as is the time duration of additional fast varying features of the FBG response which lasts until 15.00 in the afternoon, as shown in Fig. 4. (Compare FBG sensor 3 with FBG 5 in the middle of the deck and FBG sensor 7, on the other end of the deck, in Figure 2.) Our conclusion is that the changes in the data are attributable to changes in the solar radiation and not due to noise or poor sensor / installation quality.

Temperature Compensation for a FBG Patch.
Each patch consists of two Fibre Bragg Gratings. One is used to sense strain and temperature (FBGa) simultaneously and the other only senses temperature (FBGb) because it is not attached but only enclosed in the patch. The two FBGs are close enough to allow their temperatures to be considered the same. The technique for dual FBGs temperature compensation is as follows: the wavelength shift of the FBG1a which senses both temperature and strain, can be given as

$$\Delta\lambda_{F\,BGa} = k_{\varepsilon}\varepsilon + k_{1a}\Delta T \qquad\qquad\qquad (1)$$

FBG1b is sensitive only to temperature and its wavelength shift is

$$\Delta\lambda_{F\,BGb} = k_{Tb}\Delta T \qquad\qquad\qquad\qquad (2)$$

where $\Delta\lambda$ is the wavelength shift for FBGa and FBGb respectively, k_{ε} is strain coefficient and k_{Ta} and k_{Tb} are coefficients of thermal expansion which were determined experimentally in the laboratory or taken from a datasheet provided by the sensor manufacturer. In this work we calibrated in the laboratory. A typical patch was attached to the concrete block from the bridge step using the same procedure as during installation on the main bridge. Temperature coefficients were obtained experimentally by heating and cooling the specimen in the oven. Temperature compensated strain was calculated by substituting ΔT from Eq.2 into Eq.1 [3]. It is important to note that for Fibre Bragg gratings, the wavelength shift induced by 1 °C is almost equivalent to that induced by 10-20με [3].

Time Lag in Temperature Compensation Scheme and Uncertainty.
Traditionally temperature compensation is done at the same moment in time. The sensors in the shade can be reliably temperature compensated using traditional methods. However, the sensors in sunny areas show not only additional features but also a time lag between the strain and temperature responses during morning hours as clearly indicated in Fig. 6 at about 11.30. The time lag appeared to be 13 minutes for the morning data but disappeared in the afternoon data from 13.00 to 15.00.

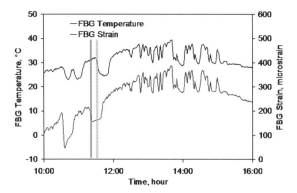

Fig. 6. FBG3 strain and temperature over 6 hours without time lag correction

If the traditional method for temperature compensation is used, strain variation of up to 300 με (peak to peak) is introduced - shown in Fig 7 by the solid line. The dotted line shows the same data but with temperature compensation including 13 minutes time lag shift correction.

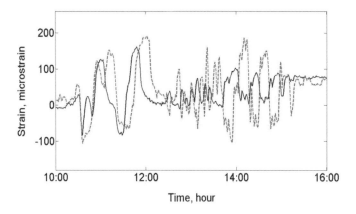

Fig. 7 FBG3 strain data after temperature compensation without time lag correction (solid line) and with 13 minutes time lag correction (dotted line).

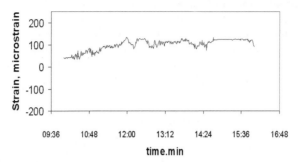

The data from other strain sensors in similar locations (see Fig 8) do not support the strain variations of that magnitude. Therefore, they are likely to be artificial and application of this simple temperature compensation (with or without time lag) could lead to misinterpretation of the data.

Fig. 8 Strain measured using a Vishay electrical resistance strain gauge located next to the sensor FBG 3 on the top of the deck.

Uncertainty in strain measurement using the temperature compensation scheme described above was determined by taking into account the uncertainty in wavelength measurements, temperature and temperature coefficients for both FBG in the patch. Preliminary calculations showed that the uncertainty associated with temperature compensation is the largest contribution to the uncertainty budget.

Summary

The observed phenomenon in strain measurements using FBG patches relate to changes in temperature due to solar radiation. This, to our knowledge, is a novel interpretation and worth bringing to attention of SHM community. This paper used FBG sensors as examples, but other sensors also show differences between sunny and shady locations. More work is required to understand fully this phenomenon.

Time lag between temperature and strain responses of two FBGs in the same patch presented an interesting challenge. It seems to depend on the rate of change in temperature and could be related to the properties of composite material used for making patches. It needs more experimental investigation in a controlled laboratory environment. At present, our recommendation would be to avoid locations exposed to direct solar radiation.

This paper is part of the continuous work on long-term performance of various sensors in outdoor environment to improve reliability of monitoring systems and interpretation of data.

References

[1] A,Karimi. Analysis of footbridge under loading, to be published

[2] Sung-Pil Chang etc . Necessity of the bridge health monitoring system to mitigate natural and man-made disasters : Structura and infrastructure Engineering, June 2009, p.173-197.

[3] Information on http://www.npl.shm

[4] Zhi Zhou, OU Jinping. Techniques of temperature compensation for FBG strain sensors: APCOM (2004), p.465-471

Applied Mechanics and Materials Vols. 24-25 (2010) pp 179-185
© *(2010) Trans Tech Publications, Switzerland*
doi:10.4028/www.scientific.net/AMM.24-25.179

Effect of loading rate on the fracture behaviour of high-strength concrete

G. Ruiz[a], X. X. Zhang[b], R. C. Yu, E. Poveda, R. Porras and J. del Viso

Universidad de Castilla-La Mancha, 13071 Ciudad Real, Spain

[a]gonzalo.ruiz@uclm.es, [b]zhangxiaoxinhrb@gmail.com

Keywords: loading rate, high-strength concrete, fracture energy.

Abstract. This research deals with the sensitivity of eight types of performance-designed high-strength concrete to the loading rate. Variations in the composition of the concrete produce the desired performance, for instance having null shrinkage or being able to be pumped at elevated heights without segregation, but they also produce variations in the fracture properties that are reported in this paper. We performed tests at five loading rates spanning six orders of magnitude in the displacement rate, from 1.74×10^{-5} mm/s to 17.4 mm/s. Load-displacement curves show that their peak is higher as the displacement rate increases, whereas the corresponding displacement is almost constant. Fracture energy also increases, but only for loading rates higher than 0.01 mm/s. We use a formula based on a cohesive law with a viscous term to study the results. The correlation of the formula to the experimental results is good and it allows us to obtain the theoretical value for the fracture energy under strictly static conditions. In addition, both the fracture energy and the characteristic length of the concretes used in the study diminish as the compressive strength of their aggregates increases.

Introduction

Compared with the extensive research into the static fracture behaviour of high-strength concrete, much less information is available on its dynamic fracture behaviour [1-4]. Schuler and Hansson [1] measured the tensile strength and the fracture energy of high-strength concretes at high strain rates between 10 s^{-1} and 100 s^{-1} with spalling tests. A three-fold increase in the fracture energy was observed at a crack opening velocity of 1.7 m/s. Nevertheless, the experimental data in scientific literature on the rate-sensitivity of the fracture behaviour of high-strength concrete is scarce.

In the case of conventional concrete however, there is abundant information on the subject. The fracture energy varies slightly under quasi-static loading if there is a variation in the loading rate, while the fracture energy changes under dynamic loading [5]. Thus some researchers deduce that the fracture energy is constant and independent of the loading rate [6-8]. However, the fracture energy increases by approximately 50% when the loading rate increases until about 1 mm/s [9]. Under high loading rates, the fracture energy greatly exceeds the static value, mainly due to structural causes, such as inertia and the geometry of the structure [10-12].

In addition, results and conclusions obtained with conventional concretes are not directly applicable to high-strength concrete, as the type and extent of the fracture process zone varies with the increasing strength of the material. In general, cracks tend to propagate around aggregates in conventional concrete, whereas in high-strength concrete, they usually go through them [13,14]. According to Carpinteri and Paggi [15], the super-singular behaviour of transgranular cracks may be a possible reason of the less pronounced effect of crack arrest by aggregates and a more brittle global behaviour of high-strength concrete. This fact implies that we should review the formulations obtained in the study of the fracture of conventional concrete when it will be used in high-strength concrete.

Therefore, in this research we want to study the variation of fracture energy in eight types of high-strength concretes designed for performance. By this we mean that besides being of high strength the concrete is required to have a special feature. Both requirements determine the mixture

features	nomenclature	Coarse aggregate	$d_{máx}$ (mm)	f_c aggregate (MPa)	E_c (GPa)	f_c (MPa)	f_t (MPa)
Conventional plant	H01	silicious	20	130	32 (2)	78 (4)	6.3 (0.4)
Pumpable	H02	andesite	12	250	30 (2)	88 (4)	5.8 (0.7)
Without retraction	H03	andesite	12	250	34 (1)	97 (6)	6.1 (0.5)
Very high strength	H05	porphyry	12	200	43 (2)	127 (11)	6.3 (0.6)
High early strength	H06	mylonite	6	150	34 (1)	71 (11)	4.9 (0.3)
Low heat of hydration	H07	andesite	12	250	29 (4)	96 (14)	5.3 (0.7)
Light	H08	arlite	10	10	20 (1)	57 (5)	3.0 (0.4)
Heavy	H09	barite	10	12	34 (1)	60 (1)	3.7 (0.4)

Table 1. Performance of design, type of aggregate used and mechanical properties of high strength concretes.

components of the concrete and its strength. The variations in the composition of the concrete, while providing the desired benefit, also produce changes in mechanical properties in fracture and, therefore, another objective of this study is to ascertain the relationship between the composition of high-strength concrete and its properties in fracture. The process of design, fabrication and the curing state of the material (for example, properties in fresh state, especially the thixotrophic changes characteristic of the use of additives) is complementary to this research and, of course, is of great technological interest. However, they are located outside of the limits of the fracture mechanical characterization of materials and therefore we have excluded this part in our work. We will only consider, as a complement to our primary objective, the influence of the composition of the concrete on the mechanical properties in fracture.

Experimental Program

High-strength concrete. The concretes studied in this investigation were designed and manufactured by Composites ID in its factory in Alpedrete (Madrid). The provision of design and nomenclature adopted for each of them are shown in Table 1 (in the nomenclature there is a gap from H03 to H05, this is because H04 was planned but not made in the end, nevertheless we respected the initial name of the concrete that Composites ID had agreed to). From a mechanical point of view its composition can be characterized by the type of coarse aggregate used, by its maximum size and compressive strength [16], this information is also included in Table 1. In addition this table provides the results of standard mechanical characterization tests (elastic modulus, Ec, compressive strength, fc, and indirect tensile strength, ft, besides the average value of these four tests, in brackets, it shows the standard deviation). For the compressive tests, we used 75 x 150 mm cylinders (diameters x height) and followed the recommendations of ASTM C39. The elastic modulus was measured using two clip strain gages centered in opposite generatices.

Tests for fracture energy. In this work we measured the fracture energy, G_F, through three-point bending tests with the procedure recommended by RILEM [17] and with the improvements proposed by Planas, Guinea and Elices [18]. In particular, the dimensions of the specimen are $100 \times 100 \times 400$ mm, the initial notch-depth ratio is approximately 0.5. The tests were conducted in position control at five loading rates: 17.4 mm/s (very fast), 0.55 mm/s (fast), 1.74×10^{-2} mm/s (intermediate), 5.5×10^{-4} mm/s (slow) and 1.74×10^{-5} mm/s (very slow). In the case of slow and very slow tests loading rates displayed were the first 0.25 mm of loading-point displacement.

This displacement always includes peak load in the first ramp. Then the test is accelerated by multiplying the rate by five to 0.75 mm of loading point displacement, and then, multiplying by five again until the end of the test. The slowest tests last about 8 hours while the fastest tests last approximately 1/5 th of a second.

Results

Figure 1 shows the typical load-displacement curves (P-δ) of the concrete H03 at each loading rate. Each curve produces a value of G_F. The average values of G_F and the loading rate for each concrete

	G_F [N/m] to loading rate v_D [mm/s]				
v_D	1.74×10^{-5}	5.5×10^{-4}	1.74×10^{-2}	0.55	17.4
H01	211 (13)	200 (41)	197 (26)	249 (35)	—
H02	121 (28)	108 (13)	119 (11)	156 (16)	177 (9)
H03	134 (14)	126 (6)	124 (8)	155 (12)	198 (23)
H05	181 (43)	156 (20)	187 (12)	215 (18)	252 (39)
H06	245 (22)	213 (9)	254 (42)	285 (24)	328 (62)
H07	149 (16)	137 (11)	157 (3)	158 (7)	204 (25)
H08	120 (3)	100 (8)	97 (5)	113 (12)	111 (2)
H09	153 (11)	138 (13)	139 (25)	169 (40)	201 (19)

Table 2: Variation of fracture energy with loading rates

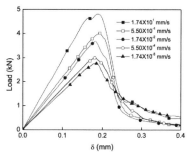

Fig. 1: Curves P-δ for concrete H03 at different loading rates (without retraction).

are provided in Table 2 (in the case of H01-mixed in a pre-cast plant-G_F was not measured for the very fast loading rate). The standard deviation of each measurement is shown between brackets.

Discussion

Variation of the mechanical properties depending on aggregate type. Figure 2 represents the relation between the concrete's mechanical properties and the compressive strength of coarse aggregates. It is clear that the compressive strength of the concretes increases almost in a linear way (except H05) with an increase in the compressive strength of aggregates. As concerns the tensile strength of the concretes, the increase is significant at first and slight later with increases in the compressive strength of aggregates. For the concrete H05 (very high strength), the increase is more pronounced for both properties. It could be that it was produced with a high quantity of additives (microsilica) [19].

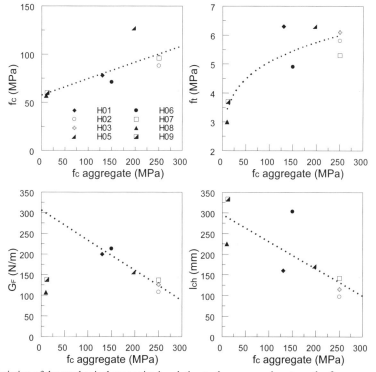

Fig. 2: Variation of the mechanical properties in relation to the compressive strength of coarse aggregates.

With regard to the variation of the elastic modulus with aggregates strength, Table 1 shows that the modulus is kept in an interval between 30 and 35 MPa approximately, except for the H08 concrete (light) and the H05 concrete (very high strength). The H08 concrete (light) is composed of an artificial aggregate, arlite, this is an expanded clay and has an elastic modulus far lower than the rest of the aggregates. The increase in the elastic modulus of the H05 concrete (very high strength) is attributable partially to the porphyry aggregate that is used in its composition and partially to the fact that the matrix is more rigid and resistant [19].

Regarding the variation of the specific fracture energy G_F with aggregate strength, Figure 2 clearly shows that G_F decreases with an increase in the compressive strength of the aggregate, except in the case of concretes with aggregates of a weaker strength (H08 and H09). Fig. 2 also shows that the characteristic length, ℓ_{ch}, decreases with increases in coarse-aggregate strength. It confirms, from a fracture mechanics view point, that the strongest concretes are the most fragile. It is important to remember that the length of fracture process and the release of fracture energy in concrete are proportional to ℓ_{ch}. Therefore, two similar structures that have the same proportion of their dimensions with ℓ_{ch} are going to generate similar crack processes with proportional dimensions and, in this respect, ℓ_{ch} characterizes the ductility or intrinsic brittleness of concrete.

Variation of fracture energy with loading rate. Figure 1 shows typical load-displacement curves, P-δ under different loading rates for the concrete H03 (without retraction). The peak load increases with an increase in the loading rate, whereas the displacement in maximum load remains practically constant. Regarding the stiffness of the beam, it is observed that, as with the peak load, it also increases with increases in the loading rate.

In Table 2 we arrange the values of G_F obtained from fracture energy tests for all of the concretes at the five loading rates next to their corresponding standard deviation in brackets, in contrast Fig. 4 represents average values graphically. It is observed that the fracture energy remains practically constant for quasi-static loading rates but increases in value for dynamic loading rates. For all of the concretes, note that the value of G_F at the lowest loading rate is slightly greater than that obtained for the following two loading rates (slow and intermediate loading rates). A possible explanation for this recovery of G_F can be in the humidity loss of the specimen during the slowest test, since the duration of the fracture process was approximately eight hours. On the other hand, the increase of G_F with increases in loading rate is attributable to a major extension of the microcraking zone around the principal fracture [20-22].

It is clear that, for static and quasi-static loading rates, from 1.74×10^{-5} mm/s up to 1.8×10^{-2} mm/s, the fracture energy scarcely changes -though the maximum load changes (See Fig. 1)-, which would explain why some researchers should have thought that fracture energy is independent from loading rate [6-8]. For higher loading rates G_F increases sensitively (with the exception of H08 -light-, probably due to the different nature of its coarse aggregate). We think that the movement of water through the network of pores of material, and the formation of a new surface of water in the way of spread of the fissures, have an influence on the variation of the peak load and the fracture energy with the loading rate that is shown in the experimental results. The aforementioned variation reproduces correctly using a cohesive model which includes a term dependent on the loading rate [23]. This approximation allows us to deduce that the adjustment to the experimental results would be of the type:

$$\frac{G_F}{G_F^S} = 1 + \left(\frac{v_D}{v_0}\right)^n. \tag{1}$$

Where G_F is the fracture energy, G_F^s is a parameter of adjustment with dimensions of fracture energy, v_D is the loading rate, v_0 is another parameter of adjustment, with rate dimensions,

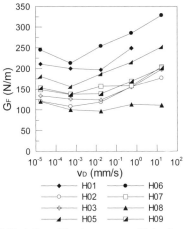

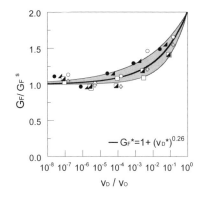

Fig. 4: Non-dimensional representation of the fracture energy with regard to the loading rate.

Fig. 3: Variation of fracture energy with loading rate.

and n is a non-dimensional exponent which describe the degree of viscosity of the material, obtained from fitting Eq. 1 to the experimental results.

In Table 3 we present the parameters that fit Eq. 1 to each type of concrete. The regression coefficients turn out to be almost 1 except in the case of H01 (mixed in a pre-cast plant), due to the fact that G_F did not measure for the very fast rate, and also in the case of H08 (light), due to its anomalous behavior; for this reason, the parameters corresponding to these two concretes are not displayed in Table 3.

It is necessary to emphasize that G_F^s can be understood as the value of the fracture energy for strictly static conditions ($v_D = 0$). It can be observed that the aforementioned parameter practically coincides, in all cases, with the fracture energy measured at slow velocity, which is habitually considered to be an intrinsic material property. On the other hand, v_0 is a parameter that represents the displacement rate that would produce a 100% increase in the fracture energy, compared with the static value; it is a parameter whose value may be between 10^2 and 10^3 mm/s and its average is 284 mm/s. The exponent n changes from between 0.2 and 0.4, and its average value is 0.27.

	G_F^s [N/m]	v_0 [mm/s]	n
H01	—	—	—
H02	106	125	0.18
H03	126	91	0.33
H05	162	317	0.20
H06	221	733	0.19
H07	145	180	0.39
H08	—	—	—
H09	142	256	0.32

Table 3: Parameters of the equation of adjustment for each concrete.

Figure 4 represents the experimental results of non-dimensional way. The x-axis corresponds to the non-dimensional velocity v_D/v_0 and the y-axis corresponds to the non-dimensional fracture energy with regard to G_F^s. In this figure, each concrete has been represented using its corresponding parameters indicated in Table 3. Two fitting curves bridged by the shaded zone are intended to cover most of the experimental data points. A new fit (the thick line) Eq. 1 to pass the center of the shaded zone gives an n of 0.26, which is very closet to the value average found above (0.27). This fit represents the non-dimensional average behaviour of the high-strength concretes in this study.

Summary and Conclusions

In this research we have investigated the loading-rate sensitivity of the fracture energy of eight high-strength concretes designed for benefits, such as being pumped or without retraction. With each type of concrete we performed tests to measure fracture energy at five different loading rates, from 1.74×10^{-5} mm/s to 17.4 mm/s. Load-displacement curves show that the maximum load increases with an increase in loading rates, yet the corresponding displacement remains almost constant. The fracture energy value increases with the loading rate from 0.01 mm/s. The results fit perfectly with a formula based on a cohesive model with a viscous term. With this adjustment we have obtained the theoretical value of fracture energy in strictly static loading conditions.

We have also investigated the changes that the variations in the composition of the concretes produced in the fracture properties.

References

[1]   H. Schuler and H. Hansson: Journal de Physique IV, Vol. 134 (2006), pps. 1145–1151.

[2]   S. Mindess, N. Banthia, and C. Yan: Cement and Concrete Research, Vol. 17 (1987), pps. 231–241.

[3]   H. Müller: CEB FIP Bulletin Vol. 42 (2008).

[4]   X. Zhang, G. Ruiz, R. C. Yu, and M. Tarifa: International Journal of Impact Engineering, Vol. 16 (2009), pps. 1204–1209.

[5]   B. Oh: Engineering Fracture Mechanics, Vol. 35 (1990), pps. 327–332.

[6]   N. Challamel, C. Lanos, and C. Casandjian: International Journal of Damage Mechanics, Vol. 14 (2005), pps. 5–24.

[7]   D. C. Jansen, S. P. Shah, and E. C. Rossow: ACI Materials Journal, Vol. 92 (1995), pps. 419–428.

[8]   J.H. Yon, N. M. Hawkins, and A. S. Kobayashi: ACI Materials Journal, Vol. 89(1992), pps. 146–153.

[9]   RILEM: Materials and Structures, Vol. 23 (1990), pps. 461–465.

[10]   Y. Lu and K. Xu: International Journal of Solids and Structures, Vol. 41 (2004), pps. 131–143.

[11]   F. Toutlemonde. PhD Thesis, E.N.P.C., Paris (1994).

[12]   J. Van Doormaal, J. Weerheijm, and L. Sluys: Journal de Physique IV, Vol. 4 (1994), pps. 501–506.

[13]   S. Choi, K. Thienel, and S. Sha: Magazine of Concrete Research, Vol. 48 (1996), pps. 103–115.

[14]   S. Caliskan, B. L. Karihaloo, and B. I. G. Barr: Magazine of Concrete Research, Vol. 54 (2002), pps. 449-461.

[15]   A. Carpinteri and M. Paggi: Engineering Fracture Mechanics, Vol. 74 (2007), pps. 59–74.

[16]   M. Alexander and S. Mindess, in: *Aggregates in Concrete*. Taylor & Francis, Madison Ave, New York (USA), (2005).

[17]   RILEM: Materials and Structures, Vol. 18 (1985), pps. 285–290.

[18] M. Elices, G. V. Guinea, and J. Planas: Materials and Structures, Vol. 30 (1997), pps. 375–376.

[19] J. Del Viso. Tesis doctoral, Universidad de Castilla-La Mancha, Ciudad Real (2008).

[20] Bischoff, P. H., and S. H. Perry: Materials and Structures, Vol. 24 (1991), pps. 425–450.

[21] Ahmed Brara and Janusz R. Klepaczko: International Journal of Impact Engineering, Vol. 34 (2007), pps. 424-435.

[22] I. Vegt, K. Van Breugel, and J. Weerheijm: *Failure mechanisms of concrete under impact loading.* Fracture Mechanics of Concrete and Concrete Structures, Vol. 1-3 (2007).

[23] G. Ruiz, X. Zhang, J. Del Viso, R. Yu, and J. Carmona, in: Anales de Mecánica de la Fractura, Vol. 25 (2008), pps. 793–798.

Session 5A: Non Linear Behaviour

Applied Mechanics and Materials Vols. 24-25 (2010) pp 189-194
© (2010) Trans Tech Publications, Switzerland
doi:10.4028/www.scientific.net/AMM.24-25.189

Numerical Prediction of the Response of Metal-to-Metal Adhesive Joints with Ductile Adhesives

N.G. Tsouvalis[1,a], K.N. Anyfantis[1,b]

[1]Shipbuilding Technology Laboratory, School of Naval Architecture and Marine Engineering, National Technical University of Athens, Athens, Greece

[a]tsouv@mail.ntua.gr, [b]kanyf@central.ntua.gr

Keywords: Adhesive bonding, Cohesive laws, Interface elements, Mixed-mode fracture

Abstract. The present work involves a numerical modelling of the Embedded Process Zone (EPZ) by utilizing the elastoplastic Mode I and Mode II fracture models for the simulation of plastically deforming adhesive joints. A traction-separation law was developed separately for Mode I and Mode II. For the analysis of the mixed-mode fracture processes, the cohesive zones in Mode I and Mode II fracture were assumed uncoupled. The experimental programme involved the fabrication and testing of Double Strap Joints (DSJs) and Single Lap Joints (SLJs). By fitting the numerical results to the experimental ones, the basic cohesive parameters of the problem were defined.

Introduction

Adhesive bonding is gaining more and more interest due to the increasing demand for joining dissimilar structural components, mostly within the framework of designing lightweight structures. During loading, the adhesive material at the crack tip experiences extreme deformation often leading to the development of extensive plasticity and microscopic cracks. This area of the adhesive is denoted as the process zone. When using ductile adhesives, this zone usually extends in a large area in front of the crack tip. During crack growth, two new surfaces are created. Before the formation of the physical crack, these two new surfaces are held together by traction within a cohesive zone. This traction is a function of the relative displacement of the surfaces which is described by a cohesive law. A cohesive law, denoted also as traction-separation law, can be defined for both Mode I and Mode II fractures [1]. Elastic-plastic fracture mechanics have been used to describe adhesive joints with the use of cohesive laws [2,3]. These laws are mostly associated with interface finite elements for the numerical modelling of the process zone. For the analysis of plastically deforming adhesive joints, the Embedded Process Zone (EPZ) approach has been proposed [4]. For applying the EPZ approach, either a coupled or an uncoupled cohesive zone modelling may be utilized [5]. A coupled constitutive relation assumes that both shear and peel tractions depend on both the peel and shear deformation of the bi-material interface. This leads to a mode dependency of the fracture behaviour. The coupled approach is based on the definition of a coupled relationship between Modes I and II [4] that does not yield the same fracture energy for all mode mixtures [5]. On the other hand, according to the uncoupled approach, the traction-separation relationships in Modes I and II are assumed to be uncoupled under mixed mode loading cases [6]. The energy release rate, J, in Modes I and II is defined as the area under the corresponding cohesive law.

The objective of this study is the development of a robust yet flexible cohesive law that can be used to simulate the constitutive behaviour of an adhesive layer under mixed mode loading. The traction-separation parameters are defined separately for Mode I and Mode II, by applying the J-integral approach in combination with the adhesive material properties. The defined cohesive laws are incorporated into interface finite elements used for modelling the adhesive behaviour of the tested single lap and double strap joints.

Experimental Procedure

All substrates (inner adherents and strap adherents) were made of normal marine grade steel, for both SLJ and DSJ configurations. The adhesive used to bond the specimens was Araldite 2015, a relatively stiff two-component epoxy adhesive manufactured by Huntsman Container Corporation Ltd. The geometry of the specimens is given in Fig. 1, showing a side view of them. The width of all specimens was equal to 25 mm. The thickness of the adhesive layer, t_a, was 0.5 mm and provisions had been taken during manufacturing to keep this value constant for all specimens. Before bonding the straps of the DSJs, the two inner 10 mm adherents were placed in contact, without any adhesive in-between them. As for the surface treatment, the common in the shipbuilding industry Sa 2½ near white grit blast cleaning was applied (approx. 2.5 mm grit size) to the bonding surfaces of all substrates. This procedure gave a measured average roughness (*Ra*) of 4.15 μm. More details can be found in [7]. After the assembly, the specimens were cured in an oven under a uniform pressure loading. According to the adhesive material manufacturer, curing procedure consisted of heating the specimens in 60°C for 4 hours, followed by a slow cooling to ambient temperature. The specimens were left in ambient temperature for 48 hours before performing the tests. Specimens were loaded by a uniaxial static tensile displacement, applied with a speed of 0.1 mm/min by an MTS hydraulic testing machine, equipped with a 100 kN load cell. During loading the applied displacement together with the reaction forces were monitored.

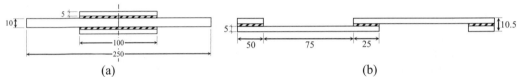

(a)   (b)

Fig. 1: Geometry and dimensions of the examined DSJ (a) and SLJ (b)

Interface Element Formulation

A 6-node interface element has been developed (see Fig. 2) based on the formulation proposed in [1]. The element has two translational degrees of freedom per node, one in each direction of a Cartesian 2D plane. The 12 x 12 element tangent stiffness matrix K and the 12 x 1 internal force vector f_I, which are required for the Newton-Raphson nonlinear solution algorithm, are given in a local curvilinear coordinate system by Eq. 1 and Eq. 2, respectively:

$$K = b \int_{-1}^{1} B^T R \bar{S} B \det J d\xi .$$ (1)

$$f_I = b \int_{-1}^{1} B^T R \bar{t} \det J d\xi .$$ (2)

where *b* denotes the element width in the out-of-plane z-direction. Matrix *B* contains the quadratic shape functions (Hermitian polynomials) $N(\xi)$, which relate the vector containing the nodal displacements *u* with the relative displacements of the elements $\bar{\delta}$, multiplied with the rotational matrix *R*. The relative displacements are calculated as the difference between the displacements of the upper (4,6,5) and lower (1,3,2) nodes and are given by:

$$\bar{\delta} = R^T N u .$$ (3)

Accounting for geometrical non-linearities, a rotational matrix *R* containing the direction cosines can be defined as

$$R = \frac{1}{\det J}\begin{bmatrix} r_1 & -r_2 \\ r_2 & r_1 \end{bmatrix} \quad \text{where} \quad \det J = \sqrt{r_1^2 + r_2^2}\,. \tag{4}$$

Vector $\bar{t}$ contains the shear and normal tractions $\{\bar{t}_s\ \bar{t}_n\}$ in the local coordinate system as described by the respective cohesive laws. A cohesive law is interpreted as the constitutive equation of the interface to be modelled, given as the function between tractions and relative displacements.

$$\{\bar{t}_s\ \bar{t}_n\} = F(\bar{\delta}_s\ \bar{\delta}_n). \tag{5}$$

Thus, the material stiffness matrix, $\bar{S}$, can be defined by:

$$\bar{S} = \frac{\partial \bar{t}}{\partial \bar{\delta}}, \quad \text{in matrix form} \quad \bar{S} = \begin{bmatrix} \bar{S}_s & \bar{S}_{sn} \\ \bar{S}_{ns} & \bar{S}_n \end{bmatrix}. \tag{6}$$

where $\bar{S}_s$ and $\bar{S}_n$ denote the derivatives of the interfacial shear and normal tractions in terms of the sliding and normal separation, respectively.

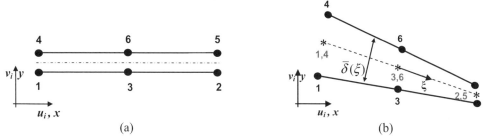

(a)   (b)

Fig. 2: 6-node 2D interface finite element in undeformed (a) and deformed state (b)

Finite Element Modelling

The finite element mesh of the SLJ and DSJ configuration are presented in Fig. 3a and Fig.3b, respectively, together with loading and boundary conditions. Because of its symmetry, only one quarter of the DSJ has been modelled and the respective symmetry boundary conditions have been applied. In order to account for geometrical and material nonlinearities, the Newton-Raphson method has been utilized together with a line search algorithm. A displacement controlled approach is utilized for aiding the convergence of the non-linear solution and avoiding numerical instability issues involved in crack growth analyses, where softening behaviour is apparent. The substrates involved in both types of joints have been modelled with CPS8 continuum 2D elements available in the element library of ABAQUS® 6.8. The width of the FE models is 25 mm in both cases. The elastic material properties of the substrates are $E = 210$ MPa and $v = 0.3$. Since all substrates are very thick with respect to the adhesive layer thickness in the present study, it can be stated with confidence that there will be no yielding of the metal substrates prior to failure of the adhesive and the consequent total failure of the joint [7]. Thus no plasticity properties of steel are needed. Experimental observations have shown that crack initiated and propagated in the area of the adhesive layer, either characterized by adhesive or cohesive failure [7]. In a numerical analysis, an adhesive layer constrained between two continua can be modelled by embedding the process zone in a cohesive law that describes the constitutive behaviour of this layer. Thus, the traditionally used continuum elements have been replaced with interface elements placed between the substrates of the SLJ and DSJ, as shown in Fig. 3a and Fig. 3b, respectively.

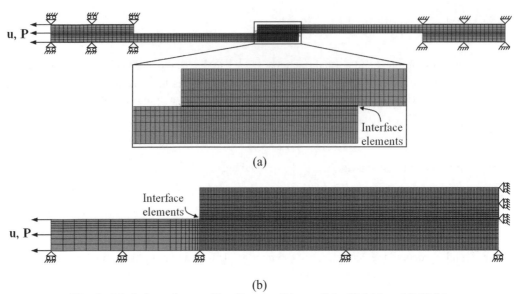

(a)

(b)

Fig. 3: Mesh, boundary and loading conditions of the SLJ (a) and DSJ (b)

For the development of cohesive laws that describe the behaviour of an adhesive layer, three sets of characteristic constitutive parameters must be defined, separately for Mode I and II [5]. The first set involves the initial elastic response of the adhesive and is described by the ratio between the Young's or shear modulus and the adhesive thickness E_a/t_a and G_a/t_a, for Mode I and Mode II respectively. In the present study, $E_a = 2$ GPa is the Young's modulus and $G_a = 0.77$ GPa is the shear modulus of the bulk adhesive [8]. The rest two sets of cohesive parameters involve the maximum attained stress (σ_0 and τ_0) together with their respective relative displacement values ($\delta_{I,0}$, $\delta_{II,0}$) and the maximum separations after which tractions vanish ($\delta_{I,C}$, $\delta_{II,C}$). Having defined the cohesive parameters, the total energy uptake can be calculated for Mode I (J_{IC}) and Mode II (J_{IIC}) respectively as [9]:

$$J_{IC} = \int_0^{\delta_{I,C}} \sigma \, d\delta_I \quad \& \quad J_{IIC} = \int_0^{\delta_{II,C}} \tau \, d\delta_{II} . \tag{7}$$

These parameters evolve together as a natural result of the interaction between the deformation of the substrates and the behaviour of the two separate cohesive laws. Thus, the total energy release rate absorbed during mixed mode loading is given by the summation of the fracture toughness of the pure cohesive laws in the following form: $J = J_I + J_{II}$.

Apart from the aforementioned sets of cohesive parameters, the well known cohesive behaviour (traction-separation function) has to be defined. In this study, an exponential function has been assumed to describe the ductile adhesive behaviour in both Modes I and II up to the point of maximum stress, followed by a linear softening behaviour after this point. The selection of the exponential function was based on the similar behaviour of the stress-strain response of the adhesive. The cohesive laws that describe the EPZ of the ductile adhesive layer used in this study are presented in Fig. 4. From the cohesive parameters shown in this figure, some magnitudes (σ_0, $\delta_{I,0}$ and $\delta_{II,C}$) were taken from data available in the literature [9,10], whereas the others (τ_0, $\delta_{I,C}$ and $\delta_{II,0}$) have been determined by an inverse method, i.e. by fitting the numerically calculated load-displacement response of the SLJ and DSJ configuration over the respective experimentally measured data.

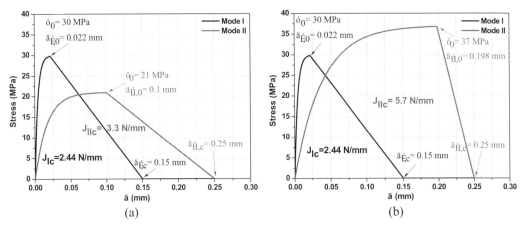

Fig. 4: Exponential cohesive laws with linear softening used for SLJ (a) and DSJ (b)

Numerical and experimental results

Fig. 5 illustrates a comparison between the load/displacement curves obtained from the SLJ and DSJ tests and the Finite Element Analysis (FEA) results based on the cohesive laws shown in Fig. 4. Both experimental curves obtained from SLJ and DSJ tests denote an initially linear response, followed by increasing non-linearities as the adhesive layer begins to substantially deform plastically. Fig. 5 shows that, although the numerical results were fitted to the experimental data by selecting appropriate values for some of the cohesive parameters (τ_0, $\delta_{I,C}$ and $\delta_{II,0}$), the proposed exponential character of the cohesive law proved adequate for modelling the elastoplastic behaviour of the ductile adhesive up to the maximum attained stress in the opening Mode I and the sliding Mode II responses. The DSJ attained much higher loads than the SLJ. This is reasonable, since the bonding surface of the DSJ is four times larger than that of the SLJ. Additionally, due to the symmetry of the DSJ configuration, the adhesive layer is stressed mostly in shear whereas the geometrical eccentricity of the SLJ leads to the development of cleavage stresses at the adhesive edge. Cleavage stresses are the most critical stresses that jeopardize the integrity of the joint by leading to premature failure. According to experimental observations and numerical predictions, in both cases and after the maximum attained load, the joints debonded in a sudden manner until complete failure. This debonding procedure is represented by the softening behaviour of the load/displacement curves after the maximum load. The sudden debonding behaviour was simulated with the use of a linear softening function which was incorporated separately in each of the cohesive laws describing pure mode failure. The FEA of the DSJ additionally captured the softening behaviour, as depicted in Fig. 5b. The SLJ experimental curve exhibits a softening response characterized by a slower rate in comparison to the rate of the respective DSJ softening behaviour. In this case, the cohesive softening law used for the simulation of the EPZ of the SLJ (see Fig. 4a) in combination with the uncoupled approach, did not capture the softening behaviour of the experimentally obtained response curve, as shown in Fig. 5a. Nevertheless it is essential to state that, instead of the shear stresses, the cleavage ones being developed at the adhesive edge are responsible for crack initiation, as cleavage stresses peak at 0.022 mm with a value of 30 MPa, whereas shear stresses peak at 0.1 mm having a value of 21 MPa. After crack initiation, total failure of the adhesive depends on the developed shear cohesive law. On the other hand, although the DSJ configuration is a shear lap type of joint in which the adhesive is loaded mostly in pure shear, the Mode I failure process must be additionaly incorporated into the mixed mode cohesive law in order to numerically capture failure initiation. Finally, the cohesive law in Mode II contributes mostly to the total energy release rate during the mixed mode fracture process of the DSJ, since J_{IIC} is almost twice as J_{IC}.

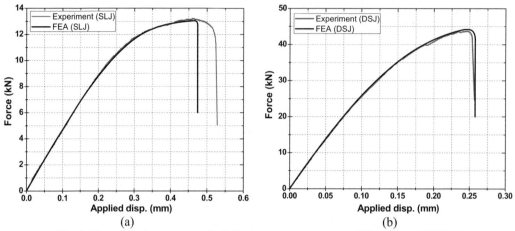

(a)   (b)

Fig. 5: Reaction force vs. applied displacement response of SLJ (a) and DSJ (b)

Conclusions

Two different cohesive laws have been developed to account for the EPZ of a ductile adhesive, one in Mode I and one in Mode II. The developed cohesive laws were used to describe the constitutive relation of user defined interface elements. These elements were utilized to model the behaviour of the adhesive and were implemented in-between the bondline of the substrates, so as to simulate the mixed-mode fracture processes of a SLJ and a DSJ. The cohesive laws were assumed uncoupled. The cohesive parameters were determined implicitly, by fitting the numerical results on the respective experimental data. The excellent agreement between the numerical and the experimental results indicate the validity of the exponential type cohesive law for modelling the elastoplastic behaviour of a ductile adhesive, at least for the present case of the very stiff adhesives where the adhesive behaviour dominates. The differences found in the softening part of the SLJ response owe probably to the linear softening law adopted herein.

References

[1] V.K. Goyal, E.R. Johnson and V.K. Goyal: Comp. Struct. Vol. 82 (2008), p. 434

[2] R.D.S.G. Campilho, M.F.S.F. de Moura, and J.J.M.S. Domingues: Int. J. Solid. Struct. Vol. 45 (2008), p. 1497

[3] Leffler, K.S. Alfredsson and U.Stigh: Int. J. Solid. Struct. Vol. 44 (2007), p. 530

[4] V. Tvergaard and W.J. Hutchinson: J. Mech. Phys. Solids Vol. 40 (1992), p. 1337

[5] L.J. Hogberg: Int. J. Fract. (2006) Vol. 141, p.549

[6] M.S. Kafkalidis and M.D. Thouless: Int. J. Solid. Struct. Vol. 39 (2002), p. 4367

[7] K.N. Anyfantis and N.G. Tsouvalis, in: Analysis and Design of Marine Structures, edited by G. Soares and P. Das, Proceedings of MARSTRUCT 2009, 2nd International Conference on Marine Structures - Analysis and Design of Marine Structures, Taylor & Francis Group, London (2009), pp. 387-392

[8] Huntsman data sheet, http://www.intertronics.co.uk/data/ara2015.pdf

[9] Q.D. Yang and M.D. Thouless: Int. J. Fract. (2001) Vol. 175, p.175

[10] M.F.S.F. de Moura, R.D.S.G. Campilho and J.P.M. Concalves: Int. J. Solid. Struct. Vol. 46 (2009), p. 1589

Applied Mechanics and Materials Vols. 24-25 (2010) pp 195-200
© (2010) Trans Tech Publications, Switzerland
doi:10.4028/www.scientific.net/AMM.24-25.195

Experimental Characterization of the Viscoplastic Material Behaviour of Thermosets and Thermoplastics

Markus Kästner, Swen Blobel, Martin Obst, Karin Thielsch and Volker Ulbricht

Technische Universität Dresden, Institute of Solid Mechanics, D-01062 Dresden, Germany

{Markus.Kaestner, Swen.Blobel, Martin.Obst, Karin.Thielsch,
Volker.Ulbricht}@tu-dresden.de

Keywords: Polymer, Composite, Viscoplasticity, Relaxation

Abstract In this contribution the mechanical behaviour of polymeric matrix materials is analysed for both thermoplastics (Polypropylene) and thermosets (RTM6, RIM935). The results obtained from tensile tests carried out at different velocities indicate a nonlinear, inelastic material behaviour with strain-rate dependence. For the clear identification and quantification of the nonlinearities, the experimental procedure has been extended to relaxation experiments and deformation controlled loading-unloading-processes with intermediate relaxations. Based on the experimental observations a small-strain viscoplastic material model is derived and material parameters are identified. The stress-strain-curves computed for different load histories are compared to the experimental results.

Introduction

The authors have applied multi-scale modelling and simulation techniques [1] to predict the effective linear elastic stiffness properties as well as the macroscopically nonlinear material behaviour for textile-reinforced polymers using only the properties of the individual constitutents and their geometrical arrangement in the composite. The observed macroscopic nonlinearities are, amongst others, driven by damage effects and the inelastic material behaviour of the polymeric matrices. Typical thermoplastics and thermosets often show strong strain-rate dependence in combination with irreversible deformation. Here the nonlinear material behaviour of different matrices is analysed experimentally and modelled numerically.

Classification of Material Behaviour

Concerning the classification of the observed phenomena we follow a proposal of Haupt [2] who suggests a classification based on the strain-rate dependence of the material behaviour and on the examination of an equilibrium relation. Examples for equilibrium points are the terminal points observed in relaxation ($\dot{\varepsilon} = 0$) or creep ($\dot{\sigma} = 0$) experiments. A set of equilibrium points for different strain states forms the equilibrium relation σ^{eq}.

If the stress-strain-curves obtained from experiments carried out at various strain-rates differ from each other, the material behaviour is named strain-rate dependent. The difference between the stress-strain-relation measured in an experiment at finite strain-rate $|\dot{\varepsilon}| > 0$ and the equilibrium relation is called overstress σ^{ov}. The second essential feature of the material behaviour is the shape of the equilibrium relation obtained in loading and unloading processes. The classification distinguishes between equilibrium relations with and without hysteresis (Fig. 1).

In the case of strain-rate independent material behaviour every stress-strain-curve is an equilibrium relation. In the case of strain-rate dependence the existence of an hysteresis in the stress-strain-curve

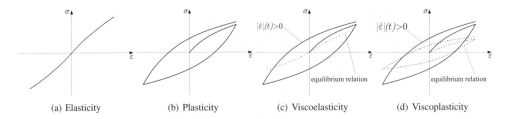

(a) Elasticity (b) Plasticity (c) Viscoelasticity (d) Viscoplasticity

Figure 1: Four types of material behaviour

does not allow for a clear classification of the material behaviour. Therefore, strain-rate dependent and independent phenomena have to be separated by a suitable experimental procedure and the shape of the equilibrium relation has to be determined. According to this, the following classification can be made:

- *Elasticity:* strain-rate independent without hysteresis (Fig. 1(a)),

- *Plasticity:* strain-rate independent with hysteresis (Fig. 1(b)),

- *Viscoelasticity:* strain-rate dependent without equilibrium hysteresis (Fig. 1(c)) ,

- *Viscoplasticity:* strain-rate dependent with equilibrium hysteresis (Fig. 1(d)).

Testing Procedure

The testing procedure applied is chosen in accordance with the proposed classification of the material behaviour. All displacement controlled experiments are carried out on a standard servo-hydraulic testing machine with a maximum force of 16 kN. The strain-state in the specimen is measured by a laser extensometer and digital image correlation (ARAMIS). An air conditioning system provides approximately constant ambient temperatures of $\vartheta = 296K$.
In the first instance, monotonic tensile tests at constant strain-rates (Fig. 2(a)) are performed. Any

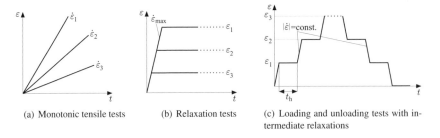

(a) Monotonic tensile tests (b) Relaxation tests (c) Loading and unloading tests with intermediate relaxations

Figure 2: Testing procedure

differences between the stress-strain-curves obtained for different strain rates $\dot{\varepsilon}_1 > \dot{\varepsilon}_2 > \dot{\varepsilon}_3$ indicate the strain-rate dependence of the material behaviour. The quantification of the strain-rate dependence is then accomplished by relaxation tests at different strain levels $\varepsilon_1 > \varepsilon_2 > \varepsilon_3$ (Fig. 2(b)).

The results obtained in tensile tests and relaxation experiments do not allow for an unambiguous classification as no unloading processes are considered. Hence, no information on the existence of an hysteresis exists. The testing procedure is therefore extended by loading and unloading processes with intermediate holding times at discrete strain levels in order to separate strain-rate dependent and independent effects (Fig. 2(c)).

Material Behaviour of Polypropylene

As one example for the proposed classification and characterization procedure, the material behaviour of Polypropylene has been investigated experimentally. In order to compare the material behaviour in monotonic tensile tests carried out at different constant velocities according to Fig. 2(a)) three engineering stress-strain-curves for distinct velocities of 1, 10 and 100 $\frac{mm}{min}$ are shown in Fig. 3.

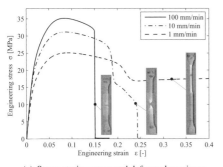

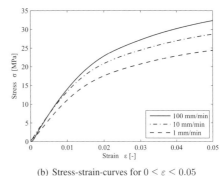

(a) Stress-strain-curves and deformed specimen　　　(b) Stress-strain-curves for $0 \leq \varepsilon \leq 0.05$

Figure 3: Comparison of characteristic stress-strain-curves for three distinct velocities

With higher velocities a clear nonproportional increase in the stress level and the stiffness can be observed. Simultaneously, the strain to failure and the amount of inelastic deformation decrease. At a velocity of 100 $\frac{mm}{min}$ the specimen fails at a maximum strain of 15 per cent. In contrast to this, no failure can be observed at a speed of 1 $\frac{mm}{min}$. In this case a neck develops and propagates through the specimen at a virtually constant load level. During this process the inelastic deformation is concentrated in two process zones at both ends of the neck.

Relaxation experiments allow for the separation of the strain-rate dependent and independent fractions of stress and are therefore of great importance for the characterization of the strain-rate dependent material behaviour. Fig. 4(a) shows the stress-time-curve of a relaxation test with a holding t_h of 48 hours. After a pronounced relaxation of stress at the beginning, the time-rate of stress decreases until the relaxation stopps after approximately 48 hours. With respect to the proposed material model this stress state is assumed to be a state of equilibrium.

So far the strain-rate dependence of the material behaviour has been demonstrated in tensile tests, relaxation experiments allow for its quantification. However, for the clear classification of the material behaviour the shape of the equilibrium relation has to be assessed. Suitable experiments are loading and unloading processes with intermediate holding times (Fig. 2(c)). For practical reasons holding times of 48 hours and more are not reasonable in these experiments. To this, a reduced holding time for intermediate relaxations $\Delta\sigma_r\left(t_{h_{ir}}\right) = 0.9 \cdot \Delta\sigma_r\left(t_h = 48\text{ h}\right)$ is defined, that is 90 percent of the overstress relaxes during $t_{h_{ir}}$. For Polypropylene the condition above leads to holding times of $t_{h_{ir}} \approx 3$ h.

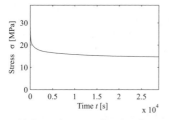
(a) Stress-time-curve for relaxation experiment

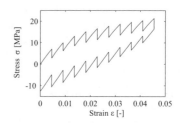
(b) Stress-strain-curve for loading and un-
loading with intermediate relaxations

Figure 4: Results for relaxation experiments and loading-unloading processes with intermediate holding times

All loading and unloading processes are carried out at a velocity of $10 \frac{mm}{min}$ and discrete points of the
equilibrium relation are determined by intermediate relaxation processes every $\Delta\varepsilon = 0.045$. Fig. 4(b)
shows the resulting stress-strain-curves for this experiment. Although the relaxation of the overstress
is not completely finished after the reduced holding time of 3 hours, the experiment indicates an equilibrium hysteresis as the terminal points at a certain strain level do not coincide for the loading and
unloading path. At least the existence of an equilibrium relation is proven by the experiment.

Material Behaviour of Thermosets

Compared to the highly inelastic behaviour of Polypropylene, the two considered thermosets – RIM935
and RTM6 – show only limited inelastic deformation. In monotonic tensile tests (Fig. 5(a), Fig. 5(c))
a minor strain-rate dependence can be detected. The strain to failure is considerably lower than for
Polypropylene. This behaviour is on the one hand due to the different microstructures of thermosets
and thermoplastics as well as due to the fact that both thermosets possess glass transition temperatures
$\vartheta_g > 373\ K$. In contrast to Polypropylene ($\vartheta_g = 273\ K$) RIM935 and RTM6 are tested below their
glass transition temperatures. In order to quantify the strain-rate dependence and to find out about the
amount of stress relaxation which can be of interest for manufacturers of fibre-reinforced polymers relaxation experiments (Fig. 5(b), Fig. 5(d)) have been performed. The procedure is currently extended
to different temperature levels below and above the glass transition temperatures of the considered thermosets.

A Material Model for Polymers

In literature there are various approaches to the simulation of the mechanical behaviour of polymers
(e.g. [3, 4, 5]). Here, in line with the presented experimental procedure, we propose constitutive
equations for modelling the inelastic material behaviour of polymers which can be characterized as a
viscoplastic material model based on an overstress formulation. The stress $\sigma = \sigma^{eq} + \sigma^{ov}$ is a combination of strain-rate independent equilibrium stress σ^{eq} and strain-rate dependent overstress σ^{ov}. Together
with the experiments performed, this allows for a clear identification of the necessary material parameter. A similar approach is presented in [6].

The material model for the equilibrium stress consists of the combination $\sigma^{eq} = \sigma^{eq,e} + \sigma^{eq,end}$ of a

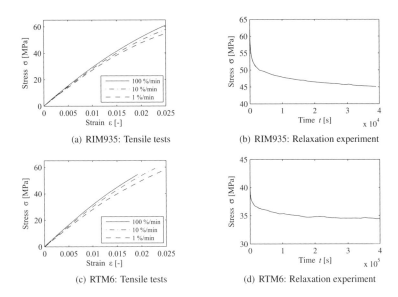

(a) RIM935: Tensile tests

(b) RIM935: Relaxation experiment

(c) RTM6: Tensile tests

(d) RTM6: Relaxation experiment

Figure 5: Experimental results obtained for two thermosets – RIM935 and RTM6

nonlinear elastic stress-strain-relation $\sigma^{\text{eq,e}}$ and an endochronic model [7, 8] of plasticity $\sigma^{\text{eq,end}}$

$$\sigma^{\text{eq,e}} = \frac{E_1}{1 + \alpha|\varepsilon|}\varepsilon \tag{1}$$

$$\sigma^{\text{eq,end}} = E_2\left(\varepsilon - q^{\text{end}}\right) \quad \text{with} \quad q'^{\text{end}}(z) = \beta\left(\varepsilon(z) - q^{\text{end}}(z)\right) \quad \text{and} \quad \dot{z}(t) = |\dot{\varepsilon}| \tag{2}$$

where q^{end} is an internal variable used to model the hysteresis in the equilibrium relation. Strain-rate independence is ensured by the use of a generalized arclength z. If no equilibrium hysteresis is observed, only the elastic part of the stress-strain-relation can be used. Linear elastic material behaviour is included in the model for $\alpha = 0$. The material properties are identified from fitting the model prediction to the terminal points of the intermediate relaxations.

The constitutive formulation for overstress is a generalized MAXWELL-model with n_{v} parallel elements

$$\sigma^{\text{ov}} = \sum_{i=1}^{n_{\text{v}}} \sigma_i^{\text{ov}} \tag{3}$$

$$\sigma_i^{\text{ov}} = c_i\left(\varepsilon - q_i^{\text{ov}}\right) \quad \text{with} \quad \dot{q}_i^{\text{ov}} = \frac{1}{\tau_i}\left(\varepsilon - q_i^{\text{ov}}\right) = \frac{c_i}{\eta_i}\left(\varepsilon - q_i^{\text{ov}}\right). \tag{4}$$

The parameters of the model c_i and τ_i form a discrete relaxation spectrum which can be determined from relaxation experiments using the window algorithm of EMRI and TSCHOEGL [9].

Nonlinear viscoelastic behaviour, which can be observed for many polymers, requires an extension of the linear overstress model. A nonlinear strain-rate dependence can be introduced by replacing the constant viscosity η_i with a function $\tilde{\eta}_i = \eta_i \exp\left(-\frac{|\sigma^{\text{ov}}|^{k_0}}{s_0}\right)$ depending on the current overstress.

The proposed material model was applied to simulate the mechanical behaviour of three different polymers – the thermoplastic Polypropylen and the thermosets RIM935 and RTM6. After the identification of the necessary material parameters the results illustrated in Fig. 6 have been obtained.

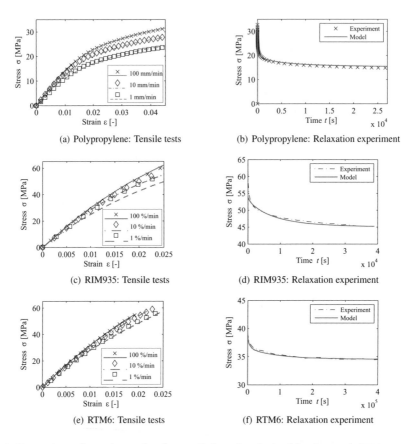

(a) Polypropylene: Tensile tests

(b) Polypropylene: Relaxation experiment

(c) RIM935: Tensile tests

(d) RIM935: Relaxation experiment

(e) RTM6: Tensile tests

(f) RTM6: Relaxation experiment

Figure 6: Comparison of experimental and numerical results obtained for three polymers

References

[1] M. Kästner, G. Haasemann, J. Brummund, V. Ulbricht, in: *Mechanical response of composites*, edited by P.P. Camanho, C.G. Davila, S.T. Pinho, J.J.C. Remmers, Computational methods in applied sciences (Springer, 2008).

[2] P. Haupt, Acta Mechanica 100 (1993), p. 129 – 154.

[3] A. Lion, Continuum Mechanics and Thermodynamics 8 (1996), p. 153 – 169.

[4] E. Krempl, F. Khan, International Journal of Plasticity 19 (2003), p. 1069 – 1095.

[5] A. D. Drozdov, J. d.C. Christiansen, Polymer 48 (2007), p. 3003 – 3012.

[6] S. Hartmann, Arch. Appl. Mech. 76 (2006), p. 349 – 366.

[7] K. Valanis, Archives of Mechanics 23 (1971), p. 517 – 533.

[8] T. Kletschkowski, U. Schomburg, A. Bertram, Mechanics of Time-Dependent Materials 8 (2004), p. 119 – 135.

[9] I. Emri, N.W.Tschoegl, Rheologica Acta 32 (1993), p. 311 – 321.

Applied Mechanics and Materials Vols. 24-25 (2010) pp 201-206
© (2010) Trans Tech Publications, Switzerland
doi:10.4028/www.scientific.net/AMM.24-25.201

Strain evolution measurement at the microscale of a Dual Phase steel using Digital Image Correlation

H. Ghadbeigi[1, a], C. Pinna[1, b], S. Celotto[2, c] and J. R. Yates[1, d]

[1]The University of Sheffield, Department of Mechanical Engineering,

Mappin Street, Sheffield, S1 3JD, UK

[2]Corus RD&T, IJmuiden, The Netherlands

[a]h.ghadbeigi@sheffield.ac.uk, [b]c.pinna@shef.ac.uk, [c]steven.celotto@corusgroup.com, j.yates@sheffield.ac.uk

Keywords: DP steel, DIC, Strain, Microstructure

Abstract. Digital Image Correlation (DIC) together with in-situ tensile testing has been used to measure in DP1000 steel the evolution of plastic strains at the microstructure scale. Interrupted tensile tests were performed on specially designed samples and scanning-electron micrographs were taken at regular applied strain intervals. Patterns defined by the microstructural features of the material have been used for the correlation carried out using LAVision software. The full field strain maps produced by DIC show a progressive localisation of deformation into bands at about $45°$ with respect to the loading direction. Plastic strains as high as 130% have been measured within the ferrite phase.

Introduction

Dual phase (DP) steels are widely used in the automotive industry since they combine high strength, high ductility as well as good formability [1]. DP steels consist of hard martensite islands dispersed throughout a soft ferrite phase. The micro-mechanical behaviour of this two-phase material is not well understood and this is a limiting factor for the development and application of future generation automotive steels.

Several attempts have been made to study the local deformation of advanced high strength steels including DP and TRIP steels [2]. Digital Image Correlation (DIC) has been used previously to measure strain distributions in DP steels [3, 4]. Tarigopula et al. [3] determined the evolution of strain localisation in DP800 steel at the macro scale during a shear test, but without any information about the contribution of each phase to the localisation process. Ososkov et al [4] used in-situ tensile tests to measure local strain partitioning in a DP600 steel before the onset of necking. They intermittently interrupted the test in a Scanning Electron Microscope (SEM) to take images and they analysed a relatively small area of the microstructure located away from the necking region. Following a small number of discrete points located either in a ferrite island or in a martensite rich region, that contained agglomerates of very fine ferrite and martensite particles, they also estimated the evolution of local deformation within each region. They reported plastic strain values inside the martensite rich regions and the ferrite phase as high as 30% and 70% respectively.

In the current study, in-situ micro tensile tests have been conducted in a SEM chamber to measure the evolution of plastic strain distributions in a DP1000 steel up to fracture. Micro-scale strain maps have been produced using DIC to study strain localisation in the material and its influence on damage.

Experiments

A commercial DP1000 steel grade with 50% ferrite and 50% martensite has been used in this study. The chemical composition of the material is given below in Table 1.

wt% C	wt% Mn	wt% Si	wt% Cr	wt% V	wt% Ni	wt% Nb
0.152	1.53	0.474	0.028	0.011	0.033	0.014

Table 1: Chemical composition of the DP1000 steel studied

The specimen geometry shown in Fig.1a has been used to obtain the required strain levels given the capacity of the in-situ SEM load frame. All the specimens were manufactured from a 1.5 mm thick sheet with the loading direction parallel to the rolling direction. The samples were mechanically polished and etched first for 5 seconds in a 2% Nital solution and then for 15 seconds in a 10% aqueous solution of sodium meta-bisulfite (SMB). The etched microstructure shows dark ferrite phase and white martensite of approximately equal volume fraction mix (Fig. 1b).

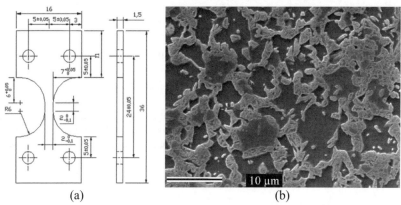

(a)   (b)

Fig. 1: (a) micro-tensile test sample with a 2mm gauge length and (b) microstructure of DP1000 steel revealed by double etching using Nital and SMB solutions

A Deben MICROTEST tensile stage with a capacity of 5 kN was used to carry out the tests. The machine was loaded in a CAMSCAN SEM, Fig. 2. The test was interrupted regularly, under loading to take images of a pre-selected area of the microstructure. This area was located as close as possible to the centre of the gauge length where necking was likely to take place. A very slow strain rate ($\dot{e} \approx 0.001s^{-1}$) was selected to give sufficient time to follow the deformation of the microstructure. The test was stopped after the start of necking. The specimen was then removed from the stage and a layer about 400µm thick was removed from the specimen by mechanical polishing to eliminate the surface roughness created by the out of plane deformation. The microstructure was then revealed following the same procedure described. Fig. 3 shows the various steps of this procedure with the specimen loaded again in the microscope at the end. Micrographs were taken again regularly until the final fracture of the sample.

DIC was used to analyse the local plastic deformation at the scale of the microstructure using micrographs taken during the test. The micrographs were analysed using the commercial software, DaVIS 7.0 from LAVision [5] to determine the in-plane displacement field from which plastic strain values were calculated. Errors due to electron beam shift [6, 7] have been neglected in this study. The magnitude of the image shift has been estimated to less than 0.5 pixels for the imaging setup used in this research. This leads to a strain error of the order of 0.1% which is very small compared to the high strain values reported in the next section.

Sequential cross correlation has been carried out between successive images and the final displacement map corresponds to the summation of the incremental displacement vectors computed for each interrogation window. The strain values were then calculated by differentiating the final displacement maps. The microstructural features of the material have been directly used to correlate the images between two successive loading steps. A multi-pass algorithm [5] leading to a 2.2 µm by 2.2 µm final interrogation window with 75% overlap between windows was used to make the correlation work. A displacement accuracy of 0.01 pixels was obtained with a strain resolution of about 0.1%. Strain maps were then produced over an area of 57 µm by 45µm and the appearance of damage was then studied in relation to local strain distributions.

Fig. 2: Tensile test stage loaded in the SEM chamber

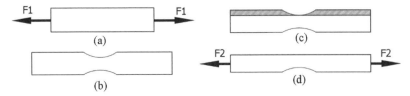

Fig. 3: (a) The specimen is loaded for the first step of the test, (b) necking occurs at the centre of the gauge section, (c) the deformed layer at the surface is removed by mechanical polishing and (d) the specimen is loaded again after etching until final fracture

Results

The engineering stress-strain curve shown in Fig. 4 is obtained from the test by measuring the elongation of the 2mm long gauge section. The elongation of the gauge section was calculated by subtracting the elastic contribution of the machine to the total displacement measured by the LVDT. The large post uniform elongation in Fig. 4 is due to the effect of the gauge section geometry, which provides an opportunity to study damage development in the material. The small stress variations along this curve correspond to the test interruptions needed to take the micrographs for subsequent DIC analysis.

Fig. 5 shows the micrographs of the selected regions at the beginning of each loading step. Distributions of the microscale engineering strain component E_{yy}, along the loading direction (which is vertical on the pictures), together with the micrographs of the deformed material at the end of each step are shown in Fig. 6. Figures 6a and 6b correspond to a macro strain of about 22% at the end of the first step and reveal the heterogeneity of the local deformation with strain values as high as 65% in the highlighted area.

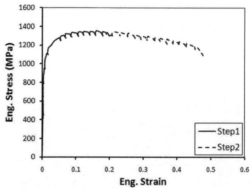

Fig. 4: Engineering stress-strain curve obtained from elongation of the gauge section

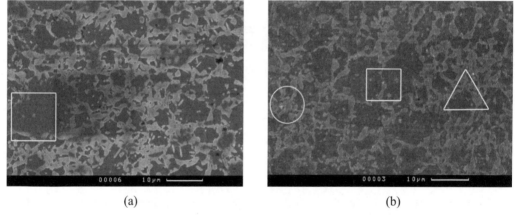

<table>
<tr><td>(a)</td><td>(b)</td></tr>
</table>

Fig. 5: Undeformed microstructure of the selected regions at the beginning of (a) the first and (b) the second loading step.

Localised bands of deformation have formed at 45° with respect to the loading direction, with the highest strain values located within a large ferrite island where slip bands can be observed. Results also show that the martensite phase experiences large strain values as high as 35% locally. Figures 6c and 6d show the deformed microstructure and corresponding strain map at a macro strain of 45% prior to the final fracture. Since the microstructure at the beginning of the second step does not correspond to the microstructure at the end of the first step, due to the material removal after polishing, the two strain maps for steps 1 and 2 cannot be added directly. The highlighted regions in Fig. 6c show some damage sites that have appeared on the surface. Plastic strain values as high as 70% have been measured locally inside the highlighted regions (Fig. 6d). The produced strain map shows again localised deformation bands that follow the path created by slip bands in ferrite islands with the appearance of damage sites at the location of high strain values in either the ferrite (square area) or the martensite phase (circle).

The measured strain values in Fig. 6 show that the material can experience very large plastic deformation (with local strain values higher than 100%), if the contributions of the two steps are added together, while the macro strain value is only 45%. Extensive plastic deformation takes place within both the ferrite and martensite for this DP1000 steel, without any observation of failure along the interface between the two phases. Further investigation is however required to understand better the damage initiation and propagation mechanisms that operate in this material using the procedure developed in this study.

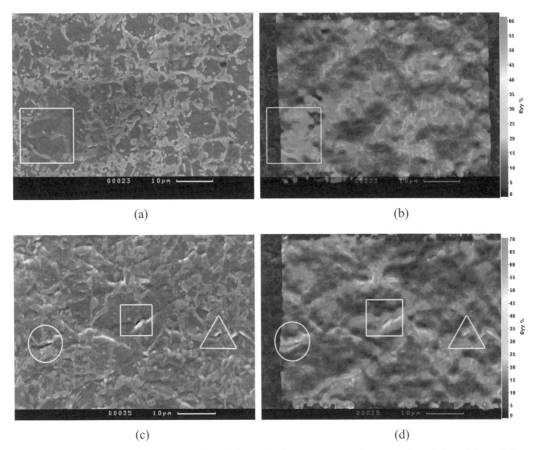

Fig. 6: (a and c) Micrographs of the deformed microstructure after steps 1 and 2, and (b and d) the corresponding strain maps superimposed onto the deformed microstructures (the loading direction is vertical on these pictures)

Conclusions

In-situ micro tensile tests have been carried out up to fracture on DP1000 steel to measure local plastic strain values at the scale of the microstructure using DIC. A progressive formation of localised bands orientated at 45° with respect to the tensile direction has been observed and provides physical insight into the localisation of deformation in this two-phase steel. Both ferrite and martensite experience large plastic deformation before the onset of local damage with strain values as high as 120% and 35% measured in the ferrite and in the martensite phases respectively. The effect of the deformability of martensite on damage initiation and evolution in various DP steels can therefore be studied using the experimental procedure developed in this work.

Acknowledgements

The authors would like to thank EPSRC (Grant number EP/F023464/1) for financial support and CORUS RD&T, IJmuiden in the Netherlands for providing the material of this study.

References

[1] T. Senuma: ISIJ INT. Vol. 41 (2001), p.520

[2] S. Oliver, T.B. Jones and G. Fourlaris: Mater. Charact. Vol. 58 (2007), p. 390

[3] V. Tarigopula, O.S. Hopperstad, M. Langseth, A.H. Clausen, F. Hild, O.-G. Lademo and M. Eriksson: Exp. Mech. Vol. 48 (2008), p. 181

[4] Y. Ososkov, D.S. Wilkinson, M. Jain and T. Simpson: Int. J. Mater. Res. Vol. 98 (2007), p. 664

[5] LAVision, Davis Strain master Software. 2005.

[6] M.A. Sutton, N. Li, D.C. Joy, A.P. Reynolds and X. Li: Exp. Mech. Vol. 47 (2007), p. 775

[7] M.A. Sutton, N. Li, D. Garcia, N. Cornille, J.J. Orteu, S.R. McNeill, H.W. Schreier, X. Li and A.P. Reynolds: Exp. Mech. Vol. 47 (2007), p. 789

Applied Mechanics and Materials Vols. 24-25 (2010) pp 207-211
© (2010) Trans Tech Publications, Switzerland
doi:10.4028/www.scientific.net/AMM.24-25.207

Biaxial Ratcheting Response of SS 316 Steel

R.Suresh Kumar [1, a], C.Lakshmana Rao [2,b] and P.Chellapandi [3,c]

[1] Scientific Officer, Nuclear Engineering Group, IGCAR, Kalpakkam-603102, India

[2] Professor, Department of Applied Mechanics, IITM, Chennai-600036, India

[3] Associate Director, Nuclear Engineering Group, IGCAR, Kalpakkam-603102, India

[a] suresh@igcar.gov.in, [b] lakshman@iitm.ac.in, [c] pcp@igcar.gov.in.

Keywords: Fast Breeder Reactor, Biaxial Ratcheting, Mechanical Behavior, Rectangular Rosette, Austenitic Stainless Steel.

Abstract. Ratcheting is one of the challenging phenomena that needs to be investigated for the *Fast breeder reactor* (FBRs), to arrive at the optimum structural dimensions that are safe and yet do not have undue redundancy. *Austenitic stainless steel* is the principal structural material for Indian FBR. Preliminary assessment indicates that there is a need to demonstrate that the main load carrying vessel made of this material can provide sufficient safety margin against ratcheting under biaxial loading conditions. This exercise calls for carrying out many simulated experiments, particularly with biaxial tension torsion specimens to generate adequate data for developing robust constitutive models to predict ratcheting. Accordingly, many *biaxial tension-torsion experiments* for austenitic stainless steel pipes were conducted and the best results have been reported here. The mechanical behavior of this material has been reported for a given axial tensile stress superimposed with a given range of cyclic shear stress for many cycles of loading. *Rectangular rosette* is used for capturing the biaxial response. Important material responses like *cyclic hardening* and *biaxial ratcheting* have been experimentally observed. Maximum accumulation of 2700 μ axial strain has been observed for a loading condition of constant axial stress of 102 MPa super imposed with a cyclic variation of shear stress amplitude of 120 MPa over 2450 cycles. The amount of progressive accumulation of axial strain was found to be directly dependent on the number of cycles. The observed rate of axial strain accumulation found decreased with increase in number of cycles. All these results are presented in detail in this paper and important conclusions that are useful in modeling the observed behavior are discussed.

Introduction

One of the innovative features that considered for future *fast breeder reactor* (FBRs) is the elimination of main load carrying vessel cooling circuit. This would result in a significant economy by saving the cost of entire cooling circuit and also by reducing the main vessel diameter. However, the challenging aspect that needs to be addressed for this cost reduction, is to arrive at the structural dimensions that are safe and yet do not have undue redundancy. This requires an investigation of thermal ratcheting failure near sodium free level. Austenitic stainless steel is the principal structural material for Indian FBR structures. Preliminary assessment indicates that there is a need to demonstrate that the vessel made of this material can provide sufficient safety margin against ratcheting under biaxial loading conditions.

Mathematical modeling of the mechanical behavior of austenitic stainless steel under cyclic loading has also seen great developments during the last 30 years. However, proper constitutive models are still not available to predict ratcheting of SS316. Literature survey indicated that the material parameters derived from biaxial experiments are able to predict the material response more accurately compared to the material parameters derived from uniaxial experiments [1,2]. Hence it is necessary to carry out many simulated biaxial experiments to generate adequate data for calibrating constitutive models. Thus, robust constitutive models can be developed to predict ratcheting phenomenon.

Many *biaxial tension-torsion experiments* for austenitic stainless steel pipes were conducted and the best results have been reported in this paper. The mechanical behavior of this material is reported for a given axial tensile stress superimposed with a given range of cyclic shear stress for many cycles of loading. All those results along with some important conclusions that are useful in mathematical modeling of some of the observed behavior is also presented in detail in this paper. The experimental results are then used to assess the shakedown limit that is prescribed in the nuclear code.

Biaxial Experiment

Experimental Facility. A photograph of the biaxial testing facility is shown in Fig. 1. In this experimental facility, axial tensile as well as cyclic torsion loading can be applied to the tubular specimen. Servo hydraulic based actuators are used in the test facility to simulate static as well as dynamic loadings. Maximum axial capacity of the test facility is 20 kN tension and that of the cyclic torsion is ±100 Nm.

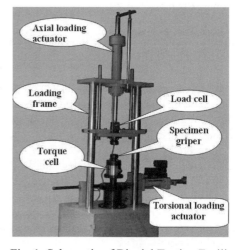

Biaxial Experiment. Biaxial state of stress can be simulated by various methods. Tension-torsion testing of tubular specimen is adopted in this paper. The cross coupling effect of shear and normal stress can be well captured by conducting such experiments. The purpose of this experiment was to provide sufficient data base under biaxial state

Fig. 1: Schematic of Biaxial Testing Facility

of stress to establish the mechanical behavior austenitic stainless steel under biaxial fatigue loading. These data are further used for validating the mathematical models [3-4].

Specimen Preparations. Austenitic stainless steel (SS316) pipes were used in these experiments. Material properties for the SS 316 at room temperature have been taken from RCCMR 2007 [5]. Pipe dimensions are arrived at based on the loading capacity of the test facility and avoiding very thin specimens that lead to buckle of the specimen. Biaxial testing specimen has been prepared as per the guide lines of ASTM E 606-04 [6]. The pipe dimensions at the guage length portion are 18.5 mm OD and 0.7 mm thickness.

Data Acquisition. Rectangular rosette was pasted over the tubular specimen to capture axial as well as shear strains. Smallest rosette from the range available in the market was selected to get proper bondage with the cylindrical surface while applying torsion. Each terminal of the rosette was connected to separate channel data acquisition channel. The tested specimen along with the schematic loading conditions is shown in Fig. 2.

Experimental Observations. The above test facility was used for various biaxial tests by applying cyclic variation of torsion super imposed with constant axial stress. The cyclic variation of shear stress has been applied in a sinusoidal manner with a frequency of 1

Fig. 2: Tested Specimen

Hz. It was observed during the course of the experiment that the specimen failed by buckling and the rosette was detached from the specimen at higher torsion. Many tests were conducted and the best results are presented below:

Biaxial Tension – Torsion Test. Constant axial stress of 102 MPa and cyclic shear stress ($\Delta\tau$) of 120 MPa has been applied. Some of the important results of these experiments are given in Fig 3. The *biaxial ratcheting* response (axial strain vs. shear strain) is shown in Fig. 3a. The axial strain is seen increasing in the direction of mean stress due to *ratcheting* and the mean shear strain is seen to be decreasing (the shift towards left) due to the *cyclic hardening* of the material. The axial stress-strain diagram is shown in Fig. 3b. For a constant stress, the strain is seen increasing in the direction of sustained stress (i.e.) *ratcheting* has been observed in the axial direction.　　The grey colour portion indicates the magnitude of the axial-shear strain during the last 200 cycles. The amount of progressive accumulation of axial strain was directly dependent on the number of cycles. The maximum accumulation of 2700 µ mean axial strain has been observed for a loading condition of constant axial stress of 102 MPa super imposed with a cyclic variation of shear stress amplitude of 120 MPa for 2450 cycles.

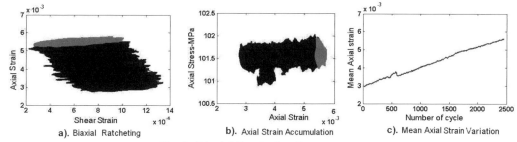

a). Biaxial Ratcheting　　　b). Axial Strain Accumulation　　　c). Mean Axial Strain Variation

Fig. 3: Biaxial Ratcheting Response

Monotonic Torsion Test. By using the above mentioned test facility monotonic torsion test was also performed. The tubular specimen was uniformly twisted and the respective torque was measured by using the torque cell attached with the specimen. Shear strain was captured using the 3 channel rectangular rosette pasted on the tubular specimen. Accordingly the *shear stress – shear strain diagram* has been plotted for the material SS316 and it is given in Fig. 4. Thus derived *shear stress –shear strain diagram* can be further used for identifying material parameters for mathematically modeling the *kinematic hardening* behavior.

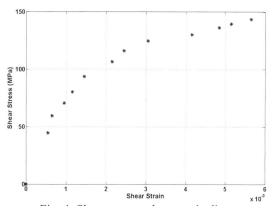

Fig. 4: Shear stress - shear strain diagram

Design Code Philosophy for the Assessment of Shakedown Strain Limit.

Design Code Procedure. RCC MR RB [7] is the design code for Indian *fast breeder reactor*. As per this, strain limit computation is based on the *efficiency index method*. Efficiency index method is used to convert the effect of biaxial stress into its equivalent uniaxial effect. Efficiency index diagram was obtained from RCC MR [7] for a given secondary stress ratio (SR). The secondary stress ratio (SR) is the ratio of the maximum range of cyclic normal stress to the axial normal stress. Secondary stress ratio for the loading conditions of axial stress (σ_{axial}) of 102 MPa and a range of cyclic shear stress of ($\Delta\tau$) 120 MPa is 1.86. Accordingly the efficiency index obtained is 0.7 as indicated in Fig. 5a. By using this efficiency index (Fig. 5a: Courtesy from RCCMR), the biaxial loading can be converted into equivalent uniaxial load (effective stress). The shakedown strain limit for this effective stress was obtained from the uniaxial reduced cyclic curve (Fig 5b.: Obtained from the cyclic test carried out based on uniaxial specimen) is 0.07%. These procedures are illustrated in Fig. 5.

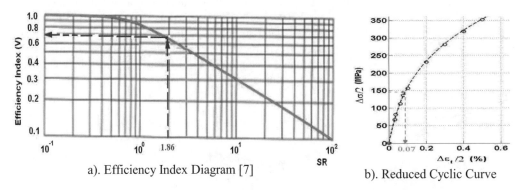

a). Efficiency Index Diagram [7]   b). Reduced Cyclic Curve

Fig 5. Shake down Strain Limit Assessment as per RCC MR Methodology

Comparison. From the above procedure, the net axial strain is obtained as 0.07 %. The comparative performance of the net strain computed by the efficiency index method as well as the net biaxial response obtained by experiment is shown in Fig. 6. In this figure, the bold straight line indicates the shake down limits as per the RCC-MR [7] and the wavy thin line indicates the biaxial ratcheting response obtained from the experiments. The wavy thin line shows that the experimentally obtained strain shakes down at higher number of cycles (say around 4000). Shakedown is the situation beyond which further

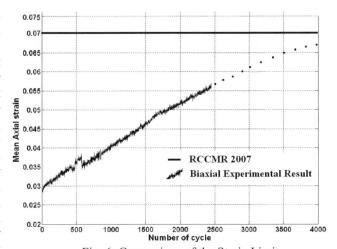

Fig. 6: Comparison of the Strain Limit

strain growth does not happen. Further repeating with multiple specimens will reveal the exact margin available in the RCC-MR ratcheting strain limits. The strain control cyclic testing will give more realistic shake down limits.

Summary

Experiments have been conducted to study the biaxial cyclic response of austenitic stainless steel, which is a primary material used in the FBR structures. Many *biaxial tension-torsion experiments* for austenitic stainless steel pipes were conducted and the best results have been reported in this paper. The mechanical behavior of this material is reported for a given axial tensile stress superimposed with a given range of cyclic shear stress for many cycles of loading. *Biaxial ratcheting* has been experimentally observed. The accumulation of 2700 μ axial strain has been observed for a loading condition of constant axial stress of 102 MPa super imposed with a cyclic variation of shear stress with a range of 120 MPa for 2450 cycles. The amount of progressive accumulation of axial strain was observed to be dependent on the number of cycles. The decrease in slope of the mean axial strain variation (Fig. 3c.) indicated that, the rate of axial strain accumulation decreased with the increase in number of cycles. Pure torsion experiment has been conducted to generate the *shear stress-shear strain curve* to compute the *cyclic hardening* material parameters so that robust mathematical model can be developed for predicting the *ratcheting* phenomenon. The experimentally observed axial strain has been further compared with the shake down limit as per RCC-MR [7]. It is found that the codes predict a shake down at higher cycles (say around 4000). In order to assess the exact margin available in the codal procedure, further experiments with higher magnitude/more number of cycles have to be performed.

References

[1] Shafiqul Bari, Tasnim Hassan, in: *An advancement in cyclic plasticity modeling for multiaxial ratcheting simulation*, Int. J. of Plasticity 18, 873–894 (2002).

[2] M.Abdel-Karim, in: *Modified kinematic hardening rules for simulations of ratcheting*, Int. J. of Plasticity, Volume 25, Issue 8, Pages 1560-1587 (2009).

[3] Laurence Portiera,b, Sylvain Callocha, Didier Marquisa, Philippe Geyerb: *Ratchetting under tension-torsion loadings: experiments and modeling*, International Journal of Plasticity 16 303-335 (2000).

[4] Guozheng Kang, Qing Gao, Lixun Cai, Yafang Sun, *Experimental study on uniaxial and nonproportionally multiaxial ratcheting of SS304 stainless steel at room and high temperatures*, Nuclear Engg. and Design 216 13–26 (2002).

[5] RCC-MR Section I, Subsection Z, "*Technical Appendix A3*", (2007).

[6] ASTM E 606-04, "*Standard practice for strain controlled fatigue testing*", (2005).

[7] RCC-MR Section I, Subsection B, "Design and construction rules for class-1 components of FBR nuclear islands", (2007).

Applied Mechanics and Materials Vols. 24-25 (2010) pp 213-218
© (2010) Trans Tech Publications, Switzerland
doi:10.4028/www.scientific.net/AMM.24-25.213

On the mutual interactions of monotonic and cyclic loading and their effect on the strength of aluminium alloys

Z.L. Kowalewski[1, a], T. Szymczak[2, b]

[1]Institute of Fundamental Technological Research,
ul. Pawinskiego 5B, 02-106 Warszawa, Poland

[2]Motor Transport Institute, ul. Jagiellonska 80, 03-301 Warszawa, Poland

[a]zkowalew@ippt.gov.pl, [b]tadeusz.szymczak@its.waw.pl

Keywords: cyclic loading, biaxial stress state, softening effect, yield surface.

Abstract. The paper investigates the interaction of monotonic cyclic loading and their effect on the yield strength of aluminium alloy. Two different loading combinations were considered, i.e. torsion-reverse-torsion superimposed on monotonic tension and monotonic torsion combined with tension-compression cycles. All strain controlled tests were carried out at room temperature using thin-walled tubular specimens. The maximum value of the total cyclic strain amplitude was less than 1%. The influence of amplitude, frequency and shape of cyclic loading signal on the proportional limit and conventional yield point was investigated. The experimental results presented in the paper may be useful to designers of structures that utilize manufacturing processes such as drawing, extrusion, forging of selected semi-finished elements or researchers working on the development of new constitutive equations.

Introduction

In many branches of industry the loading of engineering components is complex and can have an significant influence on the mechanical properties of engineering materials. A typical example of such a situation is a gas rotor turbine, for which three working periods can be distinguished i.e. start-up, operation and switching off [5]. The highest stress values and the largest phase difference between the different components of load are obtained during start-up. Therefore, the start-up is regarded as crucial with respect to variations of the mechanical properties of the material. Since the mechanical parameters of materials can be modified by complex loading many research groups are actually looking at complex loading in an effort to optimize some of the manufacturing processes [1], [2], [3], [4], [6] in order to reduce costs and prolong the lifetimes of engineering components. It can therefore be concluded that knowledge of loading history has at least a twofold role. It is important for adequate choice of materials of structural elements, and moreover, it allows the design of forming processes to guarantee optimal mechanical parameters for particular applications. Hence many mechanical laboratories are involved in such research programmes, which would be able to provide new knowledge related to the influence of complex loading on the selected material parameters of importance in industry.

Details of experimental procedure

All strain controlled tests were carried out at room temperature under biaxial stress state using thin-walled tubular specimens (Fig. 1) made of the 2024 aluminium alloy commonly used in the aircraft industry. The loading program was designed in such a way that the cyclic loading was superimposed on monotonic deformation, i.e. torsion-reverse-torsion cycles were combined with monotonic tension (Fig. 2), and tension-compression cycles were simultaneously carried out with monotonic torsion (Fig. 3). The total cyclic strain amplitude was less than 1%, since for that magnitude it can be assumed that residual stresses are negligible. After the main loading programme a yield surface concept was applied to check if the effects observed during cyclic loading have a permanent character.

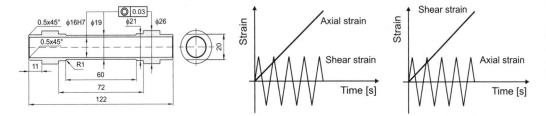

Fig. 1. Dimensions of the biaxially loaded specimen.

Fig. 2. First loading combination.

Fig. 3. Second loading combination.

Experimental results

The influence of torsion-reverse-torsion cycles on the 2024 aluminium alloy behaviour is presented in Fig 4a. The results exhibit a significant lowering of the following mechanical parameters: magnitude of hardening, proportional limit and conventional yield point. The effects are very strong and depend on the amplitude of cyclic loading. An increase of the cyclic strain amplitude led to the further decrease of the stress-strain characteristic. A reduction of the yield point was almost 250 MPa for a cyclic strain amplitude of ±0.9%.

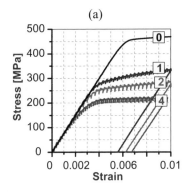
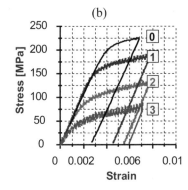

Fig. 4. Tensile (a) and torsional (b) stress-strain characteristics obtained from cyclic loading due to torsion-reverse-torsion and tension-compression, respectively. (0 – only monotonic loading, 1-4 with the cyclic loading of strain amplitude equal to ±0.3% (1), ±0.5% (2), ±0.7% (3), ±0.9% (4)).

A similar effect is observed for the tension-compression cycles superimposed on monotonic torsion, Fig. 4b. This loading regime also caused a significant essential reduction of the yield point. For example, for a cyclic strain amplitude of ±0.7% the reduction was 170MPa.

In order to confirm the variations of the mechanical parameters observed during the complex loading a yield surface was determined, after unloading of the specimens. If, for example, a reduction of the tensile yield point observed during cyclic loading would be permanent, then such a point on the subsequent yield surface should have a position much closer to the initial yield surface origin. However, as it clearly seen (Fig. 5a, b), this is not the case. The tensile yield point on the subsequent yield loci is even greater then that on the initial yield surface. This means that the reduction of strength during tension in conjunction with cyclic torsion is only temporary. It vanishes directly after interruption of cyclic loading. The same tendency can be observed for pure torsion conducted simultaneously with the tension-compression cycles. An increase of the tensile yield point can be attributed to the prior deformation induced by the previous loading history, Fig. 5a. Both diagrams in Fig. 5a, b also illustrate a residual stress evolution. All center points of the subsequent yield surfaces are moved into negative values of the axial stress. Such features identify residual stress variations due to the loading applied.

In conclusion one can say that the comparison of the subsequent yield loci with the initial yield surface exhibits only an influence of the loading history applied, and moreover, proves a transient character of the stress drop in a direction of the monotonic loading, which can be solely attributed to cycles acting in the perpendicular direction.

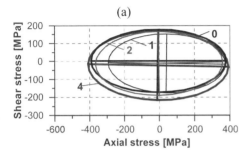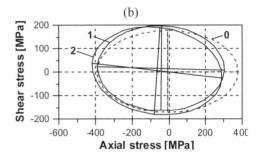

Fig. 5. An initial yield surface (0) evolution due to the presence of torsion-reverse-torsion cycles (a) or tension-compression cycles (b) during monotonic deformation by tension or torsion, respectively.
Numbers 1, 2, 4 correspond to magnitudes of strain amplitude of ±0.3%, ±0.5%, ±0.9%, respectively.

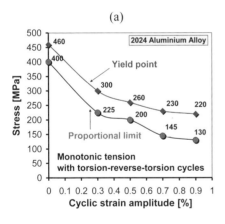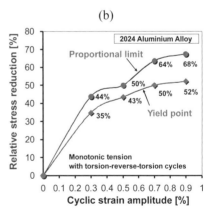

Fig. 6. Diagrams reflecting variations of the yield point and proportional limit for combined monotonic tension and torsion cycles for different strain amplitudes.

In the next stage of the experimental programme further variations of mechanical parameters were investigated. The representative results are presented in Fig. 6a, b. The figure illustrates a reduction of stress parameters due to the presence of the combined cyclic and monotonic loading shown in Fig. 2. For the highest cyclic strain amplitude (±0.9%) the reduction of the proportional limit and yield point are 70% and 50%, respectively. In the case of the second loading programme, a combination of monotonic torsion and tension-compression cycles (Fig. 3), the yield point decreases from 223MPa to 63 MPa, a 72% reduction, see Fig. 7a, b. The significant lowering of the magnitude of the hardening curve and the conventional mechanical parameters may play a role in the modification of some manufacturing processes, because it enables a reduction of the force applied during forging or extrusion processes for example. From a theoretical point of view such knowledge is essential for the development of new constitutive equations.

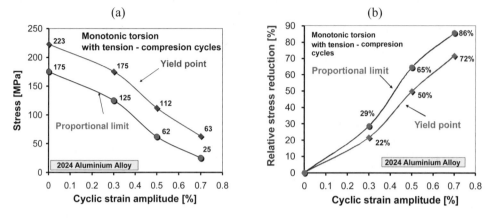

Fig. 7. Diagrams reflecting variations of the yield point and proportional limit for combined monotonic torsion and tension-compression cycles for different strain amplitudes.

From the results of these tests it is clear that cyclic loading superimposed on monotonic loading leads to significant reduction of the aluminium strength. This phenomenon could be beneficial for many industrial applications. Energy balance calculations have been performed to compare plastic strain energy dissipated during a typical monotonic test and that of a monotonic-cyclic loading combination. The results are summarized in Tab. 1 and illustrated in Fig. 8.

Table 1. Variations of the total plastic strain energy during monotonic deformation assisted by torsion-reverse-torsion cycles.

A	B	C	D	E	F
Plastic Strain Energy for Monotonic Tension [MJ/m³]	Cyclic Strain Amplitude [%]		Monotonic Tension assisted by Torsion-Reverse-Torsion for 20 cycles	Torsion-Reverse-Torsion for 20 cycles	**Total Plastic Strain Energy for Monotonic-Cyclic Loading [MJ/m³]**
	0.0				1.7
1.7	0.3	**Plastic energy**	1.0	0.0	1.0
	0.5	**[MJ/m³]**	0.9	5.0	5.9
	0.9		0.7	38.2	38.8

The total plastic strain energy dissipated during uniaxial tension is reduced by a factor of 40% in comparison to the same type of loading assisted by torsion cycles conducted for the lowest magnitude of strain amplitude considered (0.3%) in the programme. For greater cyclic strain amplitudes the total plastic strain energy increases, and therefore, these cases are not so beneficial by accounting for the energy balance. Despite of the total strain energy increasing with an increase of the cyclic strain amplitude we have to remember that superimposing cyclic loading on monotonic loading reduces the stress parameters significantly. In many industrial applications this reduction would extend the lifetimes of some engineering components. This is especially important when considering the manufacturing costs of these elements which are extremely high.

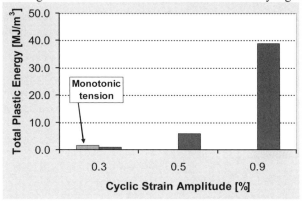

Fig. 8. Illustration of the total plastic strain energy for uniaxial tension up to a maximum strain of 1%, and a combination of monotonic tension up to a maximum strain of 1% with torsion cycles of different strain amplitudes.

Remarks and conclusions

The paper emphasizes the role of cyclic loading superimposed on monotonic loading and its effect on the strength of aluminium alloy. The results allow the following concluding remarks to be made:

- cyclic loading in combination with monotonic loading may cause a significant lowering of the proportional limit and conventional yield point,
- an increase of cyclic strain amplitude decreases the proportional limit and yield point,
- the decrease in the mechanical properties is not permanent and after termination of cyclic loading it vanishes and the material properties return to the initial values,
- cyclic loading due to tension-compression and torsion-reverse-torsion cycles in combination with monotonic loading by tension and torsion, respectively, slightly modifies the magnitude of the residual stresses,
- an initial yield surface evolution does not confirm the rapid reduction of the selected mechanical parameters during monotonic loading assisted by cyclic loading, it only points out their variations due to the loading history applied.

In order to gain a more thorough knowledge of the effects presented in this paper microscopic observations are necessary. They should be carrying out "on line" with the loading process. The stress reduction observed during tests can be treated as an important factor for modification of any manufacturing processes that are a combination of monotonic and cyclic loading. The applied loading combinations considered give a promising tool for reduction of the plastic strain energy demand, and moreover, ensure more beneficial working conditions leading to the lifetime extension of some working elements of machines used to fabricate many products in the form of rods, tubes, etc.

References

[1] W. Bochniak, A. Korbel, R. Szyndler, R. Hanarz, F. Stalony-Dobrzański, L. Błaż, P. Starski: J. Mat. Proc. Tech., Vol. 173, (2006), p. 75.

[2] W. Bochniak, A. Korbel, R. Szyndler: *Innovative solutions for metal forming*, Proc. Inter. Conf. MEFORM 2001 – Herstellung von Rohren und Profilen, Institut für Metallformung Tagungsband, 239, Freiberg/Riesa, 2001.

[3] L.X. Kong, P.D. Hodgson: Mater. Sci. Eng., A 276, (2000), p. 32.

[4] A. Korbel, W. Bochniak: J. Mater. Proc. Tech., 53, (1995), p. 229.

[5] C. Michaelsen, W. Hoffelner: The role of state of stress for the determination of the lifetime of turbine components, Fatigue under Biaxial and Multiaxial Loading, Mechanical Engineering Publication, London, 53–63, 1991.

[6] G. Niewielski, D. Kuc, K. Rodak, F. Grosman, J. Pawlicki: Journal of Achievements in Materials and Manufacturing Engineering, 17, 1-2, (2006), p.109.

Session 5B: Fracture and Damage

Applied Mechanics and Materials Vols. 24-25 (2010) pp 221-226
© *(2010) Trans Tech Publications, Switzerland*
doi:10.4028/www.scientific.net/AMM.24-25.221

Validation of Acoustic Emission (AE) Crack Detection in Aerospace Grade Steel Using Digital Image Correlation

R. Pullin[1,a], M. J. Eaton[1,b], J. J. Hensman[2,c], K. M. Holford[1,d], K. Worden[2,e] and S. L. Evans[2,f]

[1]Cardiff School of Engineering, Cardiff University, Queen's Buildings, The Parade, Cardiff, UK.

[2]University of Sheffield, Dept. of Mechanical Engineering, Mappin Street, Sheffield, U.K.

[a]pullinr@cardiff.ac.uk, [b]eatonm@cardiff.ac.uk, [c]james.hensman@sheffield.ac.uk, [d]holford@cardiff.ac.uk, [e]k.worden@sheffield.ac.uk and [f]evanssl@cardiff.ac.uk

Keywords: Acoustic emission, digital image correlation, fatigue fractures

Abstract. Acoustic Emission (AE) is a passive form of non-destructive testing that relies on the detection and analysis of stress waves released during crack propagation. AE techniques are successfully employed number of industries there remains some scepticism in aerospace engineering.

The reported investigation details a single four point bend test specimen undergoing fatigue loading. This test is part of a much larger programme designed to demonstrate a technology readiness level (TRL) of five of the use of AE to detect crack initiation and growth in landing gear structures.

The completed test required that crack growth had to be monitored to allow a comparison with the detected and located AE signals. The method of crack monitoring had to be non-contact so as not to produce frictional sources of AE in the crack region, preventing the use of crack mouth opening displacement gauges. Furthermore adhesives on the specimen surface had to be avoided to eliminate the possibility that the detected AE was from adhesive cracking, thus the use of strain gauges or foil crack gauges was not possible.

A method using Digital Image Correlation (DIC) to monitor crack growth was investigated. The test was stopped during fatigue loading at 1000 cycle intervals and a DIC image captured at peak load. The displacement due to crack growth was observed throughout the investigation and the results compared with the detected AE signals.

Results showed a clear correlation between AE and crack growth and added further evidence of TRL5 for detecting fractures in landing gears using AE.

Introduction

Messier-Dowty are world leaders in the design and manufacture of aircraft landing gear systems, structures and components. Each design requires certifying for flight by the airworthiness authorities and this process relies on information obtained from structural fatigue testing. Traditionally, non-destructive testing (NDT) is used to confirm the integrity of the landing gear structure at key stages in the fatigue test regime. The NDT inspection requires the test to be stopped for a period of time to allow the structure to be dismantled and inspected and these periods of NDT can account for 25% of the total testing time. It is proposed that Acoustic Emission (AE) can be used to monitor the landing gear during the certification test in order to reduce the down time associated with conventional NDT inspections.

Acoustic Emission (AE) is a passive NDT technique that relies on the detection of the stress waves that are released during crack propagation. Piezoelectric sensors coupled to a structure detect the surface stress waves. The sensors convert the surface displacements to a voltage that is sampled and based on user thresholds separated into individual discreet signals. Traditionally waveform parameters such as amplitude, energy and rise-time are extracted from the waveform and used for analysis.

Although AE techniques are currently employed in a number of industries, there is still some skepticism in aerospace engineering. A major aspect of implementing an AE solution for monitoring landing gears is that Messier-Dowty requires an automated system capable of identifying fractures at an early stage in what is a high-noise environment and in specimens that have complex geometry.

The reported investigation details a single specimen test that is part of a much larger programme that includes tests on real landing gears in a high noise environment, designed to demonstrate a technology readiness level (TRL) of five of using AE to detect crack initiation and growth in landing gear structures undergoing certification tests.

Methodology of Developed Technique

A methodology for automatically identifying fatigue fractures in metallic structures (Fig. 1) has been developed through an EPSRC and Messier-Dowty funded project. Full details of key aspects of the methodology have been published previously [1-6]. However, in summary, waveforms recorded using a traditional system are inputted into a novel script that extracts feature data using a fast wavelet transform. Fast wavelets are advantageous when compared to traditional AE wave descriptors which can be ambiguous. The signals are then located and clustered in the physical space (traditional location clusters). It is these clusters that form the main part of the analysis to identify fractures. Initially a measure of variance of the features of the wavelets of signals within a located cluster compared with the entire data set at a sensor is made. Signals from fractures will demonstrate a low variance in the difference in signal features as demonstrated in previous studies [5]. A further measurement of the location cluster is then made using

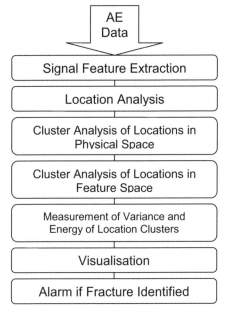

Fig. 1 Schematic of Methodology

energy rates, with a rising accumulative energy trend being an indicator of fracture growth. The rising energy trend indicator is derived from the comparison of the Paris-Erdogan equation, stress intensity factors at the crack tip and crack length [7]. As crack length increases the rate of AE emission also increases due to the greater advances in crack length per cycle, leading to a rising energy trend. It is the low variance of data in a cluster combined with a rising energy trend that signal the onset of fracture in a specimen.

Experimental Procedure

The 4PBT specimen used for this investigation is shown in Fig. 2. The specimens were manufactured by Messier-Dowty from 20mm thick aerospace grade steel. To validate the developed automated analytical techniques a digital image correlation (DIC) technique was used to monitor crack growth. During the fatigue test the cyclic loading was stopped, peak load applied and an optical measurement of the crack mouth opening was made. Using DIC offers significant advantages over foil crack gauges and a traditional crack mouth opening displacement (CMOD) gauges. Both these traditional methods can introduce acoustic emission sources into the experiment either through glue cracking in foil gauges or frictional noises from the CMOD contact point with the specimen.

The specimen was subjected to a fatigue load of 80kN / R=0.1 at 1 Hz. The specimen was instrumented with four Physical Acoustics Limited (PAL) resonant frequency sensors in a rectangular array about the notch as shown in Fig. 2. Sensors were held in position with magnetic clamps and brown grease was used as a couplant. Installed sensor sensitivity was evaluated using the pencil lead fracture technique. Waveform data was recorded using a PAL PCI-2 acquisition system at sample rate of 2 MHz. All data analysis was completed post-test.

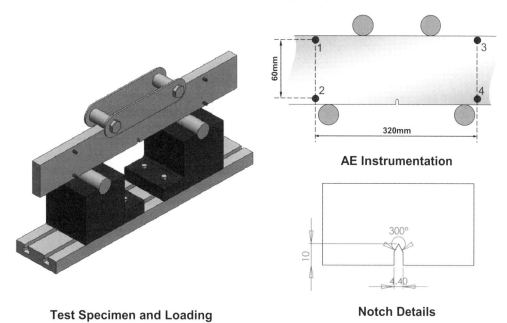

Fig 2. Specimen Details

Results and Discussion

Fig. 3 shows the digital image correlation CMOD measurement and AE data recorded during the fatigue test. The first DIC image was captured at 11k cycles and was in response to a significant increase in all detected AE energy, subsequent images were captured at 1k cycle intervals. It can be seen from the plot that the initial increase in acoustic activity coincides with an increase in crack mouth opening demonstrating that the AE is detecting the crack. This is further demonstrated by the location of the signals as shown in Fig 4.

The two results clearly demonstrate, in combination with the DIC CMOD measurements, the advantages and benefits of the acoustic monitoring technique. However the limited analysis completed does not provide an automated solution to detecting fatigue fractures.

Fig. 5 displays the first output of the developed script for the automatic detection of fractures. The plot shows the position of clusters, or groups, of over 40 signals in an area of 13 x 13 mm. The clusters are in close proximity to the notch area, acknowledging errors associated with location such as signal threshold triggering. The detected clusters all exhibit a high *crack score*, calculated from the negative log of the variance between features of waveforms associated with the cluster.

Previous investigations in this series of four point bend tests have demonstrated that clusters with crack scores above 28 are clear indicators of fractures with noise sources being considerably lower. In this example the specimen ruptured at 43.1k cycles and the initial cluster shown below occurred at approximately 8.2k cycles (the onset of crack growth). This coincides with CMOD data in Fig. 3.

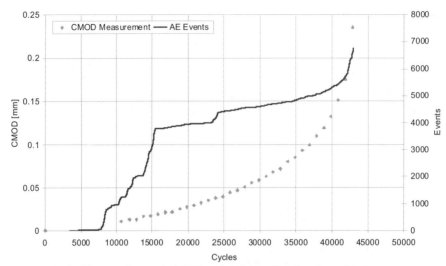

Fig 3. History of recorded DIC and AE data for duration of fatigue test

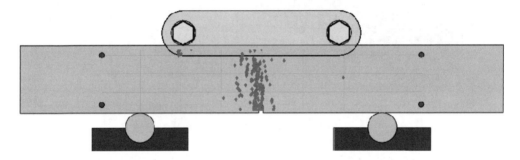

Fig 4. Location of AE signals 0-11k cycles

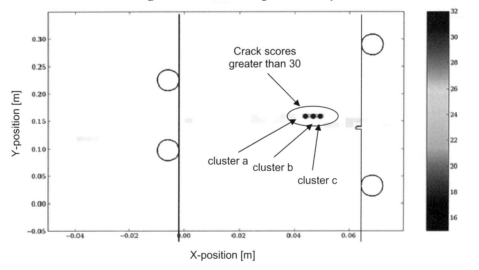

Fig. 5 Spatial and crack-score results using developed automated fracture detection script

The second aspect of the developed work examines the cumulative energy trend of the identified clusters. As previously stated in the discussed methodology a rising energy trend is indicative of fracture source and when combined with the variance indicator an automated response to the presence of the fracture can be flagged. Fig 6 shows the energy results for the three identified clusters in Fig 5.

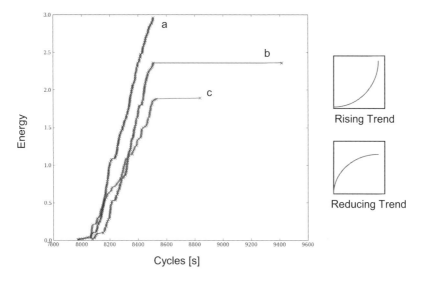

Fig. 6 Results of energy trends of spatial clusters using developed automated fracture detection script

The energy trends from the three clusters initially exhibit rising energy trends. In the script an equation for the curve is calculated mathematically and the type of curve is determined. In the example however two clusters (b and c) display a reducing trend due to signals occurring in the cluster significantly after it appears the cluster has stopped. As the fracture length increases new location clusters are created meaning earlier clusters are no longer 'live'. However if a new signal is attributed to an old 'dead' cluster the energy trend will be distorted as the cluster once again becomes 'live'. This has no bearing on the crack score. This problem is currently being investigated by introducing an online radial cluster algorithm which captures and consolidates clusters to form larger clusters which are less sensitive to additional signals being attributed to 'dead' clusters. The algorithm utilities a real time approach and reviews location clusters as they appear, hence 'online', assessing whether they are a new cluster or should be part of an existing cluster. The online radial cluster algorithm will be the focus of future publications.

The presented findings have shown that the script based on cluster variance in the feature space and energy trends would automatically indicate a fatigue fracture at 8.2k cycles in this specimen. The DIC image shows significant increase in CMOD at 11k cycles demonstrating that the fracture has started and validating the developed techniques. Results for three further investigations using four-point bend specimens are presented in Table 1 adding further validation to the developed methodology.

Further issues occur when monitoring landing gear structures undergoing certification. Landing gear structure tests are high noise environments, however noise sources have been shown to be of high variance and using an artificial source the developed techniques have been demonstrated in real tests [8]. In addition landing gears are complex structures that have a myriad of thickness changes and cutout sections which have dramatic affects on signal propagation and signal velocities which can considerably affect source location accuracy. This has been resolved using a touch and learn location system, called Delta-T, which has been demonstrated in a number of complex structures [9,10].

Table 1: Summary of results from three further four-point bend tests

Specimen Number	Position of Crack Initiation [m]	AE Location [m]	Cycles to Failure [k cycles]	Detected Failure [k cycles]	Crack Score
1	0.47	0.47	1.4	0.9	28
2	0.47	0.47	40.4	1.8	32
3	0.45	0.46	43.7	10.8	28

A series of investigations is currently being completed on a real landing gear structure with an active sliding tube creating a high noise environment to further validate the techniques. This programme of testing will be the focus of future publications.

Conclusions

The completed test has demonstrated the validity of the developed methodology. The output of the script located and identified a fracture source at approximately 8.2k cycles with final failure occurring at 43.1k cycles, furthermore this was confirmed using an optical CMOD measurement.

Acknowledgements

The authors would like to thank Messier-Dowty Ltd for supporting this research and the technical staff of Cardiff School of Engineering for their kind assistance with the testing programme.

References

[1] Hensman JJ *et al., Society for Machine Failure Prevention Technology: 5th International Conference* 15-18th July 2008

[2] Holford KM *et al., 13th International Conference on Experimental Mechanics,* Greece, July 1st-6th 2007

[3] Pullin R *et al., 13th International Conference on Experimental Mechanics,* Greece, July 1st-6th 2007

[4] Hensman JJ and Worden K, *13th International Conference on Experimental Mechanics,* Greece, July 1st-6th 2007

[5] Pullin R, *et al., Key Engineering Materials*, Vols. 347, (2007), pp. 139

[6] Roberts TM, *et al., Key Engineering Materials*, Vols 167-168 (1999), PP.142

[7] Pullin R, et al., *Fracture Mechanics of Concrete and Concrete Structures (FRAMCOS)*, Italy, June 17th-22nd 2007

[8] Hensman JJ, *et al, Mechanical Systems and Signal Processing,* Vol. 24 (2010), No. 1, pp. 211-223

[9] Baxter MG *et al., Mechanical Systems and Signal Processing*, Vols. 21 (2007), No. 3, pp. 1512

[10] Pullin R *et al., The fifth International Conference on Acoustic Emission*, Lake Tahoe, Nevada, USA, October 29 to November 2, 2007

Applied Mechanics and Materials Vols. 24-25 (2010) pp 227-232
© (2010) Trans Tech Publications, Switzerland
doi:10.4028/www.scientific.net/AMM.24-25.227

Towards the Derivation of Stress Intensity Factors by Parametric Modelling of Full-Field Thermoelastic Data

R.I Hebb[1, a], J. M. Dulieu-Barton[1, b], K. Worden[2, c], P. Tatum[3,]

University of Southampton, School of Engineering Sciences, Highfield, Southampton, SO17 1BJ, UK

*AWE, Aldermaston, Reading, Berkshire, RG7 4PR, UK

+ Department of Mechanical Engineering, University of Sheffield, Mappin Street, Sheffield, S1 3JD, UK

[a] rih103@soton.ac.uk, [b] janice@soton.ac.uk, [c] k.worden@sheffield.ac.uk

Keywords: thermoelastic stress analysis, stress intensity factors, thermoelasticity

Abstract. Thermoelastic Stress Analysis (TSA) is a well-established full-field technique for experimental stress analysis that has proved to be extremely effective for studying stress fields in the vicinity of cracks. Recently, work has focused on the observation that the stress-sum contours (isopachics) obtained from TSA take the form of a cardioid. Genetic Algorithms (GAs) and Differential Evolution (DE) have proved successful for accurate parameter estimation of the cardioids, thus allowing the SIFs to be calculated. Originally, some curve-fits indicated that a pure cardioid form is inappropriate for the base model, especially for mixed-mode cracks. The deviation from the cardioid form has been shown to be due to higher-order terms within the stress function. The objective of the current paper is to use a modified version of the original methodology (that fitted parameters to a single isopachic) to find the higher-order parameters from the entire data field obtained from the TSA.

Introduction

Effective diagnosis and prognosis algorithms for Non-Destructive Testing (NDT) and Structural Health Monitoring (SHM) require an in-depth understanding of stress/strain fields in the neighbourhood of cracks and damage. In Thermoelastic Stress Analysis (TSA) the observed isopachics (stress-sum contours) can be used to derive the Stress Intensity Factors (SIFs), which dictate the crack propagation and can be used to define the remnant life of the component. In previous work, by the authors, it was assumed that the isopachics took the form of cardioids. The approach was based on the first order Westergaard equations [1], or equivalently the singular terms in the eigenfunction expansion of Williams [2]. The effectiveness of using the cardioid form to derive the SIFs has already been demonstrated in [3, 4], where software and algorithms were developed that related the SIFs to the cardioid area and certain tangents to the curve.

Recent work [5] directly fitted the cardioid form to extracted isopachics to estimate the crack-tip SIFs. The fitting was performed using either a Genetic Algorithm (GA) [5] or Differential Evolution (DE) [6]. An added advantage of the approach is that the crack-tip is located in the data field, which can be difficult to do by visual inspection. However, some of the results indicated that a cardioid form was unsuitable for the base model. Each isopachic contour rotated by different amounts and produced a set of un-nested contours, which the first order Westergaard equations could not accommodate [7]. As a result the GA generated inaccurate values of the SIFs.

The object of the current paper is to incorporate the higher-order terms of the Williams expansion into the method and show that inclusion of these terms accommodates the rotation of the isopachics. It is demonstrated that inclusion of the higher-order terms is essential for accurate analysis. It was suggested in [7] that the omission of the higher-order terms in of the Williams expansion may have been responsible for the previous inability to explain the observed rotations and the work described in this paper clearly shows that this is the case. It has been necessary to process the entire thermoelastic data set rather than selecting individual isopachics hence the work described here significantly advances the data processing capabilities.

Theory

Thermoelastic stress analysis is based on the measurement of small temperature changes that occur in solids on the application of a cyclic stress. It can be shown that the temperature changes are proportional to the changes in the sum of principal stresses [8]. The temperature changes are measured using a highly sensitive infra-red detector. Modern infra-red detectors have been radiometrically calibrated and so are able to directly output a temperature measurement. For a homogeneous, isotropic material, the change in temperature can be related to the change in stress via:

$$\Delta T = -KT\Delta(\sigma_1 + \sigma_2) \tag{1}$$

where ΔT is the measured change in temperature, K is the thermoelastic constant for the material, T is the absolute temperature of the specimen and $\Delta(\sigma_1+\sigma_2)$ is the change in principal stresses.

The original approach [5, 6], related the SIFs to the first order Westergaard equations. In the region surrounding the crack-tip (but not the immediate vicinity where plasticity is a contributing factor and Linear Elastic Fracture Mechanics (LEFM) does not apply) the stress sum is related to the mode I and II SIFs (K_I and K_{II}) by:

$$\Delta(\sigma_1 + \sigma_2) = \Delta(\sigma_x + \sigma_y) = \frac{2K_I}{\sqrt{2\pi r}}\cos\left(\frac{\theta}{2}\right) - \frac{2K_{II}}{\sqrt{2\pi r}}\sin\left(\frac{\theta}{2}\right) \tag{2}$$

where r and θ are polar coordinates with the origin at the crack-tip, and θ measured anti-clockwise from the crack-line.

Rearranging Eq. 2 for r gives:

$$r = \frac{K_I^2 + K_{II}^2}{\pi}\left(\frac{KT}{\Delta T}\right)^2 [1 + \cos(\theta + 2\varphi_0)] \tag{3}$$

which is well-known polar representation of the cardioid curve and $\varphi_0 = tan^{-1}(K_{II}/K_I)$ is the rotation of the cardioid.

Eq. 3 shows that for the first order Westergaard case, curves of constant temperature change will form nested cardioids, all with the same rotation angle. Including the higher order terms, the Williams expansion leads to:

$$\frac{\Delta T}{KT} = \frac{2\sqrt{K_I^2 + K_{II}^2}}{\sqrt{2\pi r}}\cos\left(\frac{\theta}{2} + \varphi_0\right) + T_s + 2\sqrt{A_I^2 + A_{II}^2}\sqrt{r}\cos\left(\frac{\theta}{2} - \varphi_1\right) \tag{4}$$

where A_I and A_{II} are geometry-dependent constants, $\varphi_1 = tan^{-1}(A_{II}/A_I)$ and T_s is the T-stress.

It is clear from Eq. 4 that the $O(r^{1/2})$ term will cause a secondary rotation of the field of contours and will deform the cardioid. Physically, regions close to the crack-tip are dominated by the $O(r^{-1/2})$ term with the isopachics having a rotation of φ_0, but the $O(r^{1/2})$ term will start to dominate the field further from the tip, with a rotation of φ_1. This effect can clearly be seen in Fig. 1, where a contour plot for the lowest-order William's expansion is shown with the higher-order expansion superimposed. In order to generate Fig. 1, theoretical values have been used for K_I and K_{II} with arbitrary values set for the other parameters. The crack tip is positioned at 45° and is indicated. The contours generated when the higher-order terms are included clearly rotate towards the horizontal with increasing distance from the crack tip, showing, as expected, that the higher-order terms have more influence away from the crack-tip.

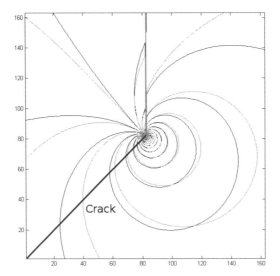

Figure 1: Contour plots of the lowest-order Williams expansion (black) and the Williams expansion including the higher-order terms (red)

Test specimen and set-up

The dimensions of the plate specimen used in this work are shown in Fig. 2. The material is Duralumin® and has dimensions 300 x 300 mm with a thickness of 1.5 mm. A slot at an angle of β = 45° and of total length 50mm has been spark eroded from a centrally located hole (in this case the slot is used to simulate a real crack). 15 holes top and bottom are used to mount the plate into the loading jig. To strengthen the area around the holes when loaded, 4 Duralumin® tabs are bonded to both sides, top and bottom, to spread the clamping load. The specimen was mounted in an Instron 8800 servo-hydraulic test machine using knuckle joints to ensure that no bending moments are introduced into the system. Fig. 3 shows a schematic of the loading jigs. Preparation of the specimen for testing required the surface of the plate to be cleaned and two passes of matt black paint to be applied. The purpose of the paint is to reduce reflections from the plate, and to increase the emissivity of the specimen. The specimen was cyclically loaded at 10Hz, with the applied stress range, $\Delta\sigma_{app}$, as 13.96 MPa.

Results

To assess the effectiveness of fitting parameters to the full field of data, the results are compared to those obtained from fitting cardioids to individual isopachics. In this case a simple (binary-coded) Genetic Algorithm (GA) was used as described in [5]. Level curves were extracted from the full-field data and the GA was used to fit the parametric form in Eq. 3 (the cardioid) to the curves. The procedure is simply an evolutionary optimisation procedure which minimises an error measure between the measured data and the parametric model. Four parameters are estimated: the crack tip location (in Cartesian coordinates, so x and y position), the rotation of the cardioid from the horizontal, γ, and a scale factor, P_1. The parameters are related to Eq. 3 via

$$P_1 = \frac{K_I^2 + K_{II}^2}{\pi}\left(\frac{KT}{\Delta T}\right)^2 = \frac{r_c}{2} \qquad \text{and} \qquad \gamma = 2\varphi_0 - \beta \tag{5}$$

where r_c is the maximum distance on the curve from the cusp.

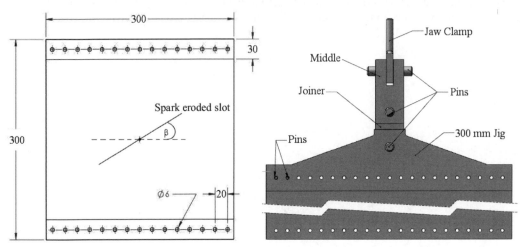

Figure 2: Schematic of the specimen   Figure 3: The loading jig

From the thermoelastic data, 10 isopachics or level curves were extracted and analysed as shown in Fig. 4; the corresponding extracted parameters are given in Table 1. The thermoelastic constant used in Eq. 5 was taken as 9.2×10^{-6} MPa^{-1}. The parameters P_1 and γ are the parameters given in Eq. 5 and the SIFs are calculated from these by simultaneously solving Eq. 5 with $\varphi_0 = tan^{-1}(K_{II} / K_I)$. The difference between the calculated cusp position and the crack tip, dr_c, is also given.

Table 1: Results from curve fitting to individual extracted isopachics

Set	x [mm]	y [mm]	$r_c/2$ [mm]	$dr_c/2$ [mm]	γ [rad]	k_I	k_{II}
12	11.75	30.33	11.70	6.41	0.393	0.79	0.53
13	10.70	27.73	8.71	3.65	0.356	0.75	0.48
14	11.26	28.65	7.01	4.72	0.669	0.64	0.57
15	9.99	26.25	5.78	2.04	0.332	0.70	0.44
16	9.92	26.25	4.97	1.99	0.381	0.68	0.45
17	9.85	26.32	4.35	2.01	0.436	0.67	0.47
18	9.57	26.11	3.81	1.67	0.405	0.66	0.45
19	9.64	26.32	3.40	1.89	0.504	0.64	0.48
20	9.26	26.04	3.06	1.46	0.424	0.66	0.45
21	9.85	26.67	2.86	2.29	0.786	0.57	0.57

The actual position of the crack-tip is at $x = 8.64$ mm and $y = 24.72$ mm. None of the cardioids accurately locate the crack-tip. As r_c increases, the accuracy deteriorates; inspection of the data in Table 1 shows that sets 12, 13 and 14 furthest from the crack-tip provide the worst values. The theoretical SIF values for a crack at angle of 45° are $k_I = 0.51$ and $k_{II} = 0.51$ [4]. The results clearly show a trend converging on the theoretical values as the data approaches the crack-tip. However, even at r_c values of much less than the slot length the values do not correspond to those given by the theory. This is because of the omission of the higher-order terms, the cardioid is not the appropriate parametric form for the level curves as the distance from the crack-tip increases. . Unfortunately, if the higher-order terms are included in the analysis, following Eq.4, the level curves (isopachics) do not have a simple parametric closed-form that makes them amenable to an analysis of the previous type. The alternative, proposed here, is to use the entire data set and use the parametric form in Eq. 4 as the basis for curve or model-fitting. This will have the added advantage of using all of the data available in the field. The parameter set for estimation now comprises 7 parameters: the x and y locations of the crack tip and the 5 parameters of the form in Eq. 4   Because of the enlarged parameter set, it is more convenient to use a real-coded evolutionary optimisation scheme so the differential evolution approach has been adopted as described in previous work [6]. As before, the objective function for the optimisation is chosen as the mean-square deviation between the model and the data.

Fig. 5 shows the experimental data with the prediction from the DE curve fit overlaid and the slot position is shown. It can be seen that the DE is providing an excellent fit to the experimental data. Table 2 gives the values of the parameters found by the DE. Here the crack-tip location is much more accurate, with the x-value in error by less than 1 mm and the y-value precisely located; this is a further indication that the DE curve fit is performing well.

The results presented here are from a preliminary study based on the full-field optimisation procedure; further refinement of the approach will be based on multiple runs of the DE algorithm as reported in previous work [6]. The finished methodology will allow estimation of confidence limits on the base parameters of the model and this will allow confidence intervals on the derived SIFs. Because of some uncertainty as to the appropriate form of the Williams expansion to use in the parametric model (it was reported in [9] that the basic reference [2] may contain errors), the SIF results for this data set are not presented here. What is clear from this study, however, is that the model form does allow a good reconstruction of the measured data as shown in Fig. 5. Setting aside the translation of the curve-fit parameters to SIFs; the results for the T-stress (as the uncontroversial constant term in the expansion) can be compared to Zanganeh *et al* [10] where the same form of the Williams expansion was used. The most similar case studied to the work presented here was biaxial loading on a cruciform with a centrally located angled slot of β equivalent to 26°. The Williams expansion up to $O(r^{1/2})$ (i.e. as Eq. (3)) gave a T-stress value of around 20 MPa which was about 50% of the applied stress. The current work shows the T-stress is about 20% of the applied stress. It should be noted that the two loading cases are not identical, as the crack angle is different, and therefore the T-stress will be different. However it is encouraging that the quantities are of the same order. Validation of the T-stress calculation will be carried out in future work. However, it is reported in [10] that an expansion to $O(r)$ provided a much better correlation with finite element analysis.

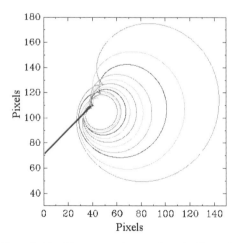

Figure 4: All of the generated curves from the first-order GA analysis.

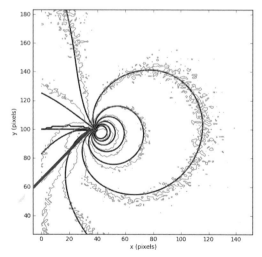

Figure 5: Experimental data with DE results over laid.

Table 2: Parameters from the DE curve fit

x [mm]	y [mm]	F_1 [MPa m$^{0.5}$]	φ_1 [rad]	T_s [MPa]	F_2 [MPa m$^{-0.5}$]	φ_2 [rad]
9.63	24.72	86.08	-0.217	1.83	0.62	0.847

As discussed above, the relationship between the full-field parametric model and the physically significant SIF parameters will be clarified before the parameters of Table 2 are used to infer the physical constants. An issue which also requires further investigation is the effect of, the plate used to collect the data not containing a real crack, but containing a slot. Analysis of actual cracks will be the object of a future paper.

Conclusions

Previous work by the authors reached a point where it was not possible to understand the discrepancies between theoretical SIF values and those inferred from the curve-fitting procedure based on an assumed cardioid form for individual isopachics. The evidence available suggested that the discrepancies were likely arising from the fact that the cardioid form was only appropriate to the lowest-order (singular) terms in the stress field expansion. In order to accommodate higher-order terms, it was necessary to adopt a different parametric modelling approach, one based on full-field data. The preliminary study presented here shows that an evolutionary optimisation approach based on an extended parametric form for the stress field is able to provide an excellent fit to measured data. A further important conclusion also follows from this work. In the previous studies, the fitting of individual isopachics did not allow an explanation as to why the fitted curves appeared to rotate with increasing distance from the crack-tip. The parametric form adopted here shows clearly that the rotation is a consequence of the higher-order terms in the stress field expansion and that the rotation can be effectively captured by the model-fitting procedure.

References

[1]   Ewalds, H.L. and Wanhill, R.J. *Fracture Mechanics*. Arnold, London, 1991

[2]   Williams, M.L. On the stress distribution at the base of a stationary crack. *Journal Applied Mechanics,* 24 (1957)

[3]   Dulieu-Smith, J.M. and Stanley, P. Progress in the thermoelastic evaluation of mixed-mode stress intensity factors. In *Proc. SEM Spring Conf. On Expt. Mech.* (Dearborn 1993), 617-629.

[4]   Dulieu-Barton, J.M., Fulton, M.C. and Stanley, P. The analysis of thermoelastic isopachic data from crack-tip stress fields. *Fatigue Fract. Engng. Mater. Struct.,* 23 (1999), 301-313.

[5]   Dulieu-Barton, J.M. and Worden, K. Genetic identification of crack-tip parameters using thermoelastic parameters. *Meas. Sci. Tech.*, 14 (2002), 176-183.

[6]   Dulieu-Barton, J.M. and Worden, K., Identification of crack-tip parameters using thermoelastic isopachics and differential evolution. *Key Engineering Materials,* 245-246 (2003), 77-86

[7]   Hebb, R.I, Dulieu-Barton, J.M., Worden, K. and Tatum, P. Curve fitting of mixed-mode isopachics. *7th International Conference on Modern Practice in Stress and Vibration Analysis, Journal of Physics*, 8-10 September, 2009

[8]   Dulieu-Barton, J.M. and Stanley, P. Development and applications of thermoelastic stress analysis. *Journal of Strain Analysis for Engineering Design,* 33 (1998), 93-104

[9]   Berto, F. And Lazzarin, P. On higher order terms in the crack tip stress field. *Int. J. Fract.,* 161 (2010), 221-226

[10]   Zanganeh, M., Tomlinson, R.A. and Yates, J.R., T-stress determination using thermoelastic stress analysis. *Journal Strain Analysis*, 43 (2008), 529-537

Applied Mechanics and Materials Vols. 24-25 (2010) pp 233-238
© *(2010) Trans Tech Publications, Switzerland*
doi:10.4028/www.scientific.net/AMM.24-25.233

A SEM-Based Study of Structural Impact Damage

M.T.H.Sultan, A.Hodzic, W.J.Staszewski and K. Worden

Department of Mechanical Engineering,

University of Sheffield,

Mappin Street,

Sheffield S13JD, United Kingdom

k.worden@sheffield.ac.uk

Keywords: CFRP Composite, Scanning Electron Microscopy, Impact Damage

Abstract
The ultimate objective of the current programme of work is to detect and quantify low-velocity impact damage in structures made from composite materials. There are many situations in the use of composites where an impact does not result in perforation of the material but causes damage that may not be visible, yet still causes a substantial reduction in structural properties. Impacts that do not cause perforation are usually termed *low-velocity*. When a composite structure undergoes such impacts, it is important to know the type and level of damage and assess the residual strength. In this study, following a systematic series of experiments on the induction of impact damage in composite specimens, Scanning Electron Microscopy (SEM) was used to inspect the topographies of the specimens at high magnification. Matrix cracking, fibre fracture, fibre pullout and delamination were the types of damage observed in the composite laminates after the low-velocity impacts. The study also conducted a (very) preliminary correlation between the damage modes and the impact energy.

Introduction
Many of the present technologies, especially among the materials used in the aerospace industry, require an unusual combination of properties that cannot be met by conventional metal alloys, ceramic and polymeric materials. For example, aerospace engineers are progressively searching for structural materials that have low densities and are strong, stiff and do not corrode over their lifetime. Therefore, since the early 1950's, composites have been used by the aerospace industry [1-3]. When composites were introduced into aircraft components, unexpected damage from in-service conditions occurred. These may have been due to impacts during flight operations [4-6] such as runway debris on composite airframes, bird-strike during flight operations or dropping of hand tools during maintenance work [5, 7]. For low-velocity impacts in composite structures, damages can not often be seen by the naked eye and are barely discernible. Such concealed damage is known as Barely Visible Impact Damage (BVID) [8] and is potentially undetectable without Non-Destructive Evaluation (NDE) techniques. Experimental investigations for identifying damages in composite laminates are habitually costly and time consuming. It is desirable to have automated systems for damage monitoring which can identify and classify damage from simply-measured response data. In order to build machine learning classifiers for such systems, one needs to acquire training data which associates measured data with given damage types and extents. This work addresses the overall aim by attempting to classify damage observed on CFRP laminates using a Scanning Electron Microscope (SEM) to inspect the topographies of impacted specimens at high magnification. The study also discusses a (very) preliminary correlation between the damage modes and the impact energy.

Fabrication of composite laminates

The composite material chosen for this work is a CFRP prepreg with a MTM57 epoxy resin system with CF2900 fabric, based on 120°C curing epoxy matrix resin. This material was fabricated to produce four large panels of 650 mm × 650 mm of which three were panels of 12 layers, and one had 11 layers on 50% area and 13 layers on the remainder. The processing used a standard vacuum-bagging procedure with the application of elevated temperature and pressure in an autoclave (cured for 30 minutes at 120°C at 5.8 bar). Each panel was cut into 8 test specimens of dimensions 250 mm × 150 mm with a diamond saw. PZT sensors of type Sonox P5 were placed on three different point of each test specimen to record the damage responses along different directions of the ply; the layout is illustrated in Figure 1.

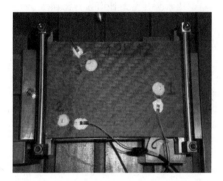

Figure 1. CFRP panel with PZT sensors as studied in this work.

Low velocity testing

An instrumented drop test rig was used to perform the tests. The rig consisted of a drop tower equipped with an impactor and a 0.0562 mV/N sensitivity force ring transducer. The impactor had a 13 mm diameter hemispherical tup nose. The height and mass of the impactor was varied in order to obtain various impact energies. For this work, the impactor was released from a chosen height and dropped freely along a guided rail. Together with the force ring transducer, the weight of the impactor was approximately 2.25 kg. The impact energy was set to range from 2.61 J to 41.72 J and the testing specimens were of 250 mm × 150 mm × 3.8 mm average thickness depending on the number of layers.

Non destructive Evaluation Techniques

X-radiography is a well-recognized non destructive evaluation technique used to assess damaged areas through images produced on film. The technique is based on differential absorption of penetrant radiation by the impacted or damaged surface; the presence of cracks and delaminations can be determined. However, difficulty may be expected in acquiring satisfactory X-ray images of adequate contrast, as there are some limitations. This is because the penetrant used only absorbs well in areas where damages are clearly visible. For damages that are not severe or only occur at the upper surface of a specimen, it is hard to produce an X-Ray image since the penetrant is not absorbed. As a result, the X-Ray radiation would not be able to image through the material. From another point of view, damage that is completely enclosed by non-damaged material will not be detected since the non-damaged material will act like a seal. Because of this constraint, X-ray is not considered the most suitable technique; although it is used here where sufficient damage occurred for penetrant absorption. X-ray analysis will generally give information about the overall area of damage, but will not allow resolution of different damage types. For this, SEM can be used with more accuracy.

SEM uses an electron beam of high intensity rather than light to produce an image. For this research work, JEOL JSM 6400 was used to determine the type of failure mechanisms observed from the composite laminates after the low-velocity impacts. Each specimen was observed at different magnification to distinguish the type of failure mechanisms. A high magnification SEM mode coupled with larger depth of focus and simplicity of sample observation makes the SEM one of the most suitable techniques for investigation of barely visible impact damage. The magnifications used for this current work were 35, 100, 500, 1000, 2500 and 5000. At 50-100 magnification, the fractographic analysis illustrates the overall view of the damage area, while at 500-750 magnification it allows one to look at the failure mechanism. Finally, at 1000-2500 magnification it gives one a chance to look at the failure mechanism in more detail. All the specimens were later used to make preliminary correlations between damage modes with the impact energy.

Results and discussion
A total of 24 specimens were used for the impact test of which 16 specimens possessed 12 layers and 4 each had 11 and 13 layers. The impact energies tested for the 12-layer specimens were set to range from 2.61 J to 41.72 J, while for the 11 and 13-layer samples, the specimens were tested at selected energy levels (10.43 J, 20.86 J, 31.29 J and 41.72 J). Due to the limited space here, only the results from the 12-layer specimens will be discussed. Two different methods were used to estimate the overall damage area; the first used the naked eye by measuring the damage size using a vernier caliper directly on the test specimen, while the second method used X-ray radiography. It was discovered from visual observation that the external damage area was circular in shape, however from the X-ray, the damage area was found to be rectangular. Table 1 summarises the different types of damage assessment for specimens with 12 layers. Note that for the damage estimation method that used X-ray, the notation ND given in the table means 'Not Detected'. This is because for damage that was not severe or that only occurred at the upper surface of a specimen, it was hard to produce an X-Ray image since the penetrant could not be sufficiently absorbed. Most of the damages are categorised as internal defects and generally consist of matrix cracking which usually cannot be detected by simply examining the surface of the specimen. The damages are only visually captured on the X-ray film for all impacts above 20.86 J. SEM was needed to determine the failure modes more precisely.

TABLE 1. ESTIMATED VALUES FOR IMPACT DAMAGE AREAS.

CFRP specimen identifier	Impact Energy (J)	Peak Force (N)	Damage Area (mm^2) from Visual Inspection	Damage Area (mm^2) from X-Ray
12LA1	41.72	3589.7	15.90	195
12LA5	39.11	3562.2	15.21	180
12LA6	36.50	3555.0	13.86	144
12LA7	33.89	3235.0	13.20	100
12LA2	31.29	3059.9	12.57	70
12LA8	28.68	3022.9	11.95	60
12LB1	26.07	2848.4	11.34	54
12LB2	23.47	2656.3	10.18	40
12LA3	20.86	2470.2	9.62	30
12LB3	18.25	2205.9	8.55	ND
12LB4	15.64	2069.0	8.04	ND
12LB5	13.04	1958.2	7.55	ND
12LA4	10.43	1927.2	7.07	ND
12LB6	7.82	1187.2	6.16	ND
12LB7	5.21	1064.2	4.52	ND
12LB8	2.60	810.0175	3.80	ND

Most of the specimens show the failure modes only at the contact area of the impactor. For low velocity impact, damage starts to propagate with the formation of a matrix crack [7,9]. The presence of the matrix crack does not affect the stiffness and toughness of the material. Delaminations, which form the critical failure mode after impact damage, propagate due to high transverse shear stresses in the location of the impacted surface [10-11]. The initiation and growth of delamination will results in progressive stiffness degradation and eventual failure of the composite structure. The damage will later propagate into other failure modes with the introduction of significant fibre damage, starting with fibre cracking, and further developing into fibre fracture and fibre pullout [12-13]. Figure 2(a-f) shows the micrograph images produced at the highest impact energy level. At this level, the highest impact energy and peak force recorded were 41.72 J and 3589.7 N respectively. Figure 2(a) illustrates the overall view of the SEM image captured at 35 magnification. At this point of enlargement, it can be seen that there is a significant damage in the test specimens. At the magnification of 100, matrix cracking and matrix breakage were observed, as shown in Figure 2(b). As the magnification increased to 500, delamination was discovered as seen in Figure 2(c), and at the magnification of 1000, there was a clear evidence of fibre pullout. Lastly, at magnifications of 2500 and 5000, there was a clear evidence of fibre breakage captured by the SEM. The threshold for damages observed at the internal sub-surface (cross sectional fractography) is still in progress and requires further study.

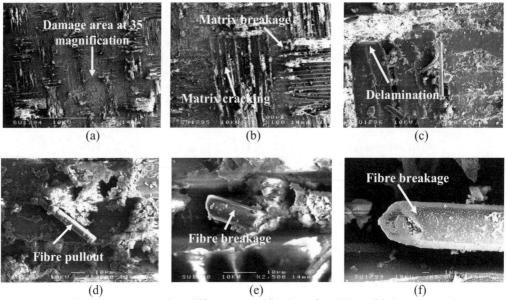

Figure 2. SEM images captured at different magnifications for 12LA1 (highest-energy impact).

Figure 3(a-f) presents micrograph images for the lowest impact energy level. As the impact energy decreases, the failure modes become different. At this level, the impact energy and peak force recorded were 2.61 J and 810.1 N, respectively. Referring to Figure 3(a), there was a clear damage detected on the surface of the plate. As the magnification increased, the damage mode became more evident; and, it could be seen that there was matrix cracking as the result of the impact. Figure 3(f) reveals that matrix cracking and matrix breakage were the only damage mechanisms discovered at this point, because the resin was still encapsulating the fibre, and this was the main difference with respect to the highest impact energy damage as shown in Figure 2.

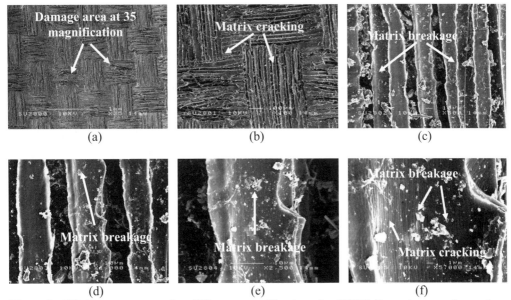

Figure 3. SEM images captured at different magnifications for 12LB8 (lowest-energy impact).

Only selected micrograph images are presented here due to the limited space; the overall failure modes observed from all the 12-layer specimens are collected and presented in Table 2. It can be seen that as the impact energy increases, the types of failure modes observed from the micrograph images proliferate. As the impact energy reaches 20.86 J, the progression of damage can be clearly observed. This is supported by the previous evidence showing that only from this energy level are damages visible on the X-ray films. From this point onwards (20.86 J), as the impact energy increases, the failure modes observed from the micrograph images progress. Damages that were observed below 20.86 J can be classified as not severe since they are in the form of matrix cracking and matrix breakage.

TABLE 2. PROGRESSION OF FAILURE MODES ON 12-LAYER SPECIMENS.

CFRP specimen identifier	Impact Energy (J)	Matrix cracking	Matrix breakage	Fibre cracking	Fibre breakage
12LA1	41.72	Yes	Yes	Yes	Yes
12LA5	39.11	Yes	Yes	Yes	Yes
12LA6	36.50	Yes	Yes	Yes	Yes
12LA7	33.89	Yes	Yes	Yes	Yes
12LA2	31.29	Yes	Yes	Yes	No
12LA8	28.68	Yes	Yes	Yes	No
12LB1	26.07	Yes	Yes	Yes	No
12LB2	23.47	Yes	Yes	Yes	No
12LA3	20.86	Yes	Yes	Yes	No
12LB3	18.25	Yes	Yes	No	No
12LB4	15.64	Yes	Yes	No	No
12LB5	13.04	Yes	Yes	No	No
12LA4	10.43	Yes	Yes	No	No
12LB6	7.82	Yes	Yes	No	No
12LB7	5.21	Yes	Yes	No	No
12LB8	2.60	Yes	Yes	No	No

Conclusions

The main objective of this study was to perform a series of low energy impacts in carbon fiber reinforced polymer composites, and then to carry out a sequence of SEM investigations to evaluate the failure modes. The objective is not to make any new observations on the nature of composite damage here, however to systematically build training data for machine learning analysis; in this respect, the SEM methods appear to have delivered a corpus of high-quality results which can form a valuable basis for further research [14]. From the work conducted, it can be concluded that (in agreement with previous fundamental research), as both impact energy and peak force detected increase, the number of distinct failure modes also increases. The experimental results allowed the identification of three critical impact energy thresholds. The threshold for matrix cracking and matrix breakage was identified as below the impact energy of 20.86 J whilst for fiber cracking it was identified between 31.29 J and 20.86 J. It is worth stating that the threshold energy for major damage was identified when the specimens were subjected to impact energy of more than 33.89 J. These thresholds will prove critical in later work when machine learning methods will be developed in order to infer damage extent and morphology from the structural dynamic response to impact.

References

[1] Williams, J.C. and Starke, E.A. Progress in structural materials for aerospace systems. *Acta Materialia*, 2003, 51(19), 5775-5799.

[2] Bannister, M. Challenges for composites into the next millennium - a reinforcement perspective. *Composites Part A: Applied Science and Manufacturing*, 2001, 32(7), 901-910.

[3] Lubin, George (Ed.), *Handbook of Composites*, New York: Van Nostrand Reinhold Company Inc., 1982.

[4] Yang JN, Jones DL, Yang SH and Meskini A. A stiffness degradation model for graphite/epoxy laminates. Journal of Composite Materials 1990; 24(7):753-769.

[5] Tai NH, Ma CCM, Lin JM and Wu GY. Effects of thickness on the fatigue-behavior of quasi-isotropic carbon/epoxy composites before and after low energy impacts. Composites Science and Technology 1999; 59(11):1753-1762.

[6] Whittingham B, Marshall IH, Mitrevski T and Jones R. The response of composite structures with pre-stress subject to low velocity impact damage. Composite Structures 2004; 66(1-4):685-698.

[7] Aslan Z, Karakuzu R and Okutan B. The response of laminated composite plates under low-velocity impact loading. Composite Structures 2003; 59(1):119-1278.

[8] Bull PH and Edgren F. Compressive strength after impact of CFRP-foam core sandwich panels in marine applications. Composites Part B: Engineering 2004; 35(6-8):535-541.

[9] Soutis, C. and Curtis, P.T. Prediction of the post-impact compressive strength of CFRP laminated composites. *Composites Science and Technology*, 1996, 56(6), 677-684.

[10]Richardson, M.O.W. and Wisheart, M.J. Review of low-velocity impact properties of composite materials. *Composites Part A: Applied Science and Manufacturing*, 1996, 27(12), 1123-1131.

[11]Cantwell, W.J. and Morton, J. The impact resistance of composite materials-a review. *Composites*, 1991, 22(5), 347-362.

[12]Li, Y., Mai, Y.-W. and Ye, L. Effects of fibre surface treatment on fracture-mechanical properties of sisal-fibre composites. *Composite Interfaces*, 2005, 12, 141-163.

[13]Iannucci, L., Dechaene, R., Willows, M. and Degrieck, J. A failure model for the analysis of thin woven glass composite structures under impact loadings. *Computers & Structures*, 2001, 79(8), 785-799.

[14]Sohn, M.S., Hu, X.Z., Kim, J.K. and Walker, L. Impact damage characterisation of carbon fibre/epoxy composites with multi-layer reinforcement. *Composites Part B: Engineering*, 2000, 31(8), 681-691.

[15]Shyr, T.W. and Pan, Y.H. Impact resistance and damage characteristics of composite laminates. *Composite Structures*, 2003, 62(2), 193-203.

Applied Mechanics and Materials Vols. 24-25 (2010) pp 239-244
© (2010) Trans Tech Publications, Switzerland
doi:10.4028/www.scientific.net/AMM.24-25.239

Application of Real-Time Photoelastic Analysis to Single Fibre Fragmentation Tests

S. Blobel[1, a], K. Thielsch[1,b], M. Kästner[1,c], V. Ulbricht[1,d]

[1]George-Bähr-Straße 3c, 01069 Dresden, Germany

[a]Swen.Blobel@tu-dresden.de, [b]Karin.Thielsch@tu-dresden.de, [c]Markus.Kaestner@tu-dresden.de

[d]Volker.Ulbricht@tu-dresden.de

Keywords: single fibre fragmentation test, micromechanical measurements, composite material, digital photoelasticity

Abstract. One of our main research areas is the trans-scale modelling of damage in composite materials, which consist of a polymer matrix and carbon or glass fibres in different material combinations and geometrical arrangements. From the local stress redistribution in the fibre-matrix interphase and in the surrounding matrix material information on the parameters of microscopic damage models for composite materials can be obtained. Owing to the difficult interface characterisation based on the properties of the single material components, a photoelastic analysis of single fibre fragmentation tests is performed. In addition to the qualitative visual interpretation in polarized light, an enhanced quantitative analysis in combination with digital photoelasticity using a four image phase shifting method will be applied [1]. As the sequential capturing of images might cause incorrect results, these four pictures are grabbed simultaneously. This allows for continuous testing. Additionally, errors due to the relaxation behaviour of the matrix material can be avoided. To this, a modular optical system consisting of a variable long distance microscope and a beam dividing module proposed by [2] was developed. It allows for the simultaneous projection of four different filtered images of one microscopic scene to the four quadrants of a CCD chip. This special equipment gives the possibility to apply quantitative photoelasticity to tensile tests performed on standard testing machines. This paper explains the measurement hardware and discusses the main problems and realised solutions from picture capturing through image processing to real-time photoelastic analysis at the present state of development. Exemplary results for the qualitative analysis of selected material combinations and different manufacturing processes are shown.

Introduction

On the one hand, modern fibre reinforced composites are more and more used in high technology applications which require a very high reliability and efficiency. On the other, the knowledge of these materials is still based on the experimental investigation of their macroscopic material behaviour. However, the effective properties result from the complex material structure at lower scales. The analysis of the characteristics on the meso and micro scale could therefore provide the understanding and prediction of the effective material behaviour on higher scales. An important role plays the understanding and the knowledge of the interface properties and behaviour between fibre and matrix under load.

The interaction between fibre and matrix can be studied by means of single fibre composites [3]. In this context the optical observation of the damage evolution was established as a special investigation method [4]. From a number of possible experiments the single fibre fragmentation test (SFFT) is often chosen to perform optical observations and measurements during the application of a tensile load to the specimen [5]. In addition to the acquisition of global load-displacement curves, the SFFT allows the qualitative optical investigation of the damage evolution and the quantitative measurement of the local stress distribution in the matrix material. Assuming a gradual destruction of the embedded fibre all stages of the micromechanical damage process can be studied. This includes adhesion of the intact fibre matrix interface, friction in the destroyed interface and stress redistribution in the vicinity of the opening break gap.

This paper describes the preparation of specimen, the experimental setup and presents the first orientating tests which will be the basis for systematic experimental investigations in future work.

Preparation of specimen for optically analysed SFFT

The SFFT specimen consists of a thermoset matrix (HT2, R&G Faserverbundwerkstoffe GmbH Waldenbuch) and an embedded commercial glass fibre with a diameter of 20 micrometer or carbon fibre with a diameter of 7 micrometer. Due to necessary cross section dimensions for SFFT a dogbone shaped specimen with a measuring length of 20 millimetres, a quadratic cross section of 4 square millimetres and a shoulder radius of 15 millimetres is chosen.

The special configuration of the SFFT specimen and the requirements to enable an optical observation necessitate a precise and reproducible manufacturing process. The specimen are produced by filling degassed uncured epoxy resin into special silicon moulds, which were fabricated out of steel forms with the shape of the specimen (fig. 1). Notches cut into the silicon moulds define the location of the fibre in the middle of the specimen cross section. Single fibres are manually extracted from commercial yarns and fixed at the notches. The preparation of the silicon moulds enables multiple fibre orientations. At first, specimens with a longitudinal fibre orientation (fig. 2) have been produced. After filling the moulds with epoxy resin the later measurement area is covered by a glass plate to guarantee a high surface quality and a uniform cross section shape. The specimen is cured inside an air pressure vessel and post-cured according to the resin specifications.

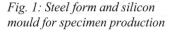

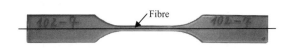

Fig. 1: Steel form and silicon mould for specimen production

Fig. 2: Specimen with longitudinal embedded carbon

Test setup for microscopically analysed SFFT

The experiments are performed on the commercial tensile testing machine "Instron 5566" with a 5kN load cell. To allow a minimal working distance for the optical observation, special specimen grips were developed and used. The microscopic investigations on the whole measurement area of the produced specimen require a translation of the optical measurement system relative to the loaded specimen. During the translation over the large measuring length features to keep the fibre in focus and to control the exposure of the image sensor are needed. To ensure an adequate frame rate at optimal image quality an automated video controlled translation unit was developed and installed. This system ensures continuous image capturing for every segment of the measuring length while every single image is focused and its brightness is adapted to obtain optimal contrast for later interpretation. The unit is controlled by a software application based on the video signal of the measurement hardware. All captured images are saved with timestamp, coordinates, force and displacement. This additional information will be of interest for the investigation and visualization of the damage evolution.

Parallel to this equipment a second camera is used to capture images with a low frame rate for a local strain measurement. The captured images of a pattern which is applied to one lateral surface of the specimen are later analysed by using the commercial image correlation software ARAMIS from GOM mbH Braunschweig. We use the whole-field strain results for the stress-strain curve of our tensile tests.

Qualitative interpretation of the damage evolution

The qualitative analysis of the SFFT provides knowledge on the local damage evolution for a selected combination of fibre and matrix. Photoelasticity shows a higher activity in matrix regions with higher stress often resulting from inhomogeneous or defect composite components. For instance, local defects of the fibre initiate a stress rearrangement near the break gap which results in higher local photoelastic activity. The intensity and dimension of these effects can give information on the local interfacial shear strength of the fibre matrix interface. Furthermore, global information like the number of fibre breaks or the average fragment length at a certain global strain or stress state can be used to characterize the selected composite material.

FEIH et al. [6] and KIM and NAIRN [7] investigated the qualitative characteristics of the SFFT by photoelasticity. Both described a typical shape of the photoelastic effect near the fibre breaks. They suppose that a fibre break has to occur in conjunction with local interfacial damage. On further investigation they found two regions of different intensity of the photoelastic effect and deduced that this is caused by different processes in the interface. Near the gap this effect is initiated through friction inside the damaged interface. The photoelastic active region far from the gap is supposed to result from stress redistributions through the intact fibre matrix interface.

Application of the realized measurement system to selected composite materials

At the time of composing this paper orientating experiments were performed to evaluate the measurement process. The first specimen were produced from epoxy resin HT2 with the stress fringe value 16.32 N/(mm·fringe) and an embedded glass or carbon fibre. In addition to the evaluation of the measurement system these experiments are performed to obtain knowledge on the influence of different manufacturing parameters on the strength of the selected composite. Different specimen types were produced, which differ in the embedded fibre, the application of air pressure during the curing process and the usage of a post curing procedure. During these first experiments the optical measurement system was still under construction. Therefore only a qualitative analysis of the resulting images was performed with a simple long distance microscope with comparable specifications (fig. 4). A manually configured set of filter plates enables the capturing of photoelastic images.

As expected, the non post-cured specimen displayed a weak interface bonding, which appeared as non-intensive photoelastic effects along the whole length of the measurement range. At the same time only some rare fibre breaks occurred until the failure of the specimen at high global strain.

Figure 3 shows a non-post-cured specimen with a tensile load applied at a constant displacement rate v=0.04mm/min. The image was captured at a tensile stress of 33.75 MPa and a global strain of 2.75 %. Referring to [7] the brighter regions are suspected to represent high shear strain at an intact interface region while the darker areas are related to friction at a failed interface.

Fig. 3: Photoelastic effects showing the debonding along the whole length of a glass fibre

Fig. 4: First experiments with tensile testing machine and a long distance microscope with polarizers

The post-cured specimen showed a contrary behaviour. The effects described by [6] and [7] could be reproduced. Intensive photoelastic effects occured in the vicinity of a fibre break gap. An increasing number of fibre breaks and saturation at higher global strain were observed. The average fragment length at damage saturation was approximately 0.5 millimetre. Figure 5 shows the photoelastic effect near the gap of the broken carbon fibre embedded in post-cured epoxy. One can see the typical shape of the photoelastic effect described by [7].

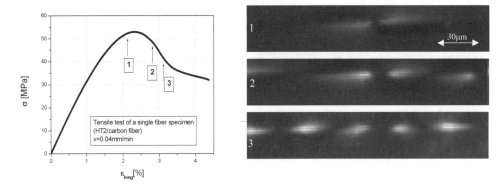

Fig. 5: Photoelastic effects in linear polarized light as a result of breaks at a carbon fibre

Optical measurement system for simultaneous capturing of four phase shifted images of one microscopic scene

The observation of photoelastic effects requires a special arrangement of optical filter plates which filter the light shining through the specimen. A simple set of polarizers allows a qualitative interpretation of the photoelastic effects. For a quantitative analysis it is necessary to take at least four phase shifted images to extract the isochromatic and isoclinic parameters. The application to a dynamic scene (like the SFFT) is only useful if a set of phase shifted images can be captured simultaneously. PATTERSON and WANG [1] first published the simultaneous capturing of four phase shifted images. Their phase stepping automated polariscope was used to quantify the shear stress in the matrix near by the free ends of fibres and near by crack tips of the matrix [8, 9].On this basis an own compact measurement system was built. It consists of a single-wavelength light source with polarizer, and quarterwaveplate and a beam splitter module with filter brackets. The beam splitter module realises the acquisition of four images of the same scene to the four quadrants of the image sensor. In every optical path there are two filter plates attached. In combination with the set of polarizer and quarterwaveplate at the light source side these polarizer and quarterwaveplates are configured analogous to [1] to produce the four differently phase shifted images (table1).

Table 1: Configuration of filter plates for phase shifting [1]

Image number	Orientation of quarter wave plate	Orientation of polarizer
1	π/4	π/4
2	0	0
3	0	-π/4
4	π/2	π/2

To be able to observe microscopic scenes, a modular long distance microscope objective is added to the described optical system. At working distances from 15 to 1430 millimetres in combination with a four mega pixel image sensor this long distance microscope offers resolutions from 1.25 to 325 micrometer per pixel. As a result of using the beam splitter a maximum effective resolution of 2.5 micrometer per pixel is possible. Long distance microscope and beam splitting module were developed by Thalheim Spezial Optik (Fig. 6).

Fig. 6: Modular image acquisition system: Light emitting unit (left), specimen, long distance microscope and beam splitter (centre), video output for an academic example of disc under diametral compression (right)

This modular system can be applied in transmission or reflection photoelastic arrangements. However, the dimension of the long distance microscope and the light source modules inhibits the usage in reflection arrangement for higher resolutions due to the necessary lower working distance.

Due to the complexity of the optical system the four phase shifted images are not absolutely geometrical congruent. Therefore, the captured images have to be corrected by image processing techniques. Based on images of a calibration object with a regular grid edge detection is performed. The detected crossings of the edges are used as reference points on which segments of the images are stretched or compressed to obtain four congruent images.

The aim of the quantitative analysis is the determination of the three-dimensional principal stress field inside the matrix material surrounding the embedded fibre which provides important information for the formulation of phenomenological damage models. This goal cannot be reached directly by photoelasticity. As the thickness of the specimen typically has to be much greater than the diameter of the embedded fibre to ensure a sufficient intensity of the photoelastic effects, the captured information can only represent integral information of the difference between the two principal stresses over the whole thickness of the specimen. This value is represented by the isochromatic value δ, which can be calculated for the used filter configuration analogous to [1] as

$$\delta = \tan^{-1}\left(\frac{i_2 - i_3}{\sin 2\theta(i_2 + i_3 - 2i_4)}\right) \qquad (1)$$

with the intensity i_n of the four images and the isoclinic value θ as

$$\theta = \frac{1}{2}\tan^{-1}\left(\frac{i_2 - i_3}{i_2 + i_3 - 2i_1}\right). \qquad (2)$$

In the case of the single fibre fragmentation test a theoretical idealization of the stress distribution around the fibre to a rotational symmetric model allows the determination of the searched principal shear stress field [10].

Summary

Single fibre specimens were manufactured with glass and carbon fibres. With these specimens first experiments were performed on a commercial testing machine using a long distance microscope. Damage evolution visualized in polarized light was investigated. A special polariscope, which permits the simultaneous capturing of four phase shifted images, was presented. Quantitative experiments will be performed in the next time and presented at the BSSM Conference 2010.

References

[1] E.A. Patterson, Z.F. Wang: Journal of Strain Analysis, Vol. 33 No1 (1998) 1-15,

[2] J. Lesniak, S.J. Zhang and E.A. Patterson: Experimental Mechanics, Vol. 44(2004)128-135

[3] M. R. Piggott: Composite Science and Technology 57(1997)965-974

[4] L. T. Drzahl, M. J. Rich, P. F. Lloyd: The Journal of Adhesion, Vol. 16(1983)1-30

[5] A. N. Netravali, R. B. Hengstenburg, S. L. Phoenix, P. Schwarz: Polymer Composites, Vol. 10(1989)226-241

[6] S. Feih, K. Wonsyld, D. Minzari, P. Westermann, H. Lilholt: *Testing Procedure for the single fiber fragmentation test*, RISO-Report 1483 (http://www.*risoe.dtu.dk/rispubl/AFM/afmpdf/ris-r-1483.pdf*)

[7] B.W. Kim, J.A. Nairn: Journal of Composite Materials, Vol. 36 (2002), p.1825

[8] F. M. Zhao, S. A. Hayes, E. A. Patterson, F.R. Jones: Composites Part A, Vol. 37(2006) 216-221

[9] F. M. Zhao, R. D. S. Martin, S. A. Hayes, E. A. Patterson, R. J. Young, F.R. Jones: Composites Part A, Vol. 37(2005) 229-244

[10] D.M. Schuster, E. Scala: Transactions of the metallurgical society of AIME, Vol. 230 (1964), p. 1635

Applied Mechanics and Materials Vols. 24-25 (2010) pp 245-250
© *(2010) Trans Tech Publications, Switzerland*
doi:10.4028/www.scientific.net/AMM.24-25.245

Characterization of Fiber Bridging in Mode II Fracture Growth of Laminated Composite Materials

K.N. Anyfantis[1,a], N.G. Tsouvalis[1,b]

[1]Shipbuilding Technology Laboratory, School of Naval Architecture and Marine Engineering, National Technical University of Athens, Athens, Greece

[a]kanyf@central.ntua.gr, [b]tsouv@mail.ntua.gr

Keywords: Delamination growth, Bridging law, Laminated composites, Interface element

Abstract. The present study involves an experimental and analytical investigation of the Mode II delamination propagation and the fibre bridging effects incorporated in the Fracture Process Zone (FPZ). End Notch Flexure (ENF) specimens from a unidirectional glass/epoxy composite material have been fabricated and tested. In order to construct the fracture resistance curve (R-curve) of the ENF tests, three different data reduction schemes have been utilized. The fibre bridging effects in the FPZ have been addressed with the use of traction-separation laws, as extracted from the corresponding calculated R-curves. These laws can be used to describe the constitutive relationship in interface finite elements, for the numerical modelling of delamination growth in laminated composite structures.

Introduction

Fibre Reinforced Polymer (FRP) laminated composite materials have attracted a wide range of uses in civil, marine, automotive and aerospace applications on account of their superior properties. Nevertheless, due to their low interlaminar strength, they are susceptible to delamination damage, hence, extensive efforts have been devoted to the development of reliable test methods that can accurately measure delamination resistance of FRP materials.

Delamination resistance, which is usually expressed by the term "interlaminar fracture toughness", is often characterized in terms of the critical strain energy release rate, G_c. Interlaminar fracture toughness for delamination in opening Mode I is known as G_{Ic} and for the sliding shear Mode II as G_{IIc}. In recent years the DCB-UBM test (Double Cantilever Beam - DCB specimen loaded by uneven bending moments) has been used often for the characterization of mixed mode fracture of composite structures [1,2]. For the determination of the Mode I interlaminar fracture toughness of composite materials, the DCB test has been accepted as a standard method (ASTM D5528, ISO/DIS 15024). On the contrary, a commonly accepted standard for the measurement of G_{IIc} has not yet been established. Despite this lack, several test methods are available for quantifying G_{IIc} and have been widely used for research purposes. Among them, the End-Notched-Flexure (ENF), the End-Loaded-Split (ELS) and the Four-Point ENF (4ENF) tests are the most popular procedures found in the literature. The ENF test [3] is most widely used due to its simple fixture based on a conventional three point bending set up and has been adopted by the Japan Industrial Standards Group (JIS) as a standard test. On the other hand the ELS [4] and 4ENF [5] tests require a more complicated test set up and present some problems related to large friction effects [6]. Therefore, in the present study focus is given on the ENF test configuration.

Mode II fracture characterization tests involve difficulties regarding monitoring of the crack tip position, as the crack tends to close with the increase of the applied load [7]. The formation of a large fracture process zone (FPZ) behind the crack tip as crack propagates is characterized by the fibre bridging effect [2], which hinders the clear visualization of the crack tip.

From a structural point of view fibre bridging is beneficial, since it results in an increasing fracture resistance as crack grows. This increase in the fracture energy as crack propagates is described by a fracture resistance curve, namely the R-curve [8]. Resistance to delamination should not be based on the R-curve when a large FPZ is present. R-curves cannot be considered as a material property, since they depend on the geometry of the specimen. Instead, fibre bridging can be expressed in the form of a traction-separation law (bridging law), which can be inferred from the experimentally obtained R-curve [2]. In the case of Mode II fracture, a bridging law expresses the dependency between the closure traction (bridging stress) and the local crack sliding. When fracture toughness reaches a plateau in the R-curve at a specific value of crack sliding (steady-state interlaminar fracture toughness, G_{ss}), closure traction vanishes, leaving the crack faces free of traction [8].

The purpose of this work is to propose a characterization method of the FPZ in Mode II fracture of unidirectional glass/epoxy composite materials using bridging laws. The experimental programme involves fabrication and testing of ENF specimens, whereas post-processing of the experimental results is based on the construction of R-curves by applying three different data reduction schemes. The R-curves were used explicitly for the definition of the bridging laws.

Experimental procedure

Six layers of 600 gr/m^2 unidirectional glass fabric were used together with epoxy resin D.E.R. 353 to manufacture a composite laminated plate. During manufacturing a PTFE film with 58 μm thickness was inserted in the mid-plane for the creation of the delamination. The plate was manufactured by the hand lay up method and was cured at 25°C for 48 hours. After curing, five ENF specimens were cut out from the plate with waterjet. The composite system had a fibre weight fraction of 50%, whereas its main Young's modulus E_1 was measured from standard tests and found to be E_1=18000 MPa. Young's modulus E_3 and shear modulus G_{13} were taken from the literature equal to 2200 and 1800 MPa, respectively.

The geometry and dimensions of the 140 mm long ENF specimens are given in Fig. 1. The thickness of the specimens is $2h$ = 4.55±0.1 mm, their width is equal to 20 mm and their unsupported length 100 mm. In order to achieve crack growth stability, the initial crack length a_o was chosen to be equal to 70% of the half unsupported length of the specimen [9], thus resulting in a value of 35 mm.

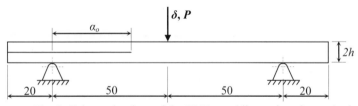

Fig. 1. Schematic view of the ENF test (dimensions in mm)

In an effort to monitor the crack length, thin vertical lines have been marked on the lateral sides of the specimens with a spacing of 2 mm. Prior to the creation of these marks, a white correction fluid was applied for improvement of the crack tip visualization. Crack propagation was monitored by visual inspection of the crack length through the use of a high resolution (1920x1080) 60 frames per second video camera equipped with 16x optical zoom lens, recording the marked side of the specimens. ENF tests were displacement controlled with a speed of 0.5 mm/min and were carried out in a 15 kN capacity hydraulic testing machine. The load-deflection curve (P-δ) was recorded during each test as a function of time.

Data reduction schemes

Post processing of the experimental results obtained from an ENF test is based on the calculation of the critical fracture energy, G_{IIc} and the corresponding R-curve. Most of the methods used for these calculations require the load-deflection (P-δ) data, together with the corresponding values of the crack length α. In order to map the crack length on the P-δ curve obtained from each test, the video was processed, measuring first the crack length as a function of time and then combining the registered time functions of P and δ data with that of the crack length.

For the purpose of this study, three different data reduction schemes have been utilized for calculating the R-curves of the ENF specimens.

Compliance Calibration Method (CCM). CCM is a classical data reduction scheme for calculating G_{IIc} [10] using the Irwin – Kies equation. According to this method, apart from loading and geometry, G_{IIc} is also given as a function of specimen's compliance C, defined as deflection per unit load ($C = \delta/P$). Thus, if a plot of C versus a is drawn for any specimen geometry, its slope at a particular point (C,a) can be measured. In order to avoid any errors occurring from fitting a continuous function to the discrete points (C,a), a different approach is followed in the present study for the calculation of dC/da. By defining crack length $a_i = a_o + i\Delta a$, where Δa is the crack increment and $i = 0,....,n$ (0 represents crack initiation and n represents maximum number of crack tip measurements), the following formulas are derived:

$$\left.\frac{dC}{d\alpha}\right|_{a=a_0,a_n} = \frac{C_i - C_{i-1}}{\alpha_i - \alpha_{i-1}}, \quad i = 1 \text{ or } n \quad , \quad \left.\frac{dC}{d\alpha}\right|_{a=a_i} = \frac{1}{2}\left(\frac{C_i - C_{i-1}}{\alpha_i - \alpha_{i-1}} + \frac{C_{i+1} - C_i}{\alpha_{i+1} - \alpha_i}\right), i = 1,...,(n-1) \quad (1)$$

Corrected Beam Theory (CBT). This is another classical method proposed in [11] and based on a beam elastic foundation model that takes into account corrections of the measured crack length. According to this method, G_{IIc} is given as a function of loading, geometry, crack length and material properties E_1, E_3 and G_{13} [11].

Compliance Based Beam Model (CBBM). The above CCM and CBT data reduction schemes require accurate crack length measurements, which present difficulties in obtaining. Thus, a novel calculation method has been proposed [12], which is based on the beam theory and according to which the results depend only on the specimen's compliance during the tests and take into account several effects that affect the behaviour of the specimen and consequently the P-δ curve, which are neglected by the classical methods. G_{IIc} is calculated by a modified Irwin – Kies equation as a function of loading, geometry and equivalent crack length that takes into account the FPZ influence.

Experimental Results

A deformed ENF specimen during the three-points bending test is shown in Fig. 2 and a magnified view of the crack tip area is shown in Fig. 3. A typical P-δ-a curve is presented in Fig. 4. The stars on the graph indicate the crack length in mm, as measured from the high resolution video camera by image processing techniques. It should be pointed out that crack initiated directly from the insert film and always propagated along the mid-plane of the specimens. As shown in Fig. 4, the experimental registered curve exhibits an initial linear response, followed by increasing non-linearities as the crack begins to propagate. After the maximum load is attained, the slope of the load-deflection curve begins to soften together with delamination progression. The crack propagated in a stable manner during the displacement control fracture testing for all five specimens tested.

According to CCM, the compliance was calculated from $C_i = \delta_i/P_i$ for each value of a_i. For the calculation of the slopes (dC/da), Eq. 1 has been utilized. The average fracture toughness values calculated with the CCM approach are listed in Table 1. The same table presents also results from the calculation of the average fracture toughness with the use of the CBT and CBBM data reduction schemes. Fig. 5 presents the corresponding R-curves obtained from the average values of Table 1.

It is noteworthy that fracture toughness values obtained with CBT and CBBM schemes compare very well between each other, exhibiting also very good repeatability, since the coefficient of variance (COV) for each crack length is rather low. On the other hand, G_{IIc} values calculated with CCM are generally higher than those of the other two schemes, having also higher COVs. This behaviour is attributed to the fact that CCM does not take under consideration any crack corrections of the measured crack length a_i for the calculation of G_{IIc}, which in contrary CBT and CBBM

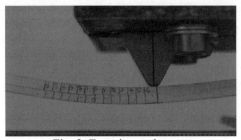
Fig. 2. Experimental setup

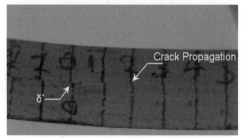
Fig. 3. Crack length measurement

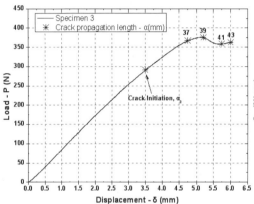
Fig. 4. Typical P-δ-α curve of an ENF test

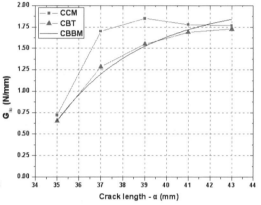
Fig. 5. Experimentally obtained R-curve

undertake. The R-curve obtained from CBBM has a continuous form, whereas the respective ones obtained from CCM and CBT have a discrete form. As Fig. 6 shows and Table 1 indicates, fracture toughness increases as the crack propagates, followed by a plateau region where it attains a steady state value, G_{ss}, for any further increase of the crack length. The same behaviour of the R-curves has been also reported from similar studies found in the literature [13]. In the R-curve obtained by CCM, steady state appears soon after crack initiation, whereas the transition is much smoother in the other two cases. In spite of the differences of the G_{IIc} values between the CCM and the other two schemes, fracture toughness at crack initiation and at the steady state indicate very good repeatability among all three schemes, with a COV being approximately equal to 6%.

Characterization of the FPZ

The extraction of a bridging law from the experimentally obtained R-curves can be done using formula

$$\tau(\delta^*) = \frac{\partial G_{II}}{\partial \delta^*} \tag{2}$$

connecting Mode II traction, τ, with fracture toughness, G_{II} and displacement δ^*, which is the crack sliding displacement at the initial pre-crack tip. For obtaining the nonlinear traction-separation law described by Eq. 2, the relationship $G_{II}(\delta^*)$ is required. In fact $G_{II}(\delta^*)$ can be implicitly defined by combining functions $G_{II}(\alpha)$ and $\delta^*(\alpha)$. Functions $G_{II}(\alpha)$ have already been defined in Fig. 5 for each data reduction scheme, whereas function $\delta^*(\alpha)$ was defined studying the recordings of the high resolution camera and using image processing techniques. Thus, the $G_{II}(\delta^*)$ discrete data sets were plotted for each data reduction scheme and an analytical function was used to fit the results for each case. This procedure is necessary for deriving the differential relationship described in Eq. 2.

Table 1. Fracture toughness G_{IIc} (N/mm)

		$\alpha_1 = 35$	$\alpha_2 = 37$	$\alpha_3 = 39$	$\alpha_4 = 41$	$\alpha_5 = 43$
CCM	Aver.	0.72	1.70	1.85	1.78	1.77
	COV (%)	14.1	15.1	12.7	9.7	8.7
CBT	Aver.	0.65	1.28	1.55	1.69	1.72
	COV (%)	7.0	5.2	3.8	2.0	2.1
CBBM	Aver.	0.66	1.21	1.55	1.74	1.85
	COV (%)	8.6	10.8	7.0	1.7	6.4

COV: Coefficient of Variation

Figures 6 to 8 present the respective G_{II}-δ^* data sets together with their corresponding fitting functions. The selected fitting functions are those yielding minimum residuals when used with the least square method for the regression of the data sets.

The discrete experimental values of G_{II}-δ^* reduced with CCM exhibit important variations. Thus, an exponential fitting function was utilized for the regression of these data sets, which exhibits a very good approximation. In contrast, G_{II}-δ^* data sets corresponding to the CBT and CBBM schemes are in very good agreement with respect to their repeatability, as shown in figures 7 and 8. CBT results (Fig. 7) exhibit a low increase rate without clearly reaching a steady state fracture energy value. This behaviour is well described with the use of a power law. In the case of CBBM results, steady state fracture toughness was reached near the final experimentally measured values of G_{II} and δ^*. Thus a second order polynomial function which exhibits a very good approximation was utilized for fitting the corresponding results, as presented in Fig. 8.

With the $G_{II}(\delta^*)$ functions known, the bridging laws $\tau(\delta^*)$ can be hereafter obtained from Eq. 2. Fig. 9 compares the bridging laws (traction-separation laws) as inferred by CCM, CBT and CBBM data reduction schemes. At crack initiation the shear traction attains a peak value followed by a linear or non-linear decay to zero, which depends on the $G_{II}(\delta^*)$ function obtained for each case. Finally after a specific value of δ^*, shear traction is released denoting the absence of the fibre bridging effect under further crack growth. Apart from the function used to describe the traction-separation laws, there are two characteristic properties that can be inferred. One is the peak value of the shear traction attained at crack initiation, τ_o and the other is the area under the $\tau(\delta^*)$ curve, I. These values have been found to be equal to 37.3 MPa and 1.04 N/mm for the CCM scheme, 48.4 MPa and 1.1 N/mm for the CBT and 13.9 MPa and 1.14 N/mm for the CBBM. A variation is observed over the peak values of shear, whereas the area integrals are in very good agreement.

Conclusions

A methodology for the characterization of delamination growth under pure Mode II fracture of laminated composite materials was proposed. Fracture energy was measured from ENF tests with the use of three different data reduction schemes. The R-curves and their respective bridging laws

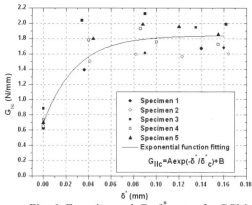

Fig. 6. Experimental G_{II}-δ^* curve for CCM

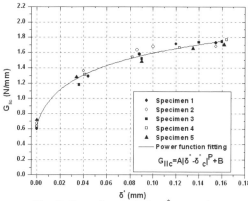

Fig. 7. Experimental G_{II}-δ^* curve for CBT

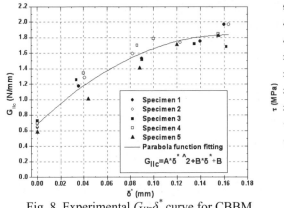
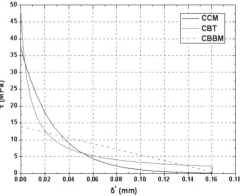

Fig. 8. Experimental G_{II}-δ^* curve for CBBM   Fig. 9. Experimental bridging laws

were obtained and compared for all three schemes. The obtained bridging laws exhibit the same softening behaviour, regardless the data reduction method used. The energy uptake of the three bridging laws obtained is in great agreement denoting the effectiveness of the proposed procedure in characterizing the FPZ. The laws defined can be used to describe the constitutive relations in interface finite elements for the numerical modelling of delamination growth.

References

[1] C. Lundsgaard-Larsen, B.F. Sørensen, C.Berggreen and R.C. Østergaard: Engineering Fracture Mechanics Vol. 75 (2008), p. 2514

[2] F.B. Sorensen and T.K. Jacobsen: Composite Science and Technology Vol. 69 (2009), p. 445

[3] J.D. Barrett and R.O. Foschi: Engineering Fracture Mechanics Vol. 9 (1977), p. 371

[4] K.R.B. Blackman, J.A. Brunner and G.J. Williams: Engineering Fracture Mechanics Vol. 73 (2006), p. 2443

[5] H.R. Martin and D.B. Davidson: Rubber and Composites Vol. 28 (1999), p. 401

[6] H. Wang, T. Vu-Khanh and N.V. Le: Journal of Composite Materials Vol. 29 (1995), p. 833

[7] T.K. O'Brien in: Composite Materials: Fatigue and Fracture: 7th Volume, ASTM STP 1330 (1998), p. 3

[8] Z. Suo, G. Bao and B. Fan: Journal of the Mechanics and Physics of Solids Vol. 40 (1992), p. 1

[9] A.L. Carlsson, W.J. Gillespie and B.R. Pipes, R.B.: Journal of Composite Materials Vol. 20 (1986), p. 594

[10] J.A. Russell and N.K. Street, in: Delamination and Debonding of Materials, edited by W.S. Johnson ASTM STP 876, Philadelphia (1985), p. 349

[11] Y. Wang and G.J Williams: Composite Science and Technology Vol. 43 (1992), p. 251

[12] M.F.S.F. de Moura and B.A. de Morais: Engineering Fracture Mechanics Vol. 75 (2008), p. 2584

[13] K. Leffler, S.K. Alfredsson and U. Stigh: International Journal of Solids and Structures Vol. 44 (2007), p. 530

Session 6A: Residual Stresses

Applied Mechanics and Materials Vols. 24-25 (2010) pp 253-259
© (2010) Trans Tech Publications, Switzerland
doi:10.4028/www.scientific.net/AMM.24-25.253

Evaluation of the impact of residual stresses in crack initiation with the application of the Crack Compliance Method Part I, Numerical analysis

G. Urriolagoitia-Sosa[a], B. Romero-Ángeles[b], L. H. Hernández-Gómez[c], G. Urriolagoitia-Calderón[d], J. A. Beltrán-Fernández[e], C. Torres-Torres[f]

[1]Instituto Politécnico Nacional, Sección de Estudios de Posgrado e Investigación,

Escuela Superior de Ingeniería Mecánica y Eléctrica, Unidad Profesional "Adolfo López Mateos"

Edificio 5, 2do Piso, Col. Lindavista, CP 07738, México, D. F. México

[a]guiurri@hotmail.com, [b]romerobeatriz97@hotmail.com, [c]luishector56@hotmail.com, [d]urrio332@hotmail.com, [e]jbeltranf@hotmail.com, [f]crstorres@yahoo.com.mx

Keywords: Crack initiation, residual stress, crack compliance method, modified SEN specimen.

Abstract. The understanding of how materials fail is still today a fundamental research problem for scientist and engineers. The main concern is the assessment of the necessary conditions to propagate a crack that will eventually lead to failure. Nevertheless, this kind of analysis tends to be more complicated, when a prior history in the material is taken into consideration and it will be extremely important to recognize all the factors involved in this process. In this work, a numerical simulation of the introduction of residual stresses, which change the crack initiation conditions, in a modified compact tensile specimen to change the condition of crack initiation is presented. Four numerical analyses were carried out; an initial evaluation was performed in a specimen without a crack and it was used for the estimation of a residual stress field produced by an overload; three more cases were simulated and a crack was introduced in each specimen (1 mm, 5 mm and 10 mm, respectively). The overload was then applied to set up a residual stress field into the component; furthermore, in each case the crack compliance method (CCM) was applied to measure the induced residual stress field. By performing this numerical simulation, the accuracy of the crack compliance method can be evaluated. On the other hand, elastic-plastic finite element analysis was utilized for the residual stress estimation. The numerical analysis was based on the mechanical properties of a biocompatible material (*AISI 316L*). The obtained results provided significant data about diverse factors, like; the manner in which a residual stress field could modify the crack initiation conditions, the convenient set up for induction of a beneficial residual stresses field, as well as useful information that can be applied for the experimental implementation of this research.

Introduction

It has been well documented that development of failure could be divided in two basic parts, initiation and spread [1]. Additionally, there is a great number of external and internal factors that contribute to the nucleation and propagation of a crack [2]. Slip bands or dislocations and surface scratches can be considered as internal effects, as external factors are considered the effect of forces and deformations. Nevertheless, when the development, performance and effect of a crack are analyzed, prior history in the material it is not considered extensively or in a sufficient manner. To consider prior history in the component raises the difficulty of the problem in a substantial way. This is why the simplest way to analyze failure and its consequences is to consider the specimen free of previous history. But on the other hand, the manufacture of components will always leave inside the material an induction of a stress or strain field and this field will interact with the development of all sorts of defects [3].

The induction of previous history into the component is based in the application of an external agent above the yielding strength of the material. The introduction of previous history can be divided in two groups, by homogenous loading or by non-homogenous loading. The consequence of a homogenous loading derives into strain hardening and Bauschinger effect, and such consequences are the introduction of residual stresses. In both cases, the consequences of the application and removal of the external agent could contribute into the material in a beneficial and/or detrimental manner. Strain hardening and Bauschinger effect can be found in the material at the same time, if the component has been strengthen by tensile strain hardening, Bauschinger effect (that is a reduction of the yield strength of the material) will be found in the compressive behavior and vice versa. In relation to residual stresses, they are also detrimental and beneficial, as tensile and compressive stresses are applied together and auto-equilibrate them self [4]. So, in the process of manufacturing pieces and components it is very important to identify the outcome that a particular process of fabrication could add to the material.

In this paper, it is presented the numerical simulation of the introduction of a residual stress field to modify the strength of the material. It can improve the mechanical resistance of the component by a tensile overload, which at the beginning of it action can propitiate the nucleation or propagation of a crack, but when the application of the external agent is ended it would leave a beneficial residual stress field.

The Crack Compliance Method [4]

In this section a brief summary of the theory relative to the CCM is presented. Let the unknown residual stress distribution in the beam be represented by the summation of an nth order polynomial series as:

$$\sigma_y(x) = \sum_{i=0}^{n} A_i P_i(x) \tag{1}$$

where A_i are constant coefficients and P_i are a power series, x^0, x^1, x^2, ... x^n etc. The strain $\varepsilon(a,s)$ (where a = crack length and s = is the distance between the location of the strain gauge and the crack plane) due to the stress fields $P_i(x)$ is known as the compliance function $C_i(a_j,s)$ and is given by:

$$C_i(a_j,s) = \frac{1}{E'} \int_0^{a_j} K_I(a)Z(a)da \tag{2}$$

$K_I(a)$ is the stress intensity factor due to the residual stress field when the crack depth in the beam is equal to a and $K_{IF}(a)$ is the stress intensity factor corresponding to the same depth due to a pair of virtual forces F applied tangentially at a position on the beam where strain measurements will be taken during the CCM cutting of the slot (where $Z(a)$ is a geometry dependant function (equation (3)):

$$Z(a) = \frac{\partial^2 K_{IF}(a,s)}{\partial F \partial s} \tag{3}$$

By following the weight function approach, $K_I(a)$ and $K_{IF}(a)$ can be expressed as:

$$K_I(a) = \int_0^a h(x,a)\sigma_y(x)dx \tag{4}$$

$$K_{IF}(a) = \int_0^a h(x,a)\sigma_{yF}(x)dx \tag{5}$$

where $\sigma_{yF}(x)$ is the stress field due to the virtual force F. Once the $C_i(a,s)$ solutions are determined the expected strain due to the stress components in equation (1) can be obtained as:

$$\varepsilon(a_j,s) = \sum_{i=0}^{n} A_i C_i(a_j,s) \tag{6}$$

The unknown terms A_i are determined so that the strains given by equation (6) match those strains measured in the experiment during cutting i.e., $\varepsilon(a_j,s)_{actual}$. In order to minimise the average error over all data points for an nth order approximation, the method of least squares is used to obtain the values of A_i. The number of cutting increments m is therefore chosen to be greater than the order of the polynomial, i.e. $m > n$.

This work used $n = 7$ with 8 constants A_i and $m = 9$, this being the number of experimental slot cutting depths at which strain readings were collected. The least square solution is obtained by minimising the square of the error relative to the unknown constant A_i, This gives a linear set of simultaneous solutions from which A_i values are determined and equation (1) is then used to determine the residual stress distribution.

Test specimen

For this research it has been chosen a keyhole specimen, which is a modification of a compact tension specimen (ASTM standard E 647-91) with special requirements (ASTM E 399) (Fig. 1) [5]. The main characteristic of the component is a hole inducted at the notch tip. When a tensile load is applied which is large enough to produce plastic deformation at the stress concentration (point A) while the surface surrounding the material remains elastic, so when removing the effects of the load a compressive residual stress remains at the concentration point?

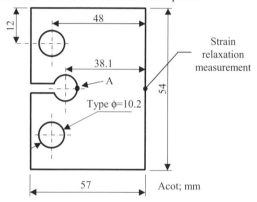

Figure 1.- Modified compact tensile specimen SEN

Statement of the problem and numerical simulation

The research presented in this paper is aimed to establish the effect of prior history in the development of a crack and how a beneficial residual stress field could increase the mechanical resistance of the material. Four cases of study are presented in this research. The mechanical properties employed for the numerical analysis were the ones obtained by four point bending tests performed on beams manufactured from a stainless steel *AISI 316L* (Fig. 2a) [6]. The elastic properties of the material were set up as follows; Young's modulus of 190 000 N/mm^2 (*E*) and Poisson ratio of 0.28 (*v*). For the elasto-plastic condition a kinematic hardening rule was applied and the mechanical properties were introduced in a tabular manner (Fig. 2b and Table 1).

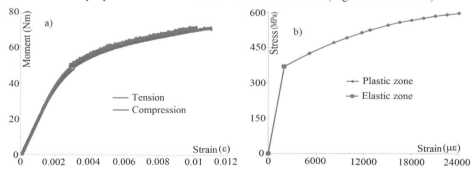

Figure 2.- Mechanical properties stainless steel *AISI* 316L
a) Bending test data used to obtain the stress-strain curve. b) *AISI 316L* stress-strain curve.

Table 1.- Stress-strain data used for the numerical simulation

Stress (MPa)	Strain
370	0.001947
425	0.005104
471	0.008171
491.7	0.009828
513.5	0.011726
525	0.012756
546	0.015000
558	0.016500
567	0.018000
575.8	0.019500
584.5	0.021000
590.5	0.022500
596.85	0.024000

The numerical simulation was performed in a symmetrical manner and all the specimens were modeled in 2D, to save computational resources (Fig. 3). Quadratic order elements (*Plane 183*) were used with 8 nodes and plane stress analysis was performed. The specimen was loaded (in all four cases) in tension and the force was uniformly distributed in 17 nodes on the loading keyhole with a magnitude of 100 N each (Fig. 3a). The base of the specimen was specially prepared with a rectangular zone (elements dimensions 0.5 mm width by 1 mm large), this zone will be employed to simulate the introduction of a slot and to evaluate the *CCM* and strain relaxation data will be obtained at the rear node of the base line (Fig. 3). Boundary conditions were applied at the bottom line in a symmetrical manner. The lack of application of boundary conditions at the near end of the bottom line is to simulate the introduction of a crack (Fig.3a) and this lack on the application of symmetry will depend on the length of the crack to be simulated. The introduction of the residual stress will be carried out by removing the effect of the load and once again resolving it (Fig. 3b).

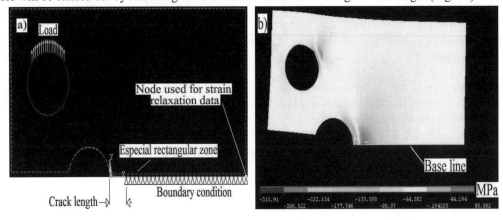

Figure 3.- Modified compact tensile specimen SEN modeled
a) Numerical model. b) Residual stress field for specimen without a crack.

Numerical case of study and results

The first case of study is the introduction of a residual stress field into a specimen free of a crack. For the next three cases of study, an analysis was performed on the effect of the introduction of a crack with different lengths (1 mm, 5 mm and 10 mm, respectively). The residual stress fields obtained for each one of the numerical simulations can be observed in Figure 4.

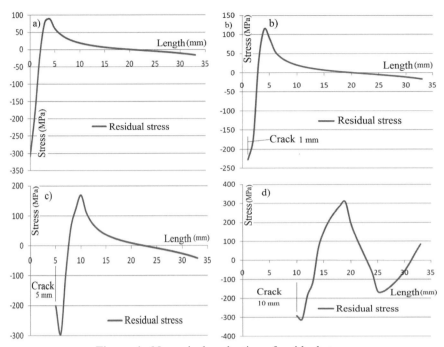

Figure 4.- Numerical evaluation of residual stress
a) Specimen without a crack. b) Specimen with a 1 mm crack length
c) Specimen with a 5 mm crack length. d) Specimen with a 10 mm crack length.

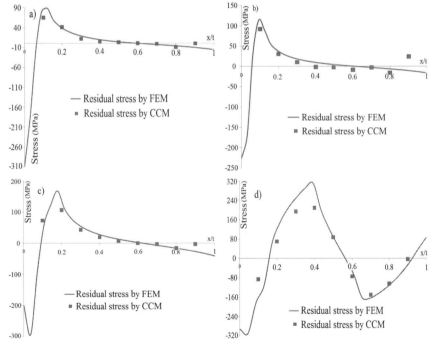

Figure 5.- Residual stress comparison between *FEM* and *CCM*
a) Specimen without a crack. b) Specimen with a 1 mm crack long
c) Specimen with a 5 mm crack long. d) Specimen with a 10 mm crack long.

CCM **numerical evaluation**

After the numerical simulation for the induction of the residual stress field was finished, a simulation of the behaviour of the *CCM* was carried out. The theory and the way to apply the *CCM* has been explained by several authors [7-9], also *Urriolagoitia-Sosa,* et al. have developed a new way to apply it [10]. The numerical evaluation of the *CCM* was done by simulating the introduction of a slot, which will cause a modification by auto-equilibrium of the residual stress on the acting base line, producing a relaxation in the material. The relaxation will produce strain data, which can be used by the CCM to determine the acting residual stress field. The residual stress results obtained by the *CCM* for each one of these cases is presented in Figure 5. In these figures there are shown as well comparisons against the numerical simulation by *FEM.*

Conclusions

This research was performed to validate the use of *FEM* in the introduction of residual stress and to validate its application to the *CCM.* The numerical data obtained will facilitate the experimental procedure for the induction of residual stress fields and the application to the *CCM.*

The modification of the compact tensile specimen *SEN,* will provide a more controllable set up for the introduction of residual stresses, this is because, if the specimen is kept with a notch tip, it cannot ensure the nucleation of a crack. The main idea of this research was to evaluate the effect of a crack in the introduction of residual stress fields, which has to be done by controlling (in certain degree) that no defects are present.

From Figure 4 it can be concluded, that after the effect of the load is removed from the specimens with a crack, a beneficial residual stress field has been induced. So, to propagate the crack it is required sufficient energy to overcome the residual stress field introduced into the component. Apparently, the combination of loading the crack and removing the load effect allows the material to raise its mechanical resistance. Nevertheless, much care has to be taken into consideration, because a crack is present in the specimen and could accumulate plastic energy due to subsequent loading cycles and then propagate.

Additionally, the exactitude on the application of the *CCM* has been evaluated by the use of *FEM.* Significant data has been obtained for the experimental procedure of the *CCM* and the proper manner to establish its application. On the other hand it can be observed in Figure 5, that similar residuals stress fields between *FEM* and *CCM* have been achieved. Nonetheless, it can be observed in Figure 5, that both ends of the curve for the calculated residual stress field are not as accurate as in the middle part of the specimen. It is thought that the mismatch observed at the opposite end of the rear location is due to the remote position in which the relaxation effect is evaluated, as the strain relaxation location cannot be fully determined. With respect to the discrepancy in residual stress results at the end near to the strain relaxation location it has been concluded, that this originates by the fact that the structural integrity of the material has been compromised by the introduction of the cut, and there is only a small part of material left.

Acknowledgement

The authors gratefully acknowledge the financial support from the Mexican government by the Consejo Nacional de Ciencia y Tecnología and the Instituto Politécnico Nacional.

References

[1] D. Broek: *Elementary engineering fracture mechanics* (Martinus Nijhoff Publishers, Netherlands, The Hague 1984), p. 24-31.

[2] M.E. Fine and D.L. Davidson: in *Fatigue Mechanisms, Advances in Quantitative Measurement of Physical Damage*, ASTM STP 811, J. Lankford, D.L. Davidson, W.L. Morris, and R.P. Wei, eds., ASTM, Philadelphia, PA, 1983, p 350–70.

[3] G. Urriolagoitia-Sosa, J.F. Durodola and N.A. Fellows: Effect of strain hardening on residual stress distribution in beams determined using the crack compliance method, *J. Strain Analysis for Eng. Design*, Vol. 42, No. 2, 2007, p. 115-121.

[4] G. Urriolagoitia-Sosa: *Analysis of prior strain history effect on mechanical properties and residual stress in beams*, PhD thesis, Oxford Brookes University.

[5] E.A. Badr: A modified compact tension specimen for the study of residual stress maintainability, *Experimental Techniques*, May/June, 2000, p. 25-27.

[6] G. Urriolagoitia-Sosa, J.F. Durodola and N.A. Fellows: A method for the simultaneous derivation of tensile and compressive behavior of materials under Bauschinger effect using bend tests, Proceeding of the I Mech E, Part C, *Journal of Mechanical Engineering Science*, Vol. 220, No. 10, 2006, p. 1509-1518.

[7] W. Chang and I. Finnie: An overview of the crack compliance method for residual stress measurement. *In Proceeding of the Fourth International Conference on Residual Stress* (Society for Experimental Mechanics, Bethel, Connecticut) 1994, p. 449-458.

[8] M.B. Prime: Residual stress measurement by successive extension of a slot; The Crack Compliance Method, *Applied Mechanics Reviews*, Vol.52, No. 2, 1999, p. 75-96.

[9] H.J. Schindler: Residual stress measurement in cracked components: Capabilities and limitations of the cut compliance method, *Materials Science Forum; Residual Stress ECRS 5*, Vol. 347-349, 2000, p. 150-155.

[10] G. Urriolagoitia-Sosa, J.F. Durodola and N.A. Fellows: Determination of residual stress in beams under Bauschinger effect using surface strain measurements, *Strain*, Vol. 39, No. 4, 2003, p. 177-185.

Applied Mechanics and Materials Vols. 24-25 (2010) pp 261-266
© *(2010) Trans Tech Publications, Switzerland*
doi:10.4028/www.scientific.net/AMM.24-25.261

Evaluation of the impact of residual stresses in crack initiation with the application of the Crack Compliance Method Part II, Experimental analysis

G. Urriolagoitia-Sosa[a], B. Romero-Ángeles[b], L. H. Hernández-Gómez[c],

G. Urriolagoitia-Calderón[d], J. A. Beltrán-Fernández[e], C. Torres-Torres[f]

[1]Instituto Politécnico Nacional, Sección de Estudios de Posgrado e Investigación,

Escuela Superior de Ingeniería Mecánica y Eléctrica, Unidad Profesional "Adolfo López Mateos"

Edificio 5, 2do Piso, Col. Lindavista, CP 07738, México, D. F. México

[a]guiurri@hotmail.com, [b]romerobeatriz97@hotmail.com, [c]luishector56@hotmail.com,

[d]urrio332@hotmail.com, [e]jbeltranf@hotmail.com, [f]crstorres@yahoo.com.mx

Keywords: Crack initiation, residual stress, crack compliance method, modified SEN specimen.

Abstract. The present work is based on a previous numerical simulation used for the introduction of a residual stress field in a modified compact tensile specimen. The main objective in that paper was to evaluate the effect that previous history has in crack initiation and to establish the new loading conditions needed to propagate a fracture. The experimental analysis presented in this paper was performed to compare and validate the numerical procedure. Several modified compact tensile specimens from a biocompatible material (*AISI 316L*) were manufactured to estimate the beneficial effect of a residual stress field. The specimens were separated in four batches; an initial group of uncracked specimens was used to establish an evaluation of the induction of a residual stress field produced by an overload; the remaining specimens were separated into three groups where a crack was introduced in each specimen (1 mm, 5 mm and 10 mm respectively) and the residual stress field caused by the application of an overload was determined. The assessment of all the residual stress fields introduced into the specimens was done by the application of the crack compliance method (*CCM*). The results obtained have provided useful information on the correlation between the numerical and experimental procedures. Furthermore, data concerning the understanding of diverse factors related to crack initiation are discussed in this paper. Finally, the beneficial aspects of the residual stresses are discussed.

Introduction

Probably one of the most complicated problems that a mechanical engineer or scientific researcher could face is the assessment of crack propagation in components with previous loading history. It is a fact, that residual stresses have been studied as a mechanical procedure to enhance the mechanical resistance of the material after the application of a manufacturing process. On the other hand, not enough effort has been drawn to investigate how a beneficial residual stress field could act after a crack has been activated by a loading procedure.

Additionally, in recent years there has been a growing interest in crack arrest problems. It has been well documented and recognized that in order to predict arrest of cracks, it is necessary a clear description of the crack growth preceding arrest [1-2]. Maybe, the principal inconvenient regarding this kind of investigation, is that most of the work done in this field has been based on the assumption of linear elastic fracture mechanics (*LEFM*); that is, the state at the crack tip is assumed to be characterized usually by the stress intensity factor [3]. Apparently the theoretical basis of *LEFM* appears to be sufficient to solve this kind of problems. However, the development of this knowledge is not sufficient to judge whether the linear theory is valid, or, rather, if it provides the desired accuracy level in a certain situation. On the other hand, it becomes more complicated when plasticity theory and previous loading history in the specimen is involved.

The investigation presented in this paper is aimed to establish the effect of prior loading history in the development of a crack and how a beneficial residual stress field could increase the mechanical resistance of the material. This paper corroborates by an experimental procedure the numerical study presented in part I of this research.

Material and test specimen

The material used in this work is stainless steel *AISI 316L*, which is one of the most utilized steel in the area of biomechanics. The chemical composition of the stainless steel *AISI 316L* is presented in Table 1 [4].

Table 1.- Chemical composition stainless steel *AISI 316L*

Element	C	Mn	Si	Cr	Ni	P	S	Mo
Weight (%)	0.03	2.0	1.0	16-18	10-14	0.045	0.03	2-3

The experimental development of this research has been performed in a keyhole specimen, which is a modification of a compact tension specimen per ASTM standard E 647-91 with special requirements as indicated in ASTM E 399 [5] (Fig. 1).

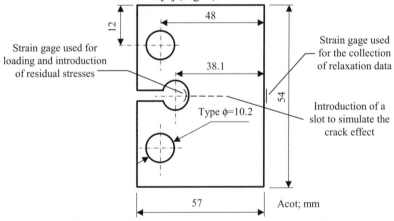

Figure 1.- Modified compact tensile specimen SEN

The main reason for the modification of the compact SEN specimen is to ensure (up to certain degree) that the manufacturing process will not produce more cracks or structural defects into the material. The main characteristic on the modification of the specimen is a hole inducted at the notch tip. A set of sixteen specimens were manufactured for this investigation and there were divided into four groups of 4 specimens each. The entire set of specimens was stress release annealed by heat treatment process of 600 °C for half an hour and slowly cold down inside a furnace [6]. The reason for applying a heat treatment process was to eliminate any kind of prior loading history acting into the material.

In the first case of study, it was included the introduction of a residual stress field into a specimen free of a crack. For the next three cases of study, an analysis was performed to evaluate the effect of an existing crack (with different lengths 1 mm, 5 mm and 10 mm, respectively) and the state of a residual stress field, which will modify the conditions for crack propagation. The generation of the crack into the specimen was performed by a wire electro discharge machine (*EDM*) in a manner as a slot (Fig. 2a). The decision to use the electric discharge machine was to produce a crack and to avoid the introduction of an additional stress field into the material. Additionally, great care was taken to induce a crack that was no thicker than 1 mm for all cases (Fig. 2b).

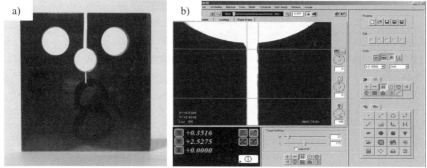

Figure 2.- Cracked modified compact tensile specimen SEN
a) Specimen with a 5 mm slot. b) Slot width evaluation.

Material characterization and preparation of specimen

For the characterization of the stainless steel *AISI 316L*, four beam specimens of 10 mm high by 6.35 mm thick by 250 mm length were prepared [7]. The beams were stress relief annealed to eliminate prior loading history [6]. The stress-strain curve for this material was obtained by four point bending tests (Fig. 3) [8]. The main advantages on this procedure are the simultaneous evaluation of tensile and compress behaviour. The Young's modulus and Poisson ratio obtained from this tests and used for the application of the *CCM* was 190 MPa and 0.28 respectively.

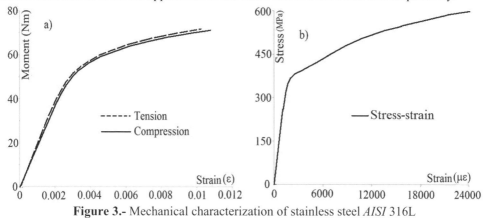

Figure 3.- Mechanical characterization of stainless steel *AISI 316L*
a) Bending test data used to obtain the stress-strain curve. b) *AISI 316L* stress-strain curve.

Experimental introduction of the residual stress field

All the specimens were prepared with a strain gauge at the rear surface (with respect to the concentration stress hole (Fig. 1)). This gauge will be used to measure the strain relaxation caused by the introduction of the slot and to be applied by *CCM* for the evaluation of the residual stress field induced into the material. Additionally, in the specimens free of an initial crack, a strain gauge was applied at the hole's border (Fig.1) to measure the strain effect related to the application of the tensile load and to correlate the experimental analysis with the numerical evaluation.

   In the numerical simulation a 100 N load was distributed in 17 nodes at the loading keyhole (Part I of this study). The resulting numerical strain obtained at the tip of the concentration hole was of 5675 $\mu\varepsilon$ and is the base for the determination of the experimental load, which was found to be 975

N. The load was applied in a tensile form by a servo-hydraulic device, with a capacity of 100 kN and was the same for all the experimental study cases. This magnitude in the load is large enough to produce a localized plasticity effect near the stress concentration zone while elasticity will remain around the specimen, so by unloading the system a residual stress field will be introduced.

Specimen preparation for the application of the crack compliance method

After the induction of the residual stress field was performed in all the study cases, the results obtained in the numerical analysis were used to assess the manner in which the specimen will relax by the introduction of the cut. With this information, it was possible to determine the best way to hold the specimen to perform the cutting and to obtain the strain relaxation for the application of the CCM. So, it was decided to clamp the specimen at one of its sides and support it at the other side, as how shown in Fig. 4.

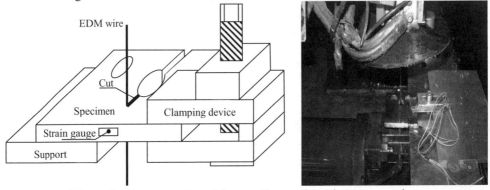

Figure 4.- Representation of the way the specimen is support and cut

The cut was done by a wire *EDM* (CHARMILLES, model ROBOFIL) in sequential steps of 1 mm of length for all cases, and the strain relaxation data was obtained using a Wheatstone bridge configuration. Also, the strain gauges were protected with M-Coat-A Air Drying Polyurethane coating as first stage and finishing with an M-Coat-B Nitride Rubber coating. The use of this procedure is to ensure the encapsulation of the strain gauge when it is submerged into the dielectric liquid used by the *EDM* machine.

Experimental cases of study and results

The theory, application and performance of the *CCM* have been extensively explained by several authors [9-13]. The first experimental case of study is the introduction of a residual stress field into a specimen free of crack. Also this first experimental case was used to set up and evaluate the accuracy of the *CCM*. For the next three experimental cases of study, an analysis was performed in the effect of the introduction of a residual stress field in a specimen with a crack of different length (1 mm, 5 mm and 10 mm, respectively). The results obtained by the application of the *CCM* and also a comparison against the numerical evaluation can be observed in Figure 5.

Conclusions

The validation for the experimental application of the *CCM* on the determination of induced residual stress fields has been presented in this paper. Additionally, in this research it has been proved that the *CCM* can be applied to components other than beams, pipes and regular plates. It has also been establish that the *CCM* can be applied in specimens wider than 10 mm, for example this investigation was applied to material thickness of 23 mm, 28 mm, 32 mm and 33 mm (which represents the specimens with crack lengths of 10 mm, 5 mm, and 1 mm, and the specimen without a crack).

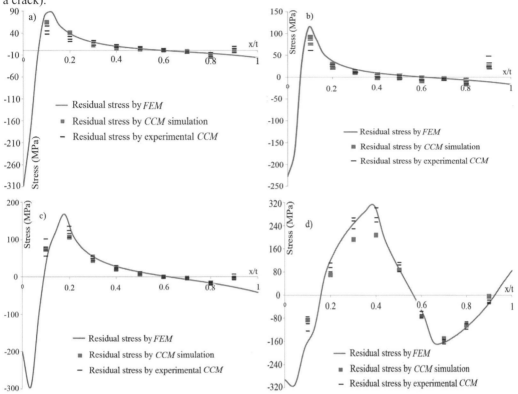

Figure 5.- Residual stress comparison between experimental evaluation against numerical analysis
a) Specimen without a crack. b) Specimen with a 1 mm crack long
c) Specimen with a 5 mm crack long. d) Specimen with a 10 mm crack long.

The development of a previous numerical analysis, which has simulated the problem applied to this experimental procedure, has provided important and significant data. The numerical study has produced results that were directly applied in the experimental testing and have been useful to simplify the *CCM* procedure. Also, it was very helpful to define the correct manner to hold and support the piece at the moment of the introduction of the slot. Furthermore, the numerical investigation is a powerful tool that has provided the expected results, which later were corroborated by the experimental procedure and have proved that the introduction of a residual stress field could raise the mechanical resistance of the material.

Once again it has been corroborated the importance in the use of the modified compact tensile specimen *SEN*, which has provided a more controllable set up for the introduction of residual stresses and the development of a methodology that effectively arrests the crack propagation. Nevertheless, the main objective in this research was to evaluate the effect of a crack with the introduction of residual stress fields. From Figure 5 it can be concluded that after the effect of the load is removed from specimens with a crack, a beneficial residual stress field has been induced. So, to propagate the crack it will be required more energy than that of the first case, because it has to overcome the residual stress field introduced into the component. On the other hand it can be observed from Figure 5, which there are similar residuals stress fields between both the numerical simulation and the experimental procedure. Nonetheless, it can be observed in Figure 5 that at the ends of the specimen the calculation for the residual stress fields are not as accurate as in the middle part of the specimen. The mismatch at the end opposite to the rear location could be related to the remote position of the effect and the relaxation cannot be read totally at the strain relaxation location. With respect to the discrepancy in residual stress results at the end near to the point of collection of relaxation data, a possible explanation could be based on the fact that the structural integrity has been compromised by the introduction of the cut and only 1 mm of material has been left. It has been corroborated that the combination of loading the crack and removing the load effect has permits the material to gain mechanical resistance. Additionally, it could be said, that a mechanical procedure (like the introduction of a residual stress field) can extend the working life of a component after a crack has been successively loaded and unloaded.

Acknowledgement

The authors gratefully acknowledge the financial support from the Mexican government by the Consejo Nacional de Ciencia y Tecnología and the Instituto Politécnico Nacional.

References

[1] J.F. Kalthoff, J. Beinert and S. Winkler: Effect of specimen geometry on crack arrest toughness, *ASME/CSME Pressure vessels and piping conference*, Montreal, Canada, 1978, p. 98-109.

[2] T. Kobayashi and J.W. Dally: Fast fracture and crack arrest, *ASTM STP 627, American Society for testing and Materials*, 1977, p. 257-273.

[3] L. Dahlberg, F. Nilsson and B. Brickstad: Influence of specimen geometry on crack propagation and arrest toughness, *Crack arrest methodology and applications*, STP 711, Baltimore Md, 1980, p. 89-108.

[4] ASM International: *Heat treater's guide; Practices and procedures for irons and steels*, 2nd Edition, 1995, p. 748-749.

[5] E.A. Badr: A modified compact tension specimen for the study of residual stress maintainability, *Experimental Techniques*, May/June, 2000, p. 25-27.

[6] Nabertherm: *Hardening guide and other heat treatments of steel*, 1997.

[7] A. Molina-Ballinas, G. Urriolagoitia-Sosa, J.M. Sandoval-Pineda, L.H. Hernández-Gómez, C. Torres-Torres, and J.A. Beltrán-Fernández: Caracterización mecánica del acero inoxidable AISI 316L bajo cargas homogéneas y no homogéneas, *IX Congreso Iberoamericano de Ingeniería Mecánica CIBIM9*, España, 2009.

[8] G. Urriolagoitia-Sosa, J.F. Durodola and N.A. Fellows: A method for the simultaneous derivation of tensile and compressive behavior of materials under Bauschinger effect using bend tests, Proceeding of the MECHE, Part C, *Journal of Mechanical Engineering Science*, Vol. 220, No. 10, 2006, p. 1509-1518.

[9] W. Chang and I. Finnie: An overview of the crack compliance method for residual stress measurement. *In Proceeding of the Fourth International Conference on Residual Stress* (Society for Experimental Mechanics, Bethel, Connecticut) 1994, p. 449-458.

[10] M.B. Prime: Residual stress measurement by successive extension of a slot; The Crack Compliance Method, *Applied Mechanics Reviews*, Vol.52, No. 2, 1999, p. 75-96.

[11] H.J. Schindler: Residual stress measurement in cracked components: Capabilities and limitations of the cut compliance method, *Materials Science Forum; Residual Stress ECRS 5*, Vol. 347-349, 2000, p. 150-155.

[12] G. Urriolagoitia-Sosa, J.F. Durodola and N.A. Fellows: Effect of strain hardening on residual stress distribution in beams determined using the crack compliance method, *J. Strain Analysis for Eng. Design*, Vol. 42, No. 2, 2007, p. 115-121.

[13] G. Urriolagoitia-Sosa, J.F. Durodola and N.A. Fellows: Determination of residual stress in beams under Bauschinger effect using surface strain measurements, *Strain*, Vol. 39, No. 4, 2003, p. 177-185.

Applied Mechanics and Materials Vols. 24-25 (2010) pp 267-272
© *(2010) Trans Tech Publications, Switzerland*
doi:10.4028/www.scientific.net/AMM.24-25.267

Mapping residual stress profiles at the micron scale using FIB micro-hole drilling

B. Winiarski [a] and P. J. Withers [b]

The University of Manchester, School of Materials, Grosvenor St, Manchester, M1 7HS, UK

[a]B.Winiarski@manchester.ac.uk, [b]Philip.Withers@manchester.ac.uk

Keywords: residual stress, micron-scale, FIB, digital image correlation, hole drilling.

Abstract. Measuring residual stress at the sub-micron scale imposes experimental challenges. We propose a new technique, namely the incremental micro-hole-drilling method (IµHM), for measurement of residual stress profiles as a function of depth with high spatial definition. Like its macroscale counterpart, it is applicable to either crystalline or amorphous materials, but at the sub-micron scale. Our method involves micro-hole milling using the focused ion beam of a dual beam FEGSEM/FIB microscope. The surface displacements are tracked by digital image correlation of SEM images recorded during milling. The displacement fields mapped around the whole are used to reconstruct the variation of the in-plane stress tensor as a function of depth. In this way the multi-axial state of residual stress has been characterised around drilled holes of 2 microns or so, enabling the profiling of the stress variation at the sub-micron scale to a depth of 2 microns. Here we demonstrate the efficacy of this method by measuring the stresses in a surface-severe-plastically-deformed (S²PD) $Zr_{50}Cu_{40}Al_{10}$ bulk metallic glass (in atomic percent, at.%) sample after failure under four-point-bending-fatigue.

Introduction

Residual stresses exist in most materials across a range of scales [1, 2]. They arise as a consequence of prior processing and/or in-service loading. Reliable information about the state of residual stress in a component or structure is often an essential part of the structural integrity assessment. At the macro scale the Incremental Hole Drilling Method [3-6] is a well established semi-destructive method for measurement of non-uniform residual stresses near the surface of a component. In essence, the method involves drilling a small shallow hole to a depth not greater than its diameter (a few millimetres say) in the specimen. The strain introduced by the partial relaxation of the stress around the hole is recorded by a strain gauge rosette stepwise over a series of drilling increments. Subsequently, the residual stress profiles are back-calculated using the Integral Method [4, 5] based on calibration data calculated by finite elements. The tendency for unstable residual stress solutions to the Integral Method, when small increments of hole-depth are selected either at very shallow or at depths approaching the hole diameter, is countered by a Tikhonov regularization scheme [7].

Destructive and semi-destructive techniques based on mechanical relaxation phenomena, such as hole/core drilling [4, 5, 8], slitting [9], and curvature methods [10] can be applied to both crystalline and amorphous materials. In principle they can be scaled down and applied to smaller structures than those to which they have been applied to-date. The advent of dual beam focused ion beam – field emission gun scanning electron microscopes (FIB-FEGSEM) in combination with digital image correlation analysis (DIC) has made it possible to make very fine holes or excisions and to record the resulting displacements with high precision. This has recently led to a number of micro-scale analogues of the mechanical stress measurement methods, principally using cantilevers or slots [11-15]. The disadvantage of these techniques is that they tend to measure one component of the in-plane stress tensor and have relatively poor lateral spatial resolution (tens of microns).

In this paper we outline the measurement principle of a new FIB micro-hole drilling method based on the Incremental Hole Drilling Method [4, 5]. We adopt a Tikhonov regularization scheme

[7] to stabilise the reconstructed residual stress profiles (σ_x, σ_y and τ_{xy}). Here this new method is used to drill a 4 µm hole to estimate the residual-stresses tensor with a depth resolution of ~200 nm and a lateral resolution of around 10 µm in a surface-severe-plastically-deformed (S^2PD) $Zr_{50}Cu_{40}Al_{10}$ BMG (in at.%) system after failure under four-point-bending-fatigue [15].

Experimental Description

Material preparation. Zr, Cu and Al melts[†] were argon arc-melted and tilt-casted to obtain $Zr_{50}Cu_{40}Al_{10}$ (atm%) BMG. Then one side of the specimen was argon shot-peened with WC/Co shots. Later the fatigue behaviour of the BMG was studied under four-point-bending-fatigue [15].

Surface Contrast Enhancement for DIC. The reliability of the DIC analysis and the scalability of the FIB-based hole drilling method strongly depend on the size, distribution, and density of the surface features that are tracked [16]. Here we have decorated the specimen surface with 20 to 30 nm equiaxed yttria-stabilized zirconia (YSZ) particles (~13% coverage) precipitated from an ethanol suspension (see Fig. 1a). We used an ultrasonic bath to break up large YSZ particle conglomerates in the suspension prior to application. To minimise surface charging and to 'protect' the surface from Ga$^+$ implantation, the surface of the specimen is coated with a carbon film of 220 Å thickness using a Gatan PECS 682 etching-coating system equipped with a Gatan 681.20000 Film Thickness Monitor.

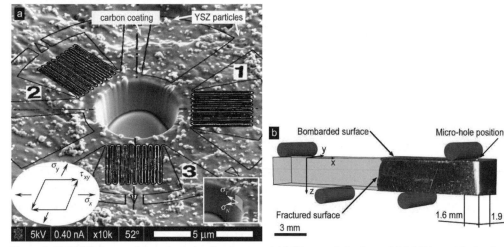

Figure 1. FEGSEM image of the microhole of radius $r_a = 4 \pm 0.07$ µm and depth, $z = 1.8 \pm 0.07$ µm: (a) showing the micro-hole at an oblique viewing angle with superimposed virtual strain-gauge rosette $r_a/r = 0.50$, where r is the radius of strain-gauge rosette; the surface is decorated with YSZ nano-particles and then carbon coated (the film thickness is about 20 nm); (b) general view showing fatigue test configuration and location of the micro-hole.

FIB Micro-hole Drilling. In order to map the stress profile in a severely plastically peen deformed BMG at the location shown in Fig. 1b, we have introduced a micro-hole of 2 µm radius to a depth of 1.8 µm (see Fig. 1a) in 11 increments using FEI Dual Beam xT Nova NanoLab 600i FEGSEM/FIB microscope. This was achieved using a focused Ga$^+$ ion beam of 280 nA accelerated by an electric field of 30 kV. At each increment the stress relaxation was recorded by FEGSEM image taken normal to the surface at a resolution of 1024×884 pixels taken with magnification of 10,000×. A focused electron beam of 0.40 nA was accelerated by an electric field of 5 kV and secondary electrons were detected using Everhart Thornley Detector (ETD). Each FEGSEM image was an average from 8 scans with a dwell time of the electron beam of 3 µs.

[†] purity better then 99.9% by weight

Digital Image Correlation Analysis. The surface displacements due to the stress relaxations are mapped from FEGSEM images at each increment using LaVision DaVis 7.2 DIC software (see Figure 2a). The DIC analysis is undertaken at locations corresponding to the positions of strain gauges of a virtual strain-gauge rosette of $r_a/r = 0.50$, where r_a is the radius of the hole and r is the radius of the strain gauge rosette, see Figs 1a and 2b. During the analysis, images are divided into smaller sub-regions (patches), which are individually correlated [16]. Before analyzing the micro-hole milling experiment, the best vector calculation parameters (VCP) were determined from two FEGSEM images captured before hole milling as described elsewhere [15]. Estimated DIC analysis accuracy is presented in Table 1. The random error is determined as the standard deviation of strain measurement determined from two FEGSEM images captured before the hole milling. Whereas the systematic error is the scatter in strain measurement using different VCPs: 32×32 pixels overlapped by 25 and 50 %, and 64×64 pixels overlapped by 50 and 75%.

Table 1. Estimated DIC uncertainty for a single vector.

Component	Random error, ε_{rnd} [$\mu\varepsilon$]	Systematic error, ε_{sys}
$\varepsilon_{G1}, \varepsilon_{G2}, \varepsilon_{G3},$	330	± 8 % of measured value

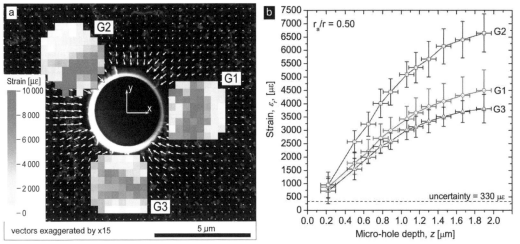

Figure 2. Digital image correlation analysis results for the final increment, $z = 1.8 \pm 0.07$ μm: (a) 2-D displacement vector field (vectors are magnified ×15) with the radial strains measured at the locations corresponding to a virtual strain-gauge rosette of $r_a/r = 0.50$ superimposed; (b) average radial strains, $G1$, $G2$ and $G3$, vs. hole depth, z.

The Integral Method. To infer the residual stress profiles (σ_x, σ_y and τ_{xy}) with depth we used directly the Integral Method proposed by Schajer [4, 5]. In essence the non-uniform stress profiles can be calculated using three pseudo-inversion equations in matrix notation

$$[P] = E/(1+v)[\bar{a}^T\bar{a}]^{-1}[\bar{a}]^T[p], \tag{1}$$

$$[Q] = E[\bar{b}^T\bar{b}]^{-1}[\bar{b}]^T[q], \tag{2}$$

$$[T] = E[\bar{b}^T\bar{b}]^{-1}[\bar{b}]^T[t], \tag{3}$$

where $E = 95 \pm 5$ GPa, and $v = 0.37$ [17]; $[\bar{a}]$ and $[\bar{b}]$ are lower triangular matrices of coefficients corresponding to the cumulative strain relaxation functions for hole drilling into a equibiaxial and pure shear stress field, respectively. We assembled these matrices using the bivariate interpolation function and tabulated cumulative strain relaxation functions from [5]. The transformed stress variables (the column matrices P, Q and T) in Eqs. (1) – (3) which are acting at corresponding hole-depths in a plane parallel to the specimen surface are related to the Cartesian stress components as

$$[P] = [(\sigma_y + \sigma_x)/2], \tag{4}$$

$$[Q] = [(\sigma_y - \sigma_x)/2],$$ (5)

$$[T] = [\tau_{xy}].$$ (6)

And the three strain relaxations measured at each hole-depth (the column matrices p, q and t) in Eqs. (1) – (3) are expressed in terms of transformed strain variables: $[p] = [(\varepsilon_{G3} + \varepsilon_{G1})/2]$, $[q] = [(\varepsilon_{G3} - \varepsilon_{G1})/2]$ and $[t] = [(\varepsilon_{G1} + \varepsilon_{G3} - 2\varepsilon_{G2})/2]$. The Cartesian stress components can be calculated from Eqs. (4) – (6)

Stabilisation of Reconstruction Algorithm. The pseudo-inversion algorithms, Eqs. (1) – (3), yield stable residual stress solutions if the entries of coefficients matrices $[\bar{a}]$ and $[\bar{b}]$ are of the same order of magnitude. However, the micro-hole milling process produces relatively small matrix entries in the diagonal bands of the coefficient matrices. Thus, Eqs. (1) – (3), become numerically ill-conditioned leading to unstable residual stress solutions. In practice this results in large oscillations of the residual stress solution with depth (see Fig. 3a). The problem of the ill-conditioned incremental calibration function matrices has been widely investigated [18]. In essence, the schemes proposed to counter this are based on careful selection of the hole-depths at which the residual stresses are calculated, which usually lead to in-depth stress profiles inferred using a reduced number of data points. In order to overcome the numerical ill-conditioning and to stabilise the residual stress solutions we used the Tikhonov regularisation scheme proposed by Tjhung & Li [7]. This scheme is significantly less sensitive to the depth increments chosen to back-calculate the residual stress profiles. We regularised Eqs. (1) – (3) into following form

$$[P_\alpha] = E/(1+v)[\alpha H_2^T H_2 + \bar{a}^T \bar{a}]^{-1} [\bar{a}]^T [p],$$ (9)

$$[Q_\alpha] = E[\alpha H_2^T H_2 + \bar{b}^T \bar{b}]^{-1} [\bar{b}]^T [q],$$ (10)

$$[T_\alpha] = E[\alpha H_2^T H_2 + \bar{b}^T \bar{b}]^{-1} [\bar{b}]^T [t],$$ (11)

where α is the regularisation parameter and H_2 is the second difference matrix operator. Within the adopted regularisation scheme we select the regularisation parameter basing on an estimate δ of the error in the measured data, e.g. $[p]$. The parameter α is chosen *a posteriori* in such way that the Euclidean norm of regularised solution discrepancy, e.g. $\|(1+v)/E \cdot [\bar{a}][P_\alpha] - [p]\|$ is equal to δ. the details are presented elsewhere [7]. The discrepancy in the measured displacements $\delta_{est} = \|[\varepsilon_{rdn}] + [\varepsilon_{sys}]\|$, the least-squares solution discrepancy $\delta_0 = \|(1+v)/E \cdot [\bar{a}][P] - [p]\|$, and the discrepancy after regularization δ are shown in Table 2.

Table 2. The discrepancy for the measurement after regularisation

δ_{est} [$\mu\varepsilon$]	δ_0 [$\mu\varepsilon$]	δ [$\mu\varepsilon$]
818	719	963

Stress Calculation Error. Several sources of measurement error arise leading to errors of different types [19]. For the IµHM method the errors originate from five main sources: (a) strain measurement errors, which include DIC calculation errors, material redeposition and additional residual stresses induced by Ga^+ ions implantation; (b) hole depth measurement errors, which include non-flatness of the bottom of the hole; (c) hole diameter measurement errors, which include tapering of the hole and deviation from roundness; (d) incorrect material constants; (e) hole eccentricity, which include possible focused ion beam drifts and the effects of possible eccentricity of DIC strain-gauge rosette from the centre of the micro-hole. In this study, following the argument in [19], we include the source (a) where the strain perturbations, e.g. $[\delta p]$ (the random error and the systematic error) result in calculated residual stress perturbation, e.g. $[\delta P]$, in Eqs. (1) – (3) and Eqs. (9) – (11).

Application of the IμHM method. Figures 3a and b shows the local in-plane residual stress profiles within the immediate vicinity of the surface of the S^2PD treated $Zr_{50}Cu_{40}Al_{10}$ bulk metallic glass as a function of depth. The same sample has been studied previously over a larger depth range in [15]. Since we used a relatively large number of data points, the matrices $[\bar{a}]$ and $[\bar{b}]$ in Eqs. (1) – (3) are ill-conditioned, which result in large oscillations of the residual stress solution with depth and extremely large uncertainty levels, in Fig. 3a. The regularization procedure has effectively reduced the oscillations of the residual stress solution, simultaneously keeping the uncertainties at acceptable level, in Fig. 3b. Our results indicate that compressive residual stress component, σ_x, in the longitudinal direction of the sample (see Figure 1c) appears to be somewhat higher (<15%) than the stress component, σ_y, in the transverse direction. This difference may arise as a result geometry of the peened bar. There is some evidence of shear stresses (~230 MPa) indicating that the principal stresses lie at around 20° to the specimen length. The longitudinal stress averages to – 920 MPa over the evaluation depth (1.8 μm) and compares to −650 MPa obtained using the FIB microslotting method to a depth of 4.1 μm [15]. Some scatter in the data could arise from material redeposition in the vicinity of the micro-hole, which is more intense than for FIB microslotting method. The virtual strain gauge rosette method uses the displacement vectors only within 3 patches; we have found that this method tends to overestimate by about 10% the inferred residual stress profiles when compared with the quasi full-field method that uses all the vectors in an annulus around the hole [20].

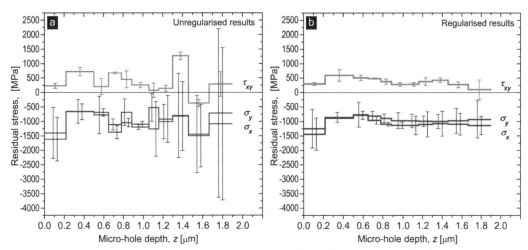

Figure 3. Inferred in-plane residual stresses profiles as a function of depth from the deformed surface: (a) unregularised stress components σ_x, σ_y and τ_{xy}; (b) regularised stress components σ_x, σ_y and τ_{xy}.

Conclusions. In summary, this short paper presents a new method based on the integral method for mapping in-plane residual or applied stresses as a function of depth at the micron-scale laterally and sub-micron scale depth-wise. Instability problems associated with the Integral Method, when small increments of hole-depth are selected near or far from the surface were minimised by application of a Tikhonov regularization scheme [7]. The results obtained by the new method reasonably agree with the results inferred by other relaxation methods [15]. The potential applications of this technique are wide-ranging including stresses in amorphous thin films, MEMS components and devices, organic electronic devices, nanostructured materials, etc. Though applicable to crystalline materials, for amorphous materials our micro-hole milling method has few competitors.

Acknowledgments. The stress measurements were made within the Stress and Damage Characterization Unit at the University of Manchester supported by the Light Alloys Towards Environmentally Sustainable Transport (LATEST) EPSRC Portfolio Project. We are grateful to P. Liaw and Y. Yokoyama for provision of the BMG.

References

[1]   P.J. Withers: Reports On Progress In Physics Vol. 70(12) (2007), p. 2211-2264.

[2]   I.C. Noyan and P.J. Withers: *Stresses in Microelectronic Circuits*, in *Encyclopedia of Materials: Science & Technology*, K.H.J. Buschow, Cahn, R.W., Flemings, M.C., et al., Editor. Elsevier: Oxford (2001).

[3]   G.S. Schajer: Experimental Mechanics (2009), DOI: 10.1007/s11340-009-9228-7.

[4]   G.S. Schajer: Journal Of Engineering Materials And Technology-Transactions Of The Asme Vol. 110(4)(1988), p. 338-343.

[5]   G.S. Schajer: Journal Of Engineering Materials And Technology-Transactions Of The Asme Vol. 110(4)(1988), p. 344-349.

[6]   P.V. Grant, J.D. Lord and P.S. Whitehead: *The Measurement fo Residual Stresses by the Incremental Hole Drilling Technique*, in *Measurement Good Practice Guide*, National Physical Laboratory: Teddington (2002).

[7]   T. Tjhung and K.Y. Li: Journal Of Engineering Materials And Technology-Transactions Of The Asme Vol. 125(2)(2003), p. 153-162.

[8]   McGinnis, M.J., S. Pessiki, and H. Turker: Experimental Mechanics Vol. 45(4)(2005), p. 359-367.

[9]   G.S. Schajer and M.B. Prime: Journal Of Engineering Materials And Technology-Transactions Of The Asme Vol. 129(2)(2007), p. 227-232.

[10]   C.A. Klein: Journal Of Applied Physics, Vol. 88(9)(2000), p. 5487-5489.

[11]   K.J. Kang, N. Yao, M.Y. He and A.G. Evans: Thin Solid Films Vol. 443(2003), p. 71-77.

[12]   N. Sabate, D. Vogel, A. Gollhardt, J. Keller, C. Cané, I. Gràcia, J.R. Morante and B. Michel: Journal Of Micromechanics And Microengineering Vol. 16(2)(2006), p. 254-259.

[13]   J. McCarthy, Z. Pei, M. Becker and D. Atteridge: Thin Solid Films Vol. 2000 (1-2)(2000), p. 146-151.

[14]   S. Massl, J. Keckes, and R. Pippan: Scripta Materialia Vol. 59(5)(2008), p. 503-506.

[15]   B. Winiarski, R.M. Langford, J. Tian, Y. Yokoyama, P.K. Liaw and P.J. Withers: Metallurgical and Materials Transactions A - In Proof (2009).

[16]   J. Quinta De Fonseca, P.M. Mummery and P.J. Withers: Journal of Microscopy Vol. 218(2004), p. 9-21.

[17]   J.M. Pelletier, Y. Yokoyama and A. Inoue: Materials Transactions, Vol. 47(2007), p. 1359-1362.

[18]   J.V. Beck, B. Blackwell and C.R. St.Clair Jr.: *Inverse Heat Conduction - Ill-Posed Problems* (New York: Wiley-Interscience 1985).

[19]   Schajer, G.S. and E. Altus: Journal Of Engineering Materials And Technology-Transactions Of The Asme Vol. 118(1)(1996), p. 120-126.

[20]   B. Winiarski and P.J. Withers: *Mapping of stress at the sub-micron scale by incremental focused ion beam hole-milling.* Paper submitted

Session 6B: Biomechanics

Applied Mechanics and Materials Vols. 24-25 (2010) pp 275-280
© *(2010) Trans Tech Publications, Switzerland*
doi:10.4028/www.scientific.net/AMM.24-25.275

Acetabular Component Deformation under Rim Loading using Digital Image Correlation and Finite Element Methods

H. Everitt[1&2, a], S. L. Evans[2, b], C. A. Holt[2, c], R. Bigsby[1, d], I. Khan[1,e]

[1]Biomet UK Ltd, Bridgend, UK

[2]Cardiff University, Cardiff, UK

[a]holly.everitt@biomet.com, [b]evansssl6@cardiff.ac.uk, [c]holt@cardiff.ac.uk,
[d]rob.bigsby@biomet.com, [e]imran.khan@biomet.com

Keywords: finite element methods, digital image correlation, deformation, acetabular cup

Abstract:
Total hip replacement is a highly successful operation; restoring function and reducing pain in arthritis patients. In recent years, thinner resurfacing acetabular cups have been introduced in order to preserve bone stock and reduce the risk of dislocation. However concerns have been raised that deformation of these cups could adversely affect the lubrication regime of the bearing; leading to equatorial and edge contact, possibly causing the implants to jam. This study aims to assess the amount of deformation which occurs due to the tight peripheral fit experienced during press-fit by applying rim loading to three different designs of acetabular cup: a clinically successful cobalt chrome resurfacing cup, a prototype composite resurfacing cup and a clinically successful polyethylene monobloc cup.

Digital Image Correlation (DIC) was used to measure the deformation and to validate Finite Element (FE) models. DIC provided a non-contacting method to measure displacement; meaning the load could be increased continuously rather than in steps as in previous studies.

The physical testing showed that the cobalt chrome cups were significantly stiffer than the composite prototype and polyethylene cups. The FE models were in good agreement with the experimental results for all three cups and were able to predict the deformation to within 10%. FE models were also created to investigate the effect of cup outside diameter and wall thickness on stiffness under rim loading. Increasing outside diameter resulted in a linear reduction in stiffness for all three materials. Increasing the wall thickness resulted in an exponential increase in cup stiffness.

Rim loading an acetabular shell does not accurately simulate the *in vivo* conditions; however it does provide a simple method for comparing cups made of different materials.

Introduction:
In the UK there has been a trend towards more cementless total hip replacement and fewer cemented procedures [1]. For a non-cemented acetabular cup, initial stability is achieved by a press fit between the cup and bone [2, 3]. The acetabulum is reamed to a slightly smaller size than the implant, typically by between 1 and 3mm, the pressure of the bone against the cup provides initial stability [3, 4]. In order to preserve bone stock and reduce the risk of dislocation, thinner resurfacing cups have been introduced [5, 6]. However the use of thinner more flexible cups has raised concerns about the potential for deformation during impaction [5, 6]. Such deformation could adversely affect the fluid film lubrication of metal-on-metal bearings [7], it could also lead to equatorial and edge contact possibly causing the implants to jam [5, 6]. Cup deformation could also influence implant stability and fixation, producing unfavourable conditions for osseointegration and possibly affect periprosthetic bone remodelling [5, 6].

On press fitting of a hemispherical acetabular cup, load transfer occurs predominantly near the rim of the cup and is greatest in the diagonal axis between the Ilium and ischial columns [6, 8] (Figure 1). This study aims to assess the amount of deformation which occurs due to the tight peripheral fit experienced during press-fit by applying rim loading to three different designs of acetabular cup: a clinically successful cobalt chrome resurfacing cup, a prototype composite resurfacing cup and clinically successful polyethylene monobloc cup. Digital image correlation (DIC) was used to measure the deformation and to validate finite element (FE) models. Various studies have used physical testing to validate FE models of implants [9-11] however to the authors' knowledge DIC has not been previously been used in this application.

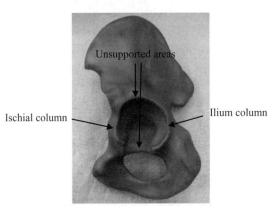

Figure 1 medial view of the pelvis

Methods:

In 3D DIC two cameras record the deformation process of a sample as it is subjected to mechanical forces (figure 2). The sample has a random dot pattern applied to the surface; this is known as a "speckled" pattern. A series of consecutive stereo images are taken throughout the test, which show the deformed speckled patterns relative to the initial pattern. Proprietary image correlation algorithms are used to analyse the digital images to produce a deformation plot [12]. In this study the speckled cups were positioned directly between the platens of a (Losenhausen) servohydraulic test machine; the displacement was controlled by a PC connected to the servohydraulic test machine, at a rate of 5 mm/min. Proprietary software (Limess Messtechnik & Software GmbH, Berlin, Germany) was used to record the camera images and simultaneously record the applied load from the test machine.

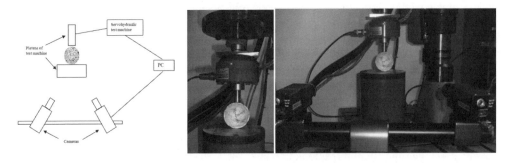

Figure 2 schematic diagram of DIC set up showing test piece, two cameras, PC and servohydralic machine (left), close up photo of speckled cup in test machine (centre), photo of camera and speckled cup (right)

Six different sized CoCr cups were tested (outside diameter 48 to 62mm). At the time of testing only one size of prototype composite cup was available. Due to availability it was not possible to test polyethylene monobloc cups; however 3 sizes of polyethylene liners were used to validate the FE model. A minimum of two cups for each size was tested for each material.

Abaqus CAE version 6.8-1 was used to create finite element models of the cups. The models were meshed using a free meshing technique with a 10 node modified quadratic tetrahedron mesh (Figure 3), the largest cups had around 20,000 elements. The test platens of the test machine were modelled as rigid bodies, a frictionless surface contact was assumed between the cup and the rigid bodies. The load was applied to the upper rigid body in the y-direction and the lower rigid body was fixed in all degrees of freedom. Finite element models were also used to investigate the effect of wall thickness and cup diameter on cup stiffness.

Figure 3 example of a typical mesh (10 node modified quadratic tetrahedron mesh) and rigid bodies

Results:

The track point function of the Vic3D software (Limess Messtechnik & Software GmbH, Berlin, Germany) was used to find the maximum diametric displacement from the digital images. The FE results were plotted as nodal displacement in the y-direction to give the diametric deformation (Table 1&2).

Material	Thickness [mm]	OD [mm]	FE predication (deformation at 300N) [mm]	DIC result (deformation at 300N) [mm]
UHMWPE	11	58	0.49	0.50
UHMWPE	5	42	2.80	2.98
UHMWPE	5	38	2.23	2.31
Composite	3	45	1.78	1.79

Table 1 details of OD and wall thicknesses of polymer cups, maximum diametric deformation FE (3mm mesh density) and DIC results taken at 300N.

Material	Thickness [mm]	OD [mm]	FE predication (deformation at 3000N) [mm]	DIC result (deformation at 3000N) [mm]
CoCr	3	48	0.33	0.36
CoCr	3	50	0.35	0.31
CoCr	3	56	0.42	0.47
CoCr	3	58	0.45	0.47
CoCr	3	60	0.47	0.50
CoCr	3	62	0.49	0.53

Table 2 details of OD and wall thicknesses of CoCr cups, maximum diametric deformation FE (3mm mesh density) and DIC results taken at 3000N.

The Finite Element models were in good agreement with the experimental results for all three materials. Figure 4 shows that refining the mesh size reduced the difference between the experimental DIC results and the FE model. The experimental results showed that increasing the cup size (with constant wall thickness) resulted in a reduction in cup stiffness; confirming that the largest cup size is the worst-case scenario in terms of deflection. As expected the cobalt chrome cups were found to be significantly stiffer than the composite prototype and the polyethylene cup. Figure 5 show the load deflection plot for the three materials tested, the cups shown had an OD of 58mm.

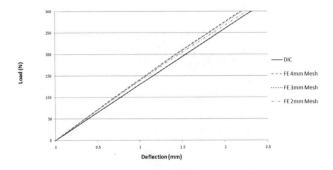

Figure 4 Load deflection plot for a 38mm UHMWPE cup showing DIC and FE results with various mesh sizes

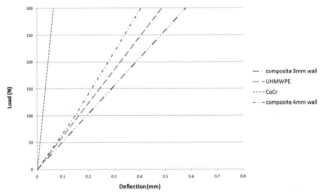

Figure 5 load deflection plot for 58mm cups of various materials

The FE models created to investigate the effect of cup outside diameter and wall thickness on stiffness showed that there was a linear decrease in cup stiffness with increasing outside diameter and an exponential decrease as wall thickness was decreased.

Discussion:
Physical testing has been used previously to validate FE models of implants [9-11, 13]. However to the authors' knowledge this is the first time digital image correlation has been used in this application. Ong et al [13] validated their FE model of acetabular cups using rim loading, cups were deformed under a load of 2kN and the deformation measured using callipers. The model was able to predict the deformations with a 7.1% difference [13]. The FE model in the current study was able to predict the deformation of an acetabular cup under rim loading to within 10%, there was a trend for the FE to overestimate the stiffness of the cups. The CoCr cups were obtained from scrap rather

than stock; therefore it is possible they are out of tolerance; meaning that the FE models are not true representations of the actual cups tested. This could explain the variation between FE predictions and DIC results for some of the cups tested.

The load cell used accurate to ±0.5%, the manufacturer of the DIC equipment reports an accuracy of 0.01 pixels, given the 1 mega pixel cameras used this should give an accuracy of 1μm for the 100mm measurement field used. However it has been shown that the properties of the speckled pattern, lighting conditions and measurement system effect the accuracy of the DIC measurement [14]. In the current study the same lighting conditions and measurement equipment was used in all cases, an initial trial run compared two methods of applying the speckled pattern and one was selected based on ease of analysis by the software. These factors may have introduced errors into the experimental measurements. Previous work has shown that the DIC equipment used in this study is accurate to within 5%.

DIC enables continuous measurement of full field displacement throughout testing, without the requirement to contact the sample. Previous studies have used callipers to measure cup deformation [13, 15], which meant that the load had to be applied in a stepwise fashion, and which also introduces the possibility of human error in the measurement. The software also has the advantage of automatically calculating strain values and the option to track the displacement of a defined point. A limitation of the system is access to the field of view required, which may be a problem for physiological loading conditions where a femoral head will be used to apply load.

Rim loading an acetabular shell does not accurately simulate the *in vivo* conditions; however it does provide a simple method for comparing cups made of different materials. A more representative test set up is required to accurately simulate loading of acetabular cups *in vivo,* further work will concentrate on this, in particular testing with cadaveric specimens.

References:
[1] NJR, *National Joint Registry for England and Wales 4th Annual Report.* (2008), The Department of Health.
[2] J.J. Callaghan, *The Clinical Results and Basic Science of Total Hip Arthroplasty with Porous-Coated Prostheses.* J Bone Joint Surg Am. 75(2) (1993), p. 299-310.
[3] C.M. Bellini, F. Galbusera, R.G. Ceroni, and M.T. Raimondi, *Loss in Mechanical Contact of Cementless Acetabular Prostheses Due to Post-Operative Weight Bearing: A Biomechanical Model.* Medical Engineering & Physics. 29(2) (2007), p. 175-181.
[4] L.D.M.D. Dorr, Z.M.D. Wan, and J.M.D. Cohen, *Hemispheric Titanium Porous Coated Acetabular Component without Screw Fixation.* Clinical Orthopaedics & Related Research. 3511998), p. 158-168.
[5] K.L. Ong, M.T. Manley, and S.M. Kurtz, *Have Contemporary Hip Resurfacing Designs Reached Maturity? A Review.* J Bone Joint Surg Am. 90(Supplement_3) (2008), p. 81-88.
[6] Z. Jin, S. Meakins, M. Morlock, P. Parsons, C. Hardaker, M. Flett, and G. Isaac, *Deformation of Press-Fitted Metallic Resurfacing Cups. Part 1: Experimental Simulation.* Proceedings of the Institution of Mechanical Engineers, Part H: Journal of Engineering in Medicine. 220(2) (2006), p. 299-309.
[7] A. Kamali, J.T. Daniel, S.F. Javid, M. Youseffi, T. Band, R. Ashton, A. Hussain, C.X. Li, J. Daniel, and D.J.W. McMinn, *The Effect of Cup Deflection on Friction in Metal-on-Metal Bearings.* J Bone Joint Surg Br. 90-B(SUPP_III) (2008), p. 552-c-.
[8] K.H. Widmer, B. Zurfluh, and E.W. Morscher, *Load Transfer and Fixation Mode of Press-Fit Acetabular Sockets.* Journal of Arthroplasty. 17(7) (2002), p. 926-935.
[9] J. Stolk, S.A. Maher, N. Verdonschot, P.J. Prendergast, and R. Huiskes, *Can Finite Element Models Detect Clinically Inferior Cemented Hip Implants?* Clinical Orthopaedics & Related Research. 4092003), p. 138.

[10]   J. Stolk, D. Janssen, R. Huiskes, and N. Verdonschot, *Finite Element-Based Preclinical Testing of Cemented Total Hip Implants*. Clinical orthopaedics and related research. 4562007), p. 138.

[11]   J.R.T. Jeffers, M. Browne, A.B. Lennon, P.J. Prendergast, and M. Taylor, *Cement Mantle Fatigue Failure in Total Hip Replacement: Experimental and Computational Testing*. Journal of Biomechanics. 40(7) (2007), p. 1525-1533.

[12]   Limess. *Digital Image Correlation* (2008)   [cited 2009 30/03/2009]; Available from: www.limess.com.

[13]   K.L. Ong, S. Rundell, I. Liepins, R. Laurent, D. Markel, and S.M. Kurtz, *Biomechanical Modeling of Acetabular Component Polyethylene Stresses, Fracture Risk, and Wear Rate Following Press-Fit Implantation*. Journal of orthopaedic research: official publication of the Orthopaedic Research Society, 2009).

[14]   D. Lecompte, A. Smits, S. Bossuyt, H. Sol, J. Vantomme, D. Van Hemelrijck, and A.M. Habraken, *Quality Assessment of Speckle Patterns for Digital Image Correlation*. Optics and Lasers in Engineering. 44(11) (2006), p. 1132-1145.

[15]   M. Squire, W.L. Griffin, J.B. Mason, R.D. Peindl, and S. Odum, *Acetabular Component Deformation with Press-Fit Fixation*. Journal of Arthroplasty. 21(6, Supplement 1) (2006), p. 72-77.

Applied Mechanics and Materials Vols. 24-25 (2010) pp 281-286
© (2010) Trans Tech Publications, Switzerland
doi:10.4028/www.scientific.net/AMM.24-25.281

Skin Thermal Effect by FE Simulation and Experiment of Laser

Ultrasonics

Chunhui Li, Sinan Li, Zhihong Huang, Wenbin Xu

Division of Mechanical Engineering and Mechatronics,

University of Dundee, Dundee DD1 4HN

C.Li@Dundee.ac.uk, S.T.Li@Dundee.ac.uk, Z.Y.Huang@Dundee.ac.uk,

X.X.Wenbin@Dundee.ac.uk

Key Words: Laser Ultrasonics, Finite Element Analysis (FEM), Thermal Effect, Skin.

Abstract. Laser ultrasonics opened possibilities to measure thermal and mechanical property of skin which occupies an essential position and is beneficial in industrial and medical applications. This paper focuses on the thermal effect in the thermal section of the laser ultrasonic technique. A transient thermal analysis is developed and promoted to simulate the interaction between the laser pulse and human skin, using a multilayered finite element model (FEM). Chicken leg had been used and irradiated by KrF laser, the thermal reactions were detected and recorded by a thermal camera. By comparison, the thermal result of experiments and simulation matches.

Introduction

The study and research of the mechanical properties of different materials occupies an essential position in many engineering and industrial applications. Laser ultrasonics uses a short laser pulse as a remote ultrasound input to excite thermal expansion, and then induces ultrasonic waves on surfaces, which contains the information of elastic properties. This technology has been widely used in industry to detect the surface condition of metallic materials and obtain relative mechanical properties of coatings [1-5].

Laser ultrasonics has potential to quantify mechanical properties of skin for diagnosis and accurate assessment of skin diseases [6-7]. In skin laser ultrasonics, the study of thermal interaction between a laser pulse and skin becomes a very essential part for the following two reasons:

Firstly, it is of great importance to understand the relationships between increased temperature and laser energy along with other parameters (e.g. shape and duration) in terms of the laser safety. Subsequently, the range of safe laser energy and its relative parameters should be defined in order to avoid skin ablation even skin damage.

Secondly, the amplitude of laser generated surface waves increases with increasing laser energy. It is important that the laser energy should be kept sufficient to produce ultrasonic waves which are readily detectable by available measurement tools such as interferometers and high frequency ultrasound transducers which allows the generated surface waves to be recorded and analysed .

The finite element model provides a clear and detailed process of the thermal effect in skin laser ultrasonics. The previous work shows that the thermal effect of laser-skin interaction is subject to the thermal properties of tested materials and the parameters of laser pulses including the energy, shape, pulse duration and focus spot [6]. However, no experiments support the results of finite element model. Thus, it is significant to compare the results of a finite element model and experiments in order to quantify the effectiveness of computer modeling, and to analyze the thermal section of skin laser ultrasonics in real conditions.

This paper mainly focuses on the thermal section of skin characterization by the laser ultrasonic technique. At the beginning, there is a discussion about the thermal results from a finite element simulation in order to provide a theoretical reference and then we introduce the experiment of thermal section of laser ultrasound to list and compare the theoretical and experiment results.

Thermal Analysis

Skin has three layers: the epidermis, the dermis and subcutaneous fat. Each of them has dramatically different structures, mechanical properties (Table 1) and physiological functions. It is assumed that the geometry of skin can be treated as constant in a small scale and mechanical and thermal properties are considered as constant in this paper.

	Epidermis	Dermis	Fat
Density (kg·m^{-3})	1200	1200	1000
Specific Heat (J·kg^{-1}·K^{-1})	3.590	3.300	1.900
Thermal conductivity (W·m^{-1}·K^{-1})	0.24	0.45	0.19
Young's Modulus (Pa)	136000	80000	34000
Poisson's Ratio	0.499	0.499	0.499
Thermal Expansion Coefficient (K^{-1})	0.0003	0.0003	0.00092

Table 1 Mechanical and thermal properties of three layers of skin [7]

In the finite element thermal analysis, PLANE55 is employed as a plane element with a two-dimensional thermal conduction capability. The element has four nodes with a single degree of freedom-temperature, at each node. An axisymmetric boundary condition is applied on the boundary of the model in the direction of the laser beam in order to simplify the solution.

In the meshing part, the element size is set up and increases gradually from epidermis to subcutaneous fat. The thickness of epidermis, dermis and subcutaneous fat are assumed as 0.08mm, 1.5mm and 10mm, in addition, the element size is 0.005mm×0.005mm, 0.02mm×0.02mm and 1mm×1mm for epidermis, dermis and subcutaneous fat respectively, and the radius of the model is 20mm. Figure 1 illustrates the schematic diagram of the laser-irradiated model.

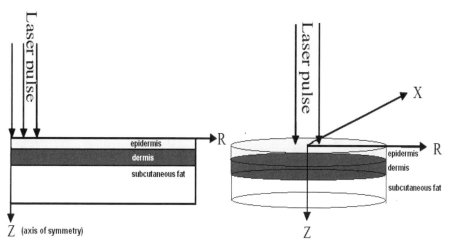

Figure 1 Schematic diagram of laser-irradiated skin model, left shows the model in simulation; right shows the real analyzed model

At the surface of the model, the laser energy is assumed to be a heat generation concentrated on the affected area. Eq.1 expresses the thermal conductive equation in three layers and the boundary condition at the surface of the skin model.

$$-k_e \frac{\partial T_e(r,z,t)}{\partial z}\Big|_{z=0} = \mu_a * I_0 * \exp(-\frac{r^2}{r_0^2}) * \frac{t}{t_0}\exp(-\frac{t}{t_0}) * \exp(-\mu_a \cdot z) \qquad (1)$$

Where K_e means the conductivity of epidermis; μ_a is the absorption coefficient of skin (mm^{-1}); I_0 is the irradiance on the surface of tissue ($W \cdot mm^{-2}$). This equation also contains the spatial and temporal distribution of the Gauss beam as we are using in the experiment, where r_0 is the beam radius and t_0 is the rise time of the Gauss beam, z is the along the axis of symmetry.

In this simulation the conditions are set as close as possible to the conditions of the experiments with the initial temperature set at 289K. The laser source in the simulation is assumed to be a KrF laser with 248nm wavelength. The laser energy is 24.7mJ, and rise time of pulse is 10ns and the spot size is 0.5mm. The time step is set up in the range of 1ns for the duration of the laser pulse interaction and it is permitted to increase after the interaction period. The temperature distribution can be plotted in Figure 2.

TEMPERATURE

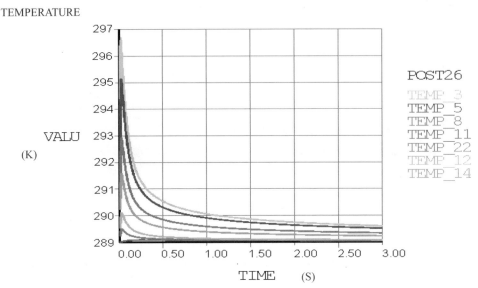

Fig. 2, Temperature in the radial direction at different distance to the center of laser spot of 24.7mJ

In Figure 2, we can get the temperature distribution along the surface of the skin model from the central laser irradiation to the points far away from the laser source. All points have different level of temperature increase, among which the central point suffers the fastest and highest temperature increase, from 289K to 296.8K. Then, the temperature declines slowly and levels off at a value of 286 Kelvin. It also shows that with the length increasing to the laser source, the peak value of temperature is cut down dramatically, however they all follow the same trend of dropping.

Experiment

In the experiment of the thermal section of skin laser ultrasonics, a chicken leg is irradiated by a laser pulse to record the temperature field by a Jade infrared thermal camera. The laser device provides the source of the laser beam, and its output energy is acquired by the computer. The optical medium in the laser generator is KrF gas, and the wavelength of the laser beam is 248nm which locates in the ultraviolet section of the electromagnetic spectrum. The laser energy is selected to be 300mJ which is maintained as a constant during the experiment and the output energy is verified by attenuation from 20% to 100%. The energy detector was used to record the energy of the laser pulse that arrived at the tissue surface which is totally different from the theoretical value set up from the computer. In this test the shape of the laser spot is rectangular with a constant energy distribution in the width direction and of Gaussian distribution in the length direction.

The thermal camera used is an infrared camera by Cedip Infrared systems. The camera can image the infrared section of the electromagnetic spectrum with windowing capabilities. The camera is a high performance, highly sensitive camera with a fast frame rate and can record the image in 320×240 pixels. Using the camera software, Altair, the frequency of imaging can be programmed and the integration time modified to an increment on 1μs.

The first step of the experimental procedure consists of setting up the laser parameters in the controlling computer. The energy of the laser beam is controlled using the PC and is varied gradually using the attenuator. The laser pulse irradiated the chicken leg and the temperature was recorded by the thermal camera and the data transferred to the PC for further processing. At the end of the test the energy detector is used to record the energy of the laser which varies from the theoretical values due to attenuation from the optical system.

Figure 3 demonstrates a typical result of the temperature distribution in the infrared wavelength and graphs of temperature evolution under irradiation of the specific laser energy, which is directly extracted from the thermal camera.

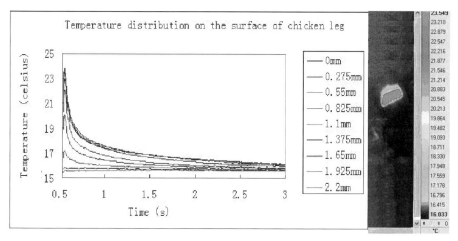

Fig. 3, Diagram of temperature distribution and temperature evolution on chicken leg with laser energy of 24.7mJ, the highest temperature happens at laser spot center and values 296.3K.

From Fig. 3, it can be concluded that the temperature evolution of chicken leg shows the exact same pattern that increases sharply due to the temporal distribution of the laser pulse and then drops down slowly as the heat diffuses into the material. The maximum value of temperature is mainly dependent on the laser energy. The highest and fastest temperature increasing point occurs in the center of the laser beam. With the length from the center increased, the temperature will rise slowly due to the diffusion of heat, with a relatively lower peak value.

Conclusions

It can be concluded that the temperature increment at the center of the laser spot is dependent on the laser pulse and is proportional with laser energy. This point is also becoming the most importance point when discussing the thermal section of skin laser ultrasonics, because it has the maximum temperature increase. It is essential to control this temperature rise to be under the level of laser damage and ablation. From comparison, the result from the simulation has a good match with the experiment data and we can assume that the skin parameters are similar to the ones from chicken skin.

In the future, mechanical section of skin laser ultrasonics will be studied, since it is of importance to relate central temperature rise with amplitude of surface wave together, and analyse the possible safe and functional range of laser energy and parameters.

Reference

1. Baiqiang Xu, Xiaowu Ni, Zhonghua Shen Numerical simulation of laser-generated ultrasound by the finite element method. Journal of Applied Physics, Vol. 95 (2004), p2116-p2122.
2. Abbate.A, Russell.W, Goldman.J, Kotidis.P, Berndt.CC Nondestructive determinination of thickness and elastic modulus of plasma spray coatings using laser ultrasonics, Review of Progress in QNDE, Vol. 18 (1999), p373-p380.
3. Bai.T.H, Pei. L.W. and Fang Q.P, Experimental studies of directivity patterns of laser generated ultrasound in neutral glasses, Ultrasonics Vol. 33 (1999), Issue. 6, p429–p436.
4. Davies.S.J, Edwards.C, Taylors.G.S, Palmer.S.B, Laser-generated ultrasound: its properties, mechanisms and multifarious applications, J.Phys.D: Appl.Phys, Vol. 26 (1993), p329-p348
5. Jijun Wang, Baiqiang Xu and Jian Lu et al. Influence of transparent coating thickness on thermalelastic force source and laser-generated ultrasound waves. Applied Surface Science (2009). p3-p10
6. Adele LEtang and Zhihong Huang. FE simulation of laser generated surface acoustic wave propagation in skin. Ultrasonics Vol. (2006) 44 p1243-p1247.
7. Jiang.S.C, Ma.N, Li.H.J, Zhang.X.X, Effects of Thermal Properties and Geometrical Dimensions on Skin Burn Injuries, Vol. 28(2002), 713-717.
8. Mansy HA, Sandler RH and Royston TJ. Excitation and propagation of surface waves on a viscoelastic half-space with application to medical diagnosis. Journal of Acoustical Society of America, Vol. 106 (1999), p3678-p3686.

Applied Mechanics and Materials Vols. 24-25 (2010) pp 287-295
© (2010) Trans Tech Publications, Switzerland
doi:10.4028/www.scientific.net/AMM.24-25.287

Biomechanics and numerical evaluation of cervical porcine models considering compressive loads using 2-D classic computer tomography CT, 3-D scanner and 3 -D Computed Tomography

J.A. Beltran-Fernández[1a], L.H. Hernández-Gómez[1b],
G. Urriolagoitia-Calderón[1c], A. González-Rebatú[2d], G. Urriolagoitia-Sosa[1e]

[1]Instituto Politécnico Nacional, Escuela Superior de Ingeniería Mecánica y Eléctrica (ESIME) Unidad Zacatenco. Sección de Estudios de Posgrado e Investigación, ESIME Zacatenco, Unidad Profesional Adolfo López Mateos (UPALM), Edificio 5, 3er Piso, 07738, D.F., México.
Tel.: 5729-6000, ext.: 54691
[2]Hospital Regional ISSSTE 1° de Octubre, Av. Instituto Politécnico Nacional. Núm. 1669, Col. Magdalena de las Salinas, 07760, México.
[a]jbeltranf@hotmail.com, [b]luishector56@hotmail.com, [c]urrio332@hotmail.com, [d]janosclub@hotmail.com, [e]guiurri@hotmail.com

Keywords: Biomechanics, cervical, 3-D Scanning, Compressive loads, Porcine vertebrae

Abstract. In this paper the biomechanical behavior and numerical evaluation results of three C3-C5 porcine cervical models created with different modeling techniques are shown. The objective of this evaluation is to know the differences between the biomechanical effects on a bone graft, which replaces a damaged C4 vertebral body, a titanium alloy (Ti-6A1-4V) cervical plate, used to isolate the C4 damaged vertebra, and the influence on the compressive loads on the complete and instrumented C3-C5 cervical model. The biomechanical integrity of the healthy C3 and C5 vertebral body after the fixation of the cervical plate using titanium alloy screws is considered. Besides, 2-D Computer Tomography classic technique, 3-D Scanner Z-Corp 700 and a CT scanning Philips Brilliance system was used to create the three FEM models. In addition, 3-D Software as Pro-E Wildfire 4.0, ScanIP 3.1, UGS NX-4 and Geomagics R 10.0 was used to create specific numerical model. Main displacements and von Misses stresses between the upper and lower surfaces of the vertebral bodies and the bone graft and the influence of the titanium alloy (Ti-6A1-4V) screws on the vertebral body of C3 and C5 were evaluated. The contribution of this study is to optimize the actual surgical technique once the numerical results on the FEM model have been analyzed. In other words, the numerical disparity between classic CT techniques versus 3-D modern techniques is established.

1.0 Introduction

At the present time, modeling techniques has changed using modern techniques which are useful in order to create complex geometries. The beginning of CAD Software and the evolution of the parametric engineering computer software has transformed the way in which the classic engineering problems are solved using structural parts. However, medical surgeries requires different solutions including complex conditions to evaluate new prosthesis or methodology using the engineering and medical analysis in order to know the mechanical and natural behavior of organic structures as bone, ceramic, titanium alloy and soft tissue, used in the cervical spine.

Diverse publications about advanced modeling techniques using CAD/CAE tools does not define the affinity between the real and the virtual model, especially in human and organic structures considering tissue, bone and the complexity of the surfaces. However, important results in orthopaedic and biomechanics studies are reported and classified in at least two substantial reviews, the first covering "the first decade" by Huiskes and Chao [1] in 1983 and the second by Prendergast [2] in 1972. Finite element modelling in cardiovascular biomechanics has not been reviewed in this way. Some of the principal aspects are discussed by Holzapfel and Ogden [3], and a review of cardiovascular stenting is given in Lally & et al [4]. Van der Meulen and Huiskes [5] have presented a survey of mechanobiology, including computational approaches, and recently Panjabi & et al. [7] using diverse biomechanics models with numerical and experimental analysis.

In this case, the present work focuses on the cervical spine, studying its mechanical behavior considering the natural state and extreme external forces caused by fracture mechanism and specific trauma evaluated by specialist medics on human spine in Mexican Hospitals as a part of medical rehabilitation programs. Medical institution is referred to ISSSTE Regional Hospital "1° de Octubre" and Academic Research Institute by the National Polytechnic Institute, and financed by The National Council for Science and Technology, CONACYT, Mexico. Numerical testing using compression loading over porcine specimens instrumented with cervical plate and bone graft were reported by Beltran [8, 9] using a FEM Model created by classic modeling technique. It represents the base of a Biomechanical reevaluation of these results, which consider the creation of an alternative cervical model created by Computer Tomography (CT) and a 3-D Scanner model ZCorp 700®.

2.0 Problem statement

The importance of the human spine as the main distribution channel of the central nervous system and a correct mechanical integrity not only influences the stability of the body, but the way in which the loads are effectively transferred through it, in order to avoid a severe disability when a cervical injury takes place in one of the cervical vertebrae subjected to a compression beyond the stress limit of the vertebra (Fig. 1). An engineering analysis of this medical effect implies to consider the understanding of medical terminology, the way in which the natural and external forces are being transferred in a composite material. In this sense the mechanical properties of cortical and trabecular bone is required in order to build a 3-D model useful and precise to evaluate diverse procedures or surgical treatments in cervical lesions. It is important to establish that the interface between the skull and the thorax is known as the cervical spine (Fig. 2) and it is composed by 7 vertebrae (C1 to C7). C1 is called Atlas, which acts as the base of the skull, C2 allows the movement of the neck system, and it is referred to as the axis. Vertebrae C3 to C6 sustain the neck, transferring load from the skull to the thoracic vertebrae T1. The study of mechanical behavior and the analysis of the parametric cervical model were published by Beltran [1, 8, 9] using classic 3-D modeling technique to get a preliminary model. The work was focused on the evaluation of a cervical plate OrtoSintese 4774/08 (55 mm) installed in the C3–C5 range of a patient, whose C4 vertebra was replaced using a bone graft taken from the iliac crest.

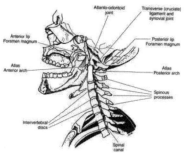
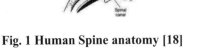
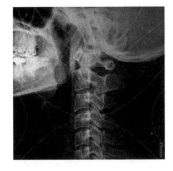

| **Fig. 1 Human Spine anatomy [18]** | **Fig. 2 Cervical Spine x-ray plate [19]** |

The plate was attached to the anterior surface of the spinal column, stabilizing and compressing the vertebrae surrounding the bone graft (Fig. 3). Finite Element Method was used for the numerical analysis. In this case, the present work shows comparative results using the same boundary conditions, using two different modeling techniques, 3-D Computed Tomography Scan and a 3-D Scanner, also a concise descriptive technique used on the developing of the models.

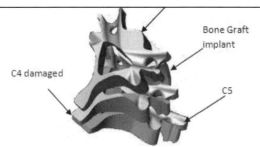

Fig. 3 CAD and FEM C3-C5 human model [8, 9]

3.0 Materials and Methods

In order to reproduce the numerical models, an instrumented porcine cervical specimen with a bone graft and a cervical titanium plate was considered. Besides, a Scanner Z-Corp 700 and a CT Scanner Philips Brilliance was used in this process. In addition 3-D software as Pro-E Wildfire 4, ScanIP 3.1, UGS NX4, Geomagics R.10 and ANSYS R.10 were the base to create and simulate each numerical model. In this sense, it was necessary to consider the differences between the materials which are implied on the cervical porcine specimen, as in the figure 3 are shown.

3.1 Modeling the cervical specimen using an Scan ZCorp 700

A 3-D Scanner ZCorp 700 was used to model the cervical specimen. It allows digitalizing specifically irregular geometries as surfaces and the possibility to print quick prototypes. In this case, the specimen represents an ideal complexity in order to reproduce the loading conditions in the numerical model in order to evaluate specific loading conditions. Some advantages over the classic 2-D CAD are the easily operation of the hardware and software. high resolution to get porosity and details in organic specimens, specifically soft tissues and the structure of the bone, and the possibility to process STL (Stereolithography) files as universal format and to be processed as a faceted solid or shrink-wrapped. Furthermore, all the instructions suggested by ZCorp for the preliminary preparations for the specimen to be scanned are described in next points:

a) Calibration of the color of the specimen. It is preferred to use matte uniform colors.(Figure 5) [10]
b) Determination of the ideal distance to acquire data.
c) Level of noise in the data (e.g. dust).

Fig. 5 Porcine Spine model with matte finishing.

There were used two methods to digitalize the specimen. The first one required a reference system in which the specimen was placed on (figure 6-a) and the second allowed to use reflective targets. A reflective effect is required on the scanning process and it was necessary to avoid that the laser ray of the scanner interacts over metallic surfaces with no painting cover, in this case in the inclusion of a titanium cervical plate (Figure 6-b). The influence of the deposition of powder on the geometry is barely perceptible (0.017 mm / 0.0007 in). **[11]** In figure 6, the 3-D Reference system allows to scan main surfaces after to colors have been calibrated.

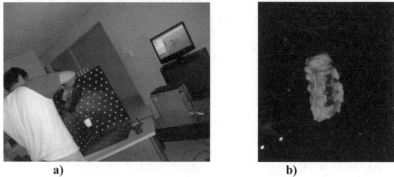

a)   b)

Fig. 6 Three Dimensional Reference System to scan vertebral model and

However, a constant scanning over the surface of the specimen was necessary in order to keep the scanner laser into the reference system. It allows avoiding the inclusion of objects with similar colors in the scanning session. Another alternative to scan different kind of tissues is by the covering using diverse colors, as in the figure 7 is shown. In this figure, the porcine spine specimen **[10]** was digitalized and some of the surfaces had to be filled.

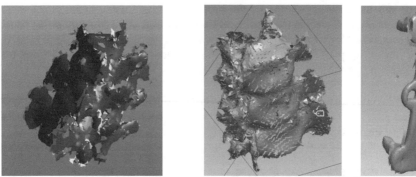

a) STL Surface   b) Faceted model (Shrink-wrapped)

Fig. 7 Partial view of the specimen with ZScan 700

3.2 Comparison between the Scan IP 3.1 Software and Computer Tomography

An ideal advantage of this method is the possibility to recreate in a digital format an in vivo specimen. In this case, for the medical team it is very important to interact with a precise and accurate model of cervical C3-C75, which was developed using computed tomography (CT) images and Scan IP 3.1 from DELCAM manufacturing. The construction of the detailed model included for each vertebra the anterior arch, odontoid process, superior and inferior articular facets, transverse processes, posterior arch and spinous process, considering a mesh with 3-D 8-node solid elements (solid 95). The scanning of the cervical specimen was done in a CT Scanner Philips Brilliance at "Hospital Regional 1° de Octubre" including 126 CT images (separately each one 0.35 mm) to be processed on ScanIP Software. Figure 8 shows an in vitro porcine specimen used in this study. Different tissues integrate the cortical and trabecular bone.

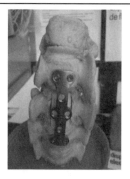 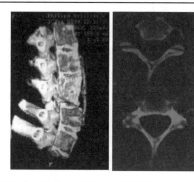

(a) In vitro porcine cervical specimen (b) Sagital and Lateral CT porcine cervical
Fig. 8 Computer Tomography image for porcine cervical specimen

Once the 126 CT images were captured by the CT scanner, the output files (*.DICOM) had to be processed on ScanIP Software using the boundary of each density zone being clarified by diverse intensity color. It was necessary to assign different colors for each component in order to identify the elements detected by the Scanner. It is very important to mention that organic and non organic elements were integrated at the same session without any kind of covering as in the last method explained. Once the boundary was closed, the rest of the area was filled automatically. It represents an advantage in order to get the precisely profile of each section and globally for the cervical specimen.

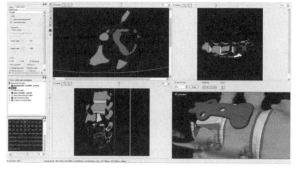

Fig. 9 ScanIP sketching – modeling process using CT image. [14]

Regardless of consider 0.35 mm between each CT image, some details were not possible to be developed efficiently, as in the figure 10 is shown. Specifically in the head of the screws (3317 Ortosintese) [13]. In base of the analysis of the specimen in the figure 10, the assigned colors for each part of the specimen are as follows: red for the cortical bone, green for the bone graft, blue for the cervical titanium plate and screws, yellow for the intervertebral disks and purple for the end disk laminate. It represents a useful tool in order to assign specific mechanical properties on FEM Software.

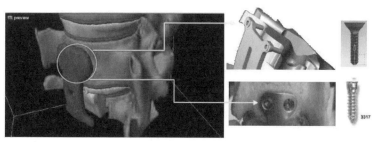

Fig. 10 Comparison between the CT image, CAD model and in vitro specimen.

The resultant file after the full 3-D model has been developed is a STL (Stereolithography format), which can be processed as parametric format (Faceted model) in diverse software as Pro-Engineer 4.0 Wildfire, UGS-NX 5 or Geomagics R10. In the figure 11 the resultant 3-D model using de described technique is shown.

Fig. 11 Optimized CT – 3D Porcine Cervical Model.

4.0 Analysis of the porcine cervical models

The analysis of the advantages and disadvantages of the described methodologies, considers the easily to process the cervical specimen, the time to develop it, the affinity to the in vivo model and the number of tetrahedric elements in the meshing FEM process. (95782 elements on 157633 nodes for the classic CAD model [1,8], 543362 elements on 891106 nodes for the ZScan 700 method [This work], and 797871 elements on 1313094 nodes for the CT image 3-D and ScanIP [This work]). In all cases the specimen had to be assembled and the nodes of each element had to be shared in order to transfer the compression loading from the upper face of the first cervical and the lower face of the last. It is very important to mention that the most affinity to the in vivo specimen was found on the model processed with CT image 3-D and ScanIP. However, the number of elements increases considerably in comparison with the first models. Besides, the time to process each model depends on the ability to process in AutoCAD R17, Mechanical Desktop 6.0 [1,8] 30 2-D tomographic slides, while in the last methods, for the Classic CAD, while for the ZCorp and CT image 3-D the process is optimized throughout the inclusion of ZScan – Pro-Engineer 4.0 Wildfire and ScanIP 3.1 – Pro-Engineer wildfire 4.0, Unigraphics NX – 5 and Geomagics R10.

5.0 Boundary and loading conditions

In order to accurately simulate the load transfer between the vertebrae, the bone graft and the cervical plate, it is highly important that the nodes, where the load is applied and transferred, coincide with each other. This validation was done on [1, 8]. However, in the figures 12 and 13 the boundary and loading conditions used for each numerical specimen are indicated. The material properties considered for the three models are shown on the table 1. Each one of the finite element model was analyzed under the same boundary and loading conditions that the Classic 2-DCAD in order to evaluate the mechanical behavior.

Table 1. Material properties for the numerical analysis [12]

	Property	Value	Units	Reference
CORTICAL BONE	Young Module	12	[GPa]	
	Poisson Ratio	0.2		
BONE GRAFT	Compression Stress Strength	5 -15	[MPa]	[12]
CERVICAL PLATE AND SCREWS 3317/04 OS (Ortosintese Titanium Ti6A14V)	Young Module	102	[GPa]	
	Poisson Ratio	0.3		
	Yielding Stress Strength	827	[MPa]	

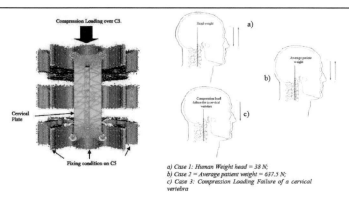

a) Case 1: Human Weight head = 38 N;
b) Case 2 = Average patient weight = 637.5 N;
c) Case 3: Compression Loading Failure of a cervical vertebra

Fig. 12. Boundary conditions [1, 8].   **Fig. 13. Loading Conditions [1, 8]**

6.0 Results

The main results about numerical testing on each specimen are reported on the table 2. It is important to take in count that preliminary publications about this research validated the surgery solution applied on the cervical range C3-C5 (Clinically called as Corporectomy), which allows to stabilize the instrumented zone after the fusion between the bone graft and the upper and lower surfaces of C3 and C5 respectively. It means that the bone graft had no significant displacements between its upper and lower surfaces. This is in accordance to the Muller criteria [15] which established that any cervical prosthesis should not be more than 3 mm in order to be considered a stable human spine after the surgery. Also the boundary conditions of the testing were described on [1, 8, and 9]. The results considering the loads, boundary conditions and FEM were solved with ANSYS 10.

Table 2. Numerical results comparison between CAD - ZCorp Scan and CT image models

	Case of Study	Location	CAD-Specimen	Zcorp – 3-D Scan model *	CT image 3-D Model	Notes
Case 1	Maximum von Mises - Stress	The centre of the cervical plate	70.26 Pa	68.9 Pa	66.46 Pa	* For these specimens, only the surface was created on the Scan session. The soft tissue elements were created on Pro-E 4.0 W.F
		Around the hole of the screws	316.2 Pa	303.7 Pa	305.3 Pa	
	Maximum Displacement between the bone graft and upper and lower C3 and C5 vertebrae	Interior surface of C3, superior surface of C5 adjacent to the surfaces of the bone graft	4.752 e-4 mm	4.397 e-4 mm	3.487 e-4 mm	
Case 2	Maximum von Mises - Stress	The centre of the cervical plate	662.8 Pa	648.7 Pa	635.7 Pa	[1,8]
		Around the hole of the screws	3351.7 Pa	3329.6 Pa	3312.2 Pa	
	Maximum Displacement between the bone graft and upper and lower C3 and C5 vertebrae	Interior surface of C3, superior surface of C5 adjacent to the surfaces of the bone graft	0.0054 mm	0.0042 mm	0.0039 mm	
Case 3	Maximum von Mises - Stress	The centre of the cervical plate	7.025 MPa	6.96 MPa	6.92 MPa	[1,8, 14]
		Around the hole of the screws	328.2 MPa	317.8 MPa	302.6 MPa	
	Maximum Displacement between the bone graft and upper and lower C3 and C5 vertebrae	Interior surface of C3, superior surface of C5 adjacent to the surfaces of the bone graft	0.0546 mm	0.0483 mm	0.0411 mm	

Conclusions

Engineering evaluation of the medical solutions on the orthopaedic area allows knowing the Biomechanical behavior of the involved structures. [17] In this case, the prosthesis of bone graft contributes to the human stability after the fusion process between the vertebral bodies of C3 and C5. The results of the numerical simulation are in accordance with experimental testing. This is one of the most recommendable methodologies that medics and hospitals have approved. Besides, the use of porcine specimens taking in count similarities between human vertebrae, specifically in mechanical properties; allows validating numerical models in order to study the mechanical behavior prior to be tested in human patients. Finally, the results of the table 2 show a convergence between the critic points using diverse modeling techniques. In each case, the von Misses stresses and the maximum displacements are lower for the CT image 3-D Model than the CAD and ZCorp – 3-D Scanner. This compliance is in accordance to the process to conform each cervical model, the accuracy of each method, the quality of each mesh for the numerical model and the similarity to the in vivo specimen, because of the inclusion of the bone density in the Computed Tomography. Therefore, the results were validated in base of experimental testing and the analysis and approbation of the medical team, in this case, specialists on spine surgery which are responsible to fix each component and support of the prosthesis.

Acknowledges

The authors kindly acknowledge the grants given by the National Council for Science and Technology, CONACYT, Mexico and the National Polytechnic Institute and the support given to this project by ISSSTE Hospital 1° de Octubre for the means and facilities for the development of this research. Project No. (IPN) SIP20091599. We also acknowledge the invitation to the 2010 BSSM International Conference on Advances in Experimental Mechanics and especially to the organizing Committee.

References

[1] J.A. Beltrán-Fernández & et. Al, "Modelling of a Cervical Plate and Human Cervical Section C3 – C5 under Compression Loading Conditions Using the Finite Element Method, Transtech publications "Applied Mechanics and Materials", Vols. 13-14-2008 p.p. 49-56, UK.

[2] Huiskes R, Chao EYS. A survey of finite element analysis in orthopaedic biomechanics: the first decade. Journal of Biomechanics 16: 385-409, 1983.

[3] Prendergast PJ. Finite element models in tissue mechanics and orthopaedic implant design. Clinical Biomechanics 12: 343-366, 1997

[4] Holzpafel GA, Ogden RW, Eds. Mechanics of Biological Tissues. Springer, Berlin, 2006

[5] Lally C, Kelly DJ, Prendergast PJ. Stents, In Wiley Encyclopedia of Biomedical Engineering, (Ed.: M.Akay), John Wiley & Sons, Vol. 5, pp. 3345-3355, 2007.

[6] Van der Meulen MCH, Huiskes R. "Why mechanobiology?" A survey article. Journal of Biomechanics 35: 401-414, 2002.

[7] Panjabi M M; Pelker R R; Friedlaender G E; Markham T C;; Moen C J Effects of freezing and freeze-drying on the biomechanical properties of rat bone. Journal of orthopaedic research: official publication of the Orthopaedic Research Society 1984; 1(4):405-11.

[8] Beltrán-Fernández JA & et.al., "Mechanical Behavior of a Calcium Phosphate Ceramic Bone Graft used in the Rehabilitation of a C4 Human Vertebra", published by Transtech in "Applied Mechanics and Materials". 5th BSSM International Conference on Advances in Experimental Mechanics, University of Manchester, UK, 4-6 September 2007.

[9] J. A. Beltrán Fernández, L. H. Hernández Gómez, G. Urriolagoitia Calderón, M. Dufoo Olvera and A. González Rebatú, "Distribución de esfuerzos por la acción de cargas de compresión en la vértebra cervical C5, empleando el Método del Elemento Finito" Científica, Vol. 9, núm. 3, Julio-Septiembre del 2005, ISSN 1665-0654, 135-142.

[10] R. Rodríguez-Cañizo, L. Hernández-Gómez, J. Beltrán-Fernández, G. Urriolagoitia-Calderón, M. Dufoo-Olvera, Análisis biomecánico de un disco intervertebral porcino lesionado - estudio experimental, 8° CONGRESO IBEROAMERICANO DE INGENIERÍA MECÁNICA, (Cusco, Perú), 23 al 25 de Octubre, (2007).

[11] Zcorp, Self-Positioning Handheld 3D Scanner, Method Sheet: Scanning Multi-colored or Multi-shaded Part, Document: MS020016 (2007).

[12] J.A Beltrán-Fernández, Análisis numérico de las cervicales C3 – C7 asociado al problema del latigazo cervical, Tesis de Doctorado, Sección de Estudios de Posgrado e Investigación, ESIME Zacatenco, México. (2007)

[13] Ortosintese (2006), Catálogo de productos 2006, Rua Friedrich Von Voith, 896 – Jaragua, Brasil.

[14] Ruíz-Muñoz, E.R, Beltrán-Fernández, J.A & Et al, "Técnicas de modelado en 3D aplicado a casos de vértebras porcinas por medio de un escáner 3D y tomografías.", XXV Congreso Nacional De Investigación Biomédica, Universidad Autónoma de Nuevo León, Facultad de Medicina, 2009.

[15] Allen, B., Ferguson, R., Et. al. (1982), A mechanistic classification of closed, indirect fractures and dislocations of the lower cervical spine, J. of Spine, vol. 7, p.p. 1 – 27.

[16] J.A. Beltran-Fernández, L.H. Hernández-Gómez, G. Urriolagoitia-Sosa, A. González-Rebatú, G. Urriolagoitia-Calderón, (2009) Biomechanical evaluation of a C3-C5 human cervical model created by computer tomography CT and 3-D scan under compression loading. 3rd International Conference on Advanced Computational Engineering and Experimenting (ACE-X-2009), Grand Hotel Palazzo Carpegna, Via Aurelia, 481 - 00165 (Rome, Italy). Organised by The Italian Association for Stress Analysis/ Department of Applied Mechanics / Seconda Universita' degli Studi di Napoli / Prof. Dr.-Ing Andreas Öchsner (chair), 22-23 June - 2009

[17] Ruíz-Muñoz, ER1, Beltrán-Fernández, JA, Rodríguez-Cañizo, RG, Hernández-Gómez LH, González-Rebatú A, (2009) "Análisis de la estabilización segmentaria en especímenes de columna lumbar porcina L3-L5, con cerclaje interespinoso empleando collarines de poliamida 6/6". IX CONGRESO IBEROAMERICANO DE INGENIERIA MECANICA, CIBIM 09. FEIBIM – Federación Iberoamericana de Ingeniería Mecánica, Escuela Técnica Superior de Ingenieros Industriales de la Universidad de las Palmas de Gran Canaria, España. 17-20 Noviembre, 2009.

[18] Bland JH. Disorders of the cervical spine, diagnosis and medical management. 2nd ed. Philadelphia: WB Saunders; 1994.

[19] Information on http://www.shutterstock.com/pic-31223248-x-ray-image-of-cervical-vertebra.html

Session 7: Manufacturing Processes

Chapter 17 Manufacturing Process

Applied Mechanics and Materials Vols. 24-25 (2010) pp 299-304
© (2010) Trans Tech Publications, Switzerland
doi:10.4028/www.scientific.net/AMM.24-25.299

Spot weld strength determination using the wedge test: in-situ observations and coupled simulations

Rémi Lacroix[1,a], Joël Monatte[1], Arnaud Lens[2,b], Guillaume Kermouche [3],
Jean-Michel Bergheau[3], Helmut Klöcker[1,c]

[1]Ecole des Mines de Saint Etienne, 158 cours Fauriel, 42023 Saint Etienne, France

[2] ArcelorMittal Maizières Voie Romaine, BP30320, 57283 Maizères-les-Metz, France

[3]Université de Lyon, ENISE, LTDS, UMR 5513 CNRS, 58 rue jean Parot,
42023 Saint-Etienne Cedex 2, France

[a]lacroix@emse.fr, [b]arnaud.lens@arcelormittal.com, [c]klocker@emse.fr

Keywords: Spot welds, High Strength Steels, wedge test, in-situ.

Abstract. This paper describes an innovative way to characterize the strength of spot welds. A wedge test has been developed to generate interfacial failures in weldments and observe in-situ the crack propagation. An energy analysis quantifies the spot weld crack resistance. Finite Element calculations investigate the stresses and strains along the crack front. A comparison of the local loading state with experimentally observed crack fronts provides the necessary data for a failure criterion in spot weld fusion zones. The method is applied to spot welds of Advanced High Strength steels.

Introduction

New steels are continuously developed in order to reduce the weight of cars, and thus their fuel consumption. The main technology used to assemble these steels is resistance spot welding. The recent development of Advanced High Strength (AHS) steels has made the determination of their weldability a crucial industrial issue. The Cross Tensile (CT) test, illustrated in figure 1.a, is widely used to evaluate the weldability of steel grades. During the CT test a crack propagates from the notch towards the weld nugget (figure 1b). Crack deviation leads to the formation of a plug. The acceptability of spot welds after CT test has been empirically related to a minimum plug diameter so defining an acceptability criterion [1] in the case of mild steels.

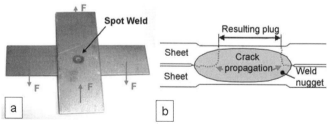

Figure 1 – (a) a spot weld and the CT test configuration, (b) a schematic representation of a cross section of weld with the resulting plug after failure.

The theoretical strength of spot welds in AHS steels submitted to the CT test has been frequently studied [2,5]. However, the experimental results of Oikawa et al [6] show that the CT-test does not lead to reproducible results for AHS steels. Moreover, experimental studies by Nait Oultit et al. [7] have highlighted inconsistencies in using the minimum plug diameter as an acceptability criterion for AHS steels. This criterion has been rejected for AHS steels by the World Auto Steel Guidelines [8].

In this study, we aim at obtaining reproducible data on the local stress and strain fields in the weldment, and at reliably quantifying the fracture properties of fusion zones. A test highlighting the properties of the fusion zone rather than the rest of the weld assembly has been developed. Reproducible fracture of the weldment through the interfacial plane has been achieved by a wedge test. The energy release rate during crack propagation has been quantified by a macroscopic analysis. A local analysis of the crack front correlated with a Finite Element model of the test investigates stresses and strains at failure. More generally the fracture properties of the fusion zone determined in this work are useful for structural integrity calculations of welded assemblies.

The Wedge Test

The Wedge Test is developed in order to allow in-situ observation of crack propagation in the weldment. It also implies an interfacial failure under limited sheet bending, such that most of the energy is used to fracture the weld rather than bend the sheets. This paragraph describes the mechanical set up of the test, the instrumentation and the automatic processing of the in-situ observations.

The test is developed for welded sheets of 2 mm thickness. A wedge test sample is obtained by cross-sectioning a spot weld, as specified on figure 2.a. The sample thus exhibits two notches and a microstructure ranging from the base material to the fusion zone, as shown in the micrography on figure 2.c. This cross section surface will be observed during the wedge test. Cross sectioning is not made exactly in the weld median plane, but is slightly offset to leave a small joint between the sheets.

One side of the sample is clamped, and a 90° wedge is driven in between the welded sheets on the opposite side, as shown in figure 2.d. The wedge is pushed along the interfacial plane of the weld assembly.

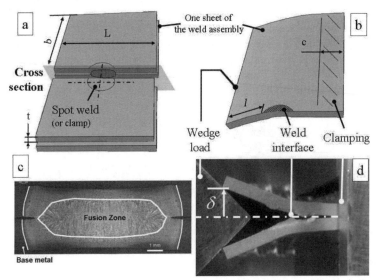

Figure 2 – (a) The wedge test sample sectioning (L=30 mm, b=40 mm, t=2mm, l=7mm, c=10 mm) (b) Mechanical loading of one sheet of the weld assembly. (c) A DP780 spot weld micrography highlighting the main weldment zones, and (d) mechanical set up of the wedge test.

The clamping ensures that the sample interface is parallel to the direction of the wedge displacement. A brushless motor applies a wedge speed of 0.01 mm/s. On the wedge crosshead guide, a Linear Variable Differential Transformer (LVDT) sensor measures the applied displacement, and a load cell records the applied load up to 10kN.

The clamping and the weld interface induce bending of one sheet of the weld assembly with a non-uniform lever arm during the wedge load, as illustrated in figure 2.b.

The weld cross-section is observed by a CDD camera which records 1280x1024 pixels grey images at a frame rate of 2s-1. A 12.5:1 optical system allows the observation of a zone of 5.5x4.4 mm². The observed surface is previously sand blasted in order to generate a speckle-like pattern, of average diameter 30μm. The recording process has been developed with the LabVIEW TM software, and simultaneously acquires the load and displacement with the associated image of the observed surface.

A Digital Image Correlation software developed in-house gives the displacement field between successive images. The crack tip is continuously located by fitting the local displacement field for zero vertical displacement. The accuracy of this measurement is 15 pixels, or 65 μm.

Local quantification of the crack resistance of the weld requires a knowledge of the complete crack front profile. Surface observation on the cross section however gives very limited information on the crack front. The complete crack front profile has been observed on samples partially fractured by interrupted Wedge Tests. These samples were later fully fractured under liquid nitrogen cooling, such that a brittle failure occurs in the remaining joint. The profile of the crack front after the interrupted wedge test is the limit between the all-brittle zone and the mixed (brittle and ductile) zone. The limit between both zones can be easily observed by SEM.

Fracture analysis

Various studies of the wedge test have produced macroscopic analyses of the test, particularly in adhesive [9,10] or brazed joints [11]. However, these tests are all close to a two dimensional configuration, and well suited to an elastic Double Cantilever Beam (DCB) analysis, as described by Kanninen [12]. The variation of the local loading state along the crack tip is neglected in most studies [9,11]. In the case of spot welds, the wedge test has a fully three dimensional geometry. A macroscopic analysis of the wedge test is first investigated, and then a 3D FE analysis is carried out to investigate the stresses and strains along the crack front.

Macroscopic analysis. A macroscopic analysis of the wedge test is based on the load - displacement curve. Figure 3.a illustrates these data recorded for a spot weld of Dual Phase 780 (DP780) steel. The spot welds are loaded in 4 successive cycles. The displacement $u_{start}^{(i)}$ at the beginning of the ith cycle gives information about the energy dissipated during the previous cycles (1,..i-1). We assumed in this study that this energy is dissipated only in the sheet plastic bending and in the weld failure.

The elastic energy $W^{e(i)}$ stored in the sample during cycle (i) can be determined from the load - displacement curves. The linear part of the loading cycle n° i gives $W^e(i)=P.(u_{end}^{(i)}-u_{start}^{(i)})/2$, P and $u_{end}^{(i)}$ being the load and displacement at the end of the linear behaviour, $u_{start}^{(i)}$ the displacement at the beginning of the linear behaviour, as illustrated on the figure 3.a for one cycle. Assuming a linear evolution of the displacements u_{start} between the measured locations for two cycles ($u_{start}^{(i)}$ and $u_{start}^{(i+1)}$) it is possible to estimate W^e for any displacement u of the wedge : $W^e(u)=P(u-u_{start})/2$.

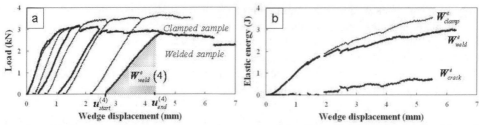

Figure 3 – (a) Load - displacement curve for repeated loading of a DP780 spot weld, and for clamped DP780 steel sheets. The hatched area corresponds to the elastic energy. (b) Elastic energy stored in these two samples.

In a second test, the (elastic) energy (W^e_{clamp}) stored during deformation of the sample without crack propagation is determined. A specific sample made out of the same material with the same geometry is used. One bolt clamps two steel sheets in the same manner as the weld interface joins the sheets in the welded sample. A comparison of the energy W^e_{clamp}, stored in the sample without crack propagation, with that of the energy in the sample with crack propagation (W^e_{weld}), allows an estimation of the energy released during crack propagation (W^e_{crack}).

This analysis neglects some small sliding between the clamping bolt and the sheets. Figure 3.a shows a typical result of this test. W^e_{crack} is thus estimated by $W^e_{crack} = W^e_{clamp} - W^e_{weld}$, as illustrated in figure 3.b.

Finite Element analysis. The FE model of the Wedge Test has been carried using standard ABAQUS according to the procedures described in the manual [13], and is shown in figure 4. This model investigates the stresses and strains along the crack front before propagation, as crack propagation is not yet implemented. The weld interfacial plane is a plane of symmetry. Thus only one sheet is studied, assuming it is clamped in the weldment interface and the sample clamping (figure 2.b). The wedge movement into the sample is modelled assuming a frictionless hard contact. The current model is limited to two zones: the base metal and the fusion zone. The latter is extended through the HAZs to meet the base metal zone. The mechanical behaviour of the base metal and the fusion zone have been characterised independently by tensile and compressive tests, as described in [14]. The mesh is made of 7400 C3D20R elements, with an element edge size of about 150 µm at the weld interface.

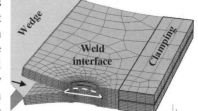

Figure 4 – FE model of the test

Results

The wedge test has been applied to DP780 steel spot welds. Figure 3.a illustrates the load evolution and Figure 5.c the recorded images during a test. The load increases quasi-linearly until a crack appears, and grows up to full interfacial failure. After weld failure, the load drops abruptly and then stays roughly constant during bending of the clamped sheets.

Macroscopic analysis. Figure 5.a illustrates the crack advance based on the in-situ observations, as described above. The crack has advanced 7.4 mm when full interfacial failure occurs, corresponding to the distance between the two notches on the observed surface.

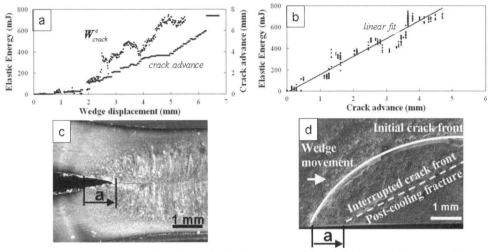

Figure 5 – (a) Crack advance and estimated elastic energy released during crack propagation as a function of the wedge displacement. (b) Estimated elastic energy release as a function of crack advance. (c) In-situ observation of crack advance a. (d) Weld interface after fracture on liquid nitrogen showing the initial and current crack front.

The evolution of the estimated energy W^e_{crack} as a function of the crack advance is plotted in figure 5.a. A linear fit of W^e_{crack} gives an estimate of the derivative dW^e_{crack}/da, the energy release rate on the crack front, as shown in figure 5.b. It is nevertheless very difficult to conclude on the energy release rate per unit crack front length. Considering the geometry of the test, the loading state along the crack front is not expected to be perfectly homogeneous. However, a first order assumption of homogeneity leads to $G = 1/c.dW/da = 34\ kJ/m^2$, for a crack front length c of 5 mm at crack initiation. This is also equivalent to a mode I toughness $K_{Ic} = \sqrt{G.E} = 84\ MPa\ \sqrt{m}$, with E the Young modulus of 210GPa.

Stresses and strains along the crack front. Observations of the crack front on an interrupted wedge test are illustrated in the figure 5.c and 5.d. The original circular crack front is shown by the solid line and the interrupted crack front by the dashed line. The latter is straight and makes an angle between 60 and 70 degrees with the direction of wedge movement. The crack front length varies between 0 and 5 mm during crack propagation.

The loading state along the crack front for a wedge displacement of 1 mm has been evaluated by the model. Figure 6.a illustrates respectively the mesh of the model in the weld interface, and figures 6.b, 6.c and 6.d, illustrate the maximum principal stress, the stress triaxiality and the plastic deformation at the weld interface.

A qualitative comparison between the simulated stresses and strains along the original crack front with figure 5.d shows that a failure criterion based on the maximum principal stress would be more consistent with the observed crack propagation, than a deformation based criterion.

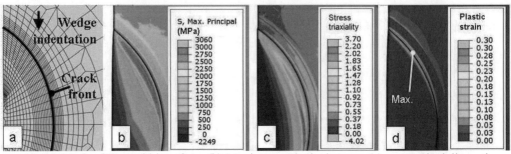

Figure 6 – (a) Mesh of the weld interface. (b) Maximum principal stress, (c) stress triaxiality ratio, (d) equivalent plastic strain in the weld interface.

Conclusions

The wedge test is able to load a spot weld in a quasi-rigid way up to crack initiation and propagation through the weld interface. A simple macroscopic analysis of the test allows the energy released during crack propagation to be estimated. This model has qualitatively shown that a failure criterion based on the maximum principal stress would generate a crack front of a similar profile to the experimentally observed ones.

Further investigations will adapt the FE model to the real crack geometry. The stresses and strains along the evolving crack front during the test could thus be evaluated. The determination of a specific volume of the fracture process zone will finally provide quantitative information for an appropriate failure criterion. The observation of the crack front by means of the interrupted test is a major requirement of this method.

References

[1] NF EN ISO 18278-2, Resistance welding — Weldability.

[2] H Lee, J Choi, Mechanics of Materials 37, 19–32, 2005.

[3] DJ Radakovic, M Tumuluru, Welding Journal 87, 96s-104-s, 2008.

[4] MN Cavalli et al., Fatigue Fract Engng Mater Struct 28, Issue: 10, 861-874, Oct. 2005.

[5] S Dancette et al., New Developments on Metall and Apps of HS steels Buenos Aires 2008.

[6] Oikawa H. et al., Nippon Steel Technical Report No. 95 January 2007.

[7] B Naït Oultit, A Lens, H Klöcker, SMW Conference XIII, AWS Detroit Section, 2008.

[8] AHSS Application Guidelines - Version 3, World Steel Association - September 2006.

[9] J Cognard, J Adhesion 22, 97-108, 1987.

[10] JP Sargent, Int J Adhes Adhes 25, 247-256, 2005.

[11] NR Philips, MY He, AG Evans, Acta Materiala 56, 4593-4600, 2008.

[12] MF Kanninen, Int J Fracture 9, No 1, 83-92, 1973.

[13] ABAQUS, Hibbitt, Karlsson and Sorensen, 2000.

[14] R Lacroix, A Lens, JM Bergheau, G Kermouche, H Klöcker, CFM Conference XIX, 2009.

Applied Mechanics and Materials Vols. 24-25 (2010) pp 305-310
© *(2010) Trans Tech Publications, Switzerland*
doi:10.4028/www.scientific.net/AMM.24-25.305

Improving fatigue performance of alumino-thermic rail welds

M. Jezzini-Aouad[1, 2, a], P. Flahaut[1, 2, b], S. Hariri[1, 2, c], D. Zakrzewski[1, 2, d] and L. Winiar[3, e]

[1]Univ. Lille Nord de France, F-59500 Lille, France

[2]EMDouai, MPE-TPCIM, F-59508 Douai, France

[3]Railtech International, Zone industrielle du Bas Pré, 59590 Raismes, France

[a]jezzini@ensm-douai.fr, [b]flahaut@ensm-douai.fr, [c]hariri@ensm-douai.fr, [d]zakrzewski@ensm-douai.fr, [e]lwiniar@railtech.fr

Keywords: alumino-thermic weld / residual stress / fatigue / microstructure / rail.

Abstract. Rail transport development offers economic and ecological interests. Nevertheless, it requires heavy investments in rolling material and infrastructure. To be competitive, this transportation means must rely on safe and reliable infrastructure, which requires optimization of all implemented techniques and structure. Rail thermite (or alumino-thermic) welding is widely used within the railway industry for in-track welding during re-rail and defect replacement. The process provides numerous advantages against other welding technology commonly used. Obviously, future demands on train traffic are heavier axle loads, higher train speeds and increased traffic density. Thus, a new enhanced weld should be developed to prevent accidents due to fracture of welds and to lower maintenance costs.

In order to improve such assembly process, a detailed metallurgical study coupled to a thermomechanical modeling of the phenomena involved in the rail thermite welding process is carried out. Obtained data enables us to develop a new improved alumino-thermic weld (type A). This joint is made by modifying the routinely specified procedure (type B) used in a railway rail by a standard gap alumino-thermic weld. Joints of type A and B are tested and compared. Based on experimental temperature measurements, a finite element analysis is used to calculate the thermal residual stresses induced. Besides, experimental investigation was carried out in order to validate the numerical model. Hence, X-Ray diffraction has been used to map the residual stress field that is generated in welded rail of types A and B. In the vicinity of the weld, the residual stress patterns depend on the thermal conditions during welding. Their effect on fatigue crack growth in rail welds is studied. In the web region, both longitudinal and vertical components of residual stresses are tensile, which increases the susceptibility of that region to crack initiation and propagation from internal material defects. Indeed, weld fracture in track initiates at the web fillet. Thus, to be closer to real issue, fatigue tests specimens has been defined within the split-web area. Fatigue tests was performed on the defined specimens, welded by conventional and improved processes and obtained results adjudicates on the new advances.

Introduction

Rail joints are the weakest units in the track structure. Railway rails are made in various length up to about 100m. In the past the most common method of joining sections was by bolting. In traditional bolted-joint track, there is differential movement of the running surface at the rail ends that creates severe wear and batter conditions under train traffic [1,2]. The dynamic interaction between wheels and rail joints not only contributes to the rail end damage, but also degrades the cross ties, fasteners, ballast and even the subgrade. Wherever possible, rails are now joined by welding which is a cheaper process and produces a superior joint. Probably, the most common technique for joining the rail in track on-site, for various reasons including easier alignment and other on-site factors (simplicity, portability and economy), is alumino-thermic welding [2,3]. A large number of alumino-thermic rail welds are made in track each year using standard procedures and a very high proportion are reliable [3], however, some do fail. Hence, this process should continue to be studied for its optimization and improvement. Different residual stress patterns generated by alumino-thermic welding may contribute to the failures and they must be controlled. Compressive residual stresses are generally beneficial in that it would inhibit the initiation and propagation of fatigue cracks from surface defects. In contrast, tensile stresses increase the susceptibility to crack initiation and propagation from internal material defects. Whether the residual stresses effect is beneficial or detrimental to the structure reliability, we will seek either to develop them or to avoid them.

Improving the behavior in service of welded track structure must start through the analysis and understanding of metallurgical, physical and mechanical phenomena involved in the welding process. Thus, this work focuses on the reliability of alumino-thermic welded rails. After an overview of the method used, we will try to understand the phenomena involved in the generation of residual stresses. These phenomena once mastered, it can then act on different parameters to obtain a distribution and levels of residual stresses that increase resistance to service. This improvement is visible through the results of fatigue tests performed. In this investigation, two types of thermite welds (A and B) are studied. The welding-type B made with the standard PLR process used in France and type A is one developed for this study.

Weld Fabrication

A full-scale laboratory thermite welding was conducted in the RailTech International laboratory (Raismes- France). The parent rail used in welding was a 60 kg/m rail (profil 60E1). A RailTech standard PLR thermite welding kits and procedures, using a one-shot crucible, were used in this study. The chemical analyses for the rail and weld metal are given in Table 1. Depending on the tests, two pieces of approximately 350 to 650 mm long new rail were lined up and placed on a support. The rails were carefully aligned, with the ends raised about 1 mm to allow drop on cooling, and the gap between them was adjusted to 25 mm. A standard mold was assembled and fitted, sealed with luting sand and preheated for 4-5 minutes (3 bars) using the induction air torch to ensure that the assembly was dry. After preheating, the loaded welding crucible was put over the mold and the thermite charge was ignited. The exothermic reaction produced caused the charge to melt and the liquid metal was poured into the weld cavity. After pouring, the welds were allowed to cool before the mould shoe, the weld top, and the risers were removed.

For the weld type A, a new step is added to the standard procedure. In fact, in this step the welded rail is subjected to rapid cooling in order to reduce the post-weld time elapsed before to reach 350°C (temperature at which the train is allowed to roll).

Specimen	C[%]	Si[%]	Mn[%]	Cr[%]	Al[%]
Weld	0,721	0,722	0,909	0,100	0,265
Rail	0,735	0,310	1,115	0,030	0,002

Table 1: Chemical compositions of parent rail and completed thermite weld (wt %)

Weld Microstructure

In order to reveal the weld microstructure and the extent of the melted zone (MZ) and heat-affected zone (HAZ), alumino-thermic welds were sectioned, polished and etched with 3% nital for metallurgical observation. Sections reveal that the MZ varies from 40 mm (head & foot) to 50 mm (web), with a 'vase' shape (Fig. 1). In fact, in the head and foot regions, where the rail sections are thickest, there is least melting of the adjacent rail metal. The MZ is surrounded by a heat-affected zone that varies from 15 to 25 mm. Samples were then examined using a scanning electron microscopy (SEM). Whatever the weld type is A or B, there are three successive zones starting from the weld centerline to the molten / base metal interface: a zone with equiaxed grains in size up to 1 mm, then a columnar region and finally, an area composed of small equiaxed crystals at the cold interface.

Indeed, when the temperature drops locally a few degrees below the liquidus temperature, solidification starts with nucleation of solid metal on various heterogeneities. Then, the weld solidification leads to pearlite lamellae formation with ferrite at grains boundaries (Fig. 2). These microstructures are responsible, at least partially, of the brittle fracture of welds [4,5,6]. Evidence of columnar growth that occurred during the cooling process is apparent in directions roughly from the boundary of the weld to near the centerline (Fig. 3). The MZ/HAZ boundary is strongly marked by a refinement of pearlitic grains at the HAZ (Fig. 3).

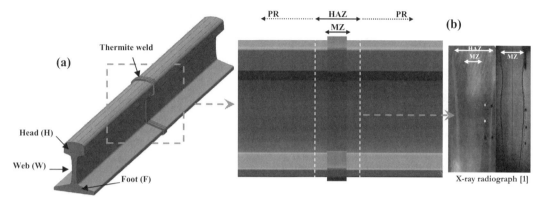

Fig. 1: (a) Geometry of a aluminothermite welded rails showing its Head, Web and foot; (b) Vertical central section of a thermite weld, polished and etched by Nital, showing the extent of the melted (MZ) and the heat-affected zones (HAZ) [1].

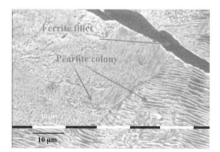

Fig. 2: A typical microstructure of thermite weld metal showing pearlite colonies and ferrite on the grain boundary

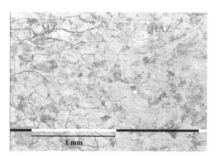

Fig. 3: Interface between weld metal (MZ) and the HAZ where pearlite structure is finer

Tensile Tests

Uniaxial tensile tests were performed on cylindrical specimens, machined from the rails head, web and foot, of alumino-thermic welds type A and B. In order to characterize the different zones produced by thermite welding, the melted zone (MZ), heat affected zone (HAZ) and parent rail (PR) within the tensile specimens were instrumented by strain gauges (Fig. 4).

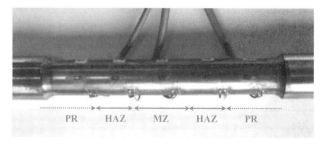

Fig. 4: Tensile test specimen within MZ, HAZ and PR instrumented by strain gauges

The mechanical properties are given in Fig. 5 and 6. It is noted that either for type A or B, the yield strength of parent metal (PR) is not homogeneous in the head (H), web (W) and foot (F) of the rail (Fig 1 (a)). This is may be due to manufacturing processes, in particular, rail kneading. The level of Young's modulus and the yield strengths of the HAZ and MZ for types A and B remained equivalent. These parameters are more pronounced in the HAZ compared to other zones. The yield strength decreases slightly in the MZ for type A compared with type B.

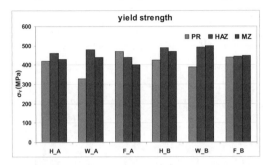

Fig. 5: Comparison between yield strength of welds types A & B. Where, σ_Y is determined for the MZ, HAZ and PR from tensile tests on specimens machined from the H, W and F of thermite weld

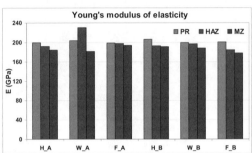

Fig. 6: Comparison between Young's modulus of welds types A & B. Where, E is determined for the MZ, HAZ and PR from tensile tests on specimens machined from the H, W and F of thermite weld

Residual Stresses

Within the melted zone (MZ), the heat-affected zone (HAZ) and the adjacent parent material where the thermal softening and thermal strains caused by the heat flow from the weld are sufficient to cause yielding during welding, the residual stress field are dominated by the weld-induced residual stresses. At greater distances from the weld, the residual stresses after welding are a function of the superimposition of the weld-induced residual stresses and any pre-existing residual stresses in the parts being joined. This superimposition may be in the linear elastic or nonlinear elastic plastic or creep range, depending on combined magnitude of the residual stresses and the mechanical properties of the parent material [7].

Residual stresses before welding may be caused by thermal or mechanical processes during materials manufacture or fabrication operations. Material or product manufacturing operations that cause significant residual stresses include casting, forging, rolling, heat treatments, quenching, straightening and carburisation.

X-ray diffraction technique is used to investigate the distribution of residual stresses across the alumino-thermic welded rail. In crystalline materials such as metals and their alloys, the constituent atoms are regularly arranged on the crystal lattice, which has dimensions precisely defined and characteristic for a particular alloy. The principle of diffraction is to measure the spacings between atomic planes in materials. When a beam of X-rays is incident upon a polycrystalline material the rays are scattered in specific directions defined by the Bragg equation (Eq. 1):

$$\lambda = 2d\sin\theta. \tag{1}$$

Where λ is the X-rays wavelength, d is the interplanar spacing and θ is the incidence angle. If the material is stressed, the lattice dimension is changed by δd. Measurements were made on thermite welded rails using parameters given in Table 2.

Cryst. Latt.	Plane {hkl}	$\lambda_{K\alpha}$	$2\theta_0$	Pen (ψ=0)
CC	{211}	Cr 2,29 [nm]	156,41°	5,8 [µm]

Table 2: Ferritic steel parameters used for X-ray diffraction

After welding and cooling, the residual stresses are determined at different locations and their distribution is compared to temperature evolution throughout the welding operation. To do so, holes were drilled in the rail at regular distances (50, 100 and 150 mm) from the weld centerline to accept thermocouple probes. K-type thermocouples were used for the rail temperature measurement. A National Instrument multi-channel system was used to collect data during welding and cooling. Typical temperature measurements in the rail head for types A and B are shown (Fig. 7). Similar temperature history curves were also obtained in the rail web and base. Distribution of longitudinal residual stresses is given in Fig. 8.

Within the MZ and the HAZ, residual stress field is dominated by the weld-induced residual stresses. These stresses are maximal at the MZ (35-35mm) and HAZ (50-50mm) where temperatures exceeded 650°C before cooling (Fig. 7, Fig. 8). The level of these stresses decreases moving away from the MZ.

Initial results show that levels, signs and distributions of stresses depend on the thermal history. For example, the compressive residual stresses are highest in welded rails where the temperature gradients were more pronounced (Fig. 7, Fig. 8). Fig. 8 shows obtained results for both types A & B, a higher compression stress level is observed for type A.

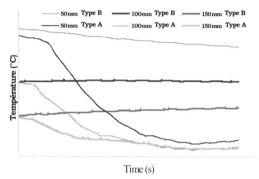

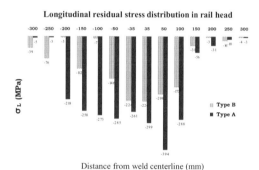

Fig. 7: Comparisons between measured temperature in the rail head of welds types A & B at distances of 50, 100 and 150 mm from the weld centreline

Fig. 8: Comparisons of residual stresses distribution in the rail head, determined using X-ray diffraction, between thermite welded rails of types A & B

Fatigue Tests

Fatigue tests were performed on type A and B welds in order to compare the two process and to verify whether type A provide improvements of the rails fatigue life or not. First, tests were carried out according to the European four point bending fatigue test draft standard EN 14730-1 [8]. Specimens were full scale 25 mm gap thermite rail welds, 1200 mm rail long, and applying the load on the running surface i.e. the rail head. In this case, the head is in compression and the foot is in tension. The fatigue fractures sites were located at the web-base fillet of the rail. Notwithstanding, observed in-track failures, for this study case, always initiate at the HAZ head-web fillet. Hence, to approximate in-track fracture conditions, fatigue tests were performed in reverse order, i.e. foot in compression and head in traction. Using this new method, failures initiate at the head-web fillet, which is similar to those obtained in-track.

Preliminary results performed with the reversed fatigue tests on full-scale specimens, support assumptions on the beneficial influence of compressive residual stresses on fatigue life. Welds of type B tested failed after about two million cycles where type A welds stand the test up to about ten million cycles.

The validation of a fatigue test result requires a large number of specimens. However, tests on full-scale specimens are very expensive, so it is necessary to set up a scale-down specimen. This sample should keep the main characteristics of a full-scale welded rail. Given that in-track failures initiate from the head-web fillet, the scale-down specimens are made by machining while keeping untouched this zone (Fig. 9).

Reversed fatigue tests were performed on scale-down specimens of types A and B. Results confirm those obtained on full-scale specimens. An improvement of about 30 - 40% is obtained for the welding-type A developed in this work.

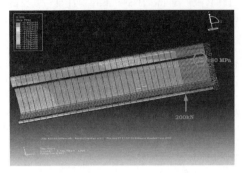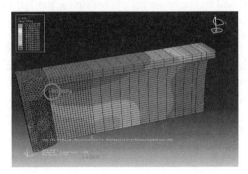

Fig. 9: Full-scale (left figure) / scale-down (right figure) specimens design basis using finite element method

Summary

This work aims to improve the service performance of alumino-thermic rail welds. A metallurgical and thermal study enabled to validate a new thermite welding process. The standard and new methods were compared.

The residual stresses associated with the two methods have been determined. Results show an increase of maximal longitudinal compressive stresses associated with the new process.

Results of fatigue four points bending tests performed, on full and scale-down specimens, confirm the improved resistance of the new weld type.

References

[1] Y. Chen, A heat transfer modelling study of rail thermite welding, PhD Dissertation, University of Illinois at Urbana-Champaign (2004).

[2] W.W. Hay, *Railroad Engineering*, edited by John Wiley & Sons, New York (1982).

[3] P.J. Webster, G. Mills, X.D. Wang, W.P. Kang and T.M. Holden: Journal of Strain Analysis Vol. 32 (1997), pp. 389-400.

[4] F.T. Lee: Welding Journal Vol. 85 n°1 (2006), pp. 24-29.

[5] L.C. Schroeder and D.R. Poirier: Proc. of Railroad Rail Welding (1983), pp. 21-59.

[6] J. Myers, G.H. Geiger and D.R. Poirier: Welding Journal Research Supplement Vol 61 (1982), pp. 258-268.

[7] R.H. Leggatt: Int. J. Pressure Vessels Piping Vol 85 (2008), pp. 144-151.

[8] EN 14730-1: *European standard* (2006).

Applied Mechanics and Materials Vols. 24-25 (2010) pp 311-316
© *(2010) Trans Tech Publications, Switzerland*
doi:10.4028/www.scientific.net/AMM.24-25.311

The Effect of Ultrasonic Excitation in Metal Forming Tests

S. Abdul Aziz[1,a], M. Lucas[1,b]

[1]School of Engineering, University of Glasgow, Glasgow G12 8QQ, UK

[a]s.aziz@eng.gla.ac.uk, [b]m.lucas@eng.gla.ac.uk

Keywords: ultrasonics; metal forming; ultrasonic forming tests

Abstract. The use of ultrasonic excitation of tools and dies in metal forming operations has been the subject of ongoing research for many years. However, the lack of understanding about the effects of ultrasonic vibrations on the forming process has resulted in difficulties in maximising the benefits and applications of this technology. In particular, experimental characterisations of the effects of superimposing ultrasonic oscillations have largely relied on interpretations of measurements of the mean forming load and have ignored the oscillatory forces. Previous research [1] has shown that by applying ultrasonic vibrations to the lower platen in compression tests on pure aluminium specimens, the resulting stress-strain relationship can be characterised by a temporary effective softening of the material during intervals of ultrasonic excitation. The current research investigates this effect in a series of simple forming tests using a number of different metal specimens. In this research, the forming tests are conducted using a piezoelectric force transducer to measure the oscillatory force data during ultrasonic excitation of the die. It is shown that the benefits of superimposing ultrasonic excitation of the die are highly dependent on the material being formed and that, in many cases, the maximum oscillatory force exceeds the static forming load even where the mean forming load is reduced significantly during the interval of ultrasonic excitation.

Introduction

Research in the use of ultrasonic vibration in metal forming and joining processes has been ongoing since the middle of the 20th century. In particular, for metal forming processes, extensive research has reported significant benefits of superimposing ultrasonic vibrations on the forming tools in terms of forming load reductions. In early studies [2] into the use of ultrasonic vibration, many researchers focussed on the influence of ultrasonic oscillations on the internal stresses during plastic flow of metal and also on interfacial friction effects, known as the volume and surface effects respectively. Many of these studies have been associated with the development of ultrasonic metalworking processes related to industrial applications such as die forming, wire drawing and extrusion [3]. In all of these studies, the evaluation of the benefits of ultrasonic excitation relied on measurements of the mean forming load only and not on measurement of the oscillatory force during ultrasonic excitation.

This paper presents the results of a simple forming test where a flat sheet sample of metal is forced into a shaped female die by a round nose male die on a test machine. The die forms part of a tuned ultrasonic horn, so that ultrasonic excitation can be applied during the tests and results can be compared with tests performed without ultrasonic excitation. The design and tuning of the ultrasonic horn is achieved using finite element modelling and experimental modal analysis using a 3D laser vibrometer. For the forming tests, the plunger is attached to the cross-head of the test machine and is also attached to a piezoelectric force transducer for measurement of the static-oscillatory forming force. The results of this study illustrate how ultrasonically assisted metal forming can result in a lowering of the mean forming load during ultrasonic excitation of the die and, further, investigates the oscillatory force during ultrasonic forming.

Design of the Ultrasonic Die

The ultrasonic excitation system is shown in Fig. 1 and consists of a Langevin piezoelectric transducer, a booster and an ultrasonic horn, all tuned to their first longitudinal mode of vibration at 20 kHz. The tuned booster was included to allow a flange to be incorporated between the transducer and the die horn to provide a nodal mounting to the test machine. This ensures that the mounting rig does not affect the vibratory motion of the horn, booster or transducer.

The forming die in this study constituted the output end of the ultrasonic horn. The die horn and booster were designed using finite element (FE) analysis and the modal frequencies and associated mode shapes were subsequently confirmed using experimental modal analysis (EMA). The FE software Abaqus was used in these studies to determine the modal parameters of the horn and booster. The ultrasonic booster was designed using the five-element horn configuration as reported by Peshkovsky [4]. The transducer can only provide ultrasonic amplitudes up to 10 μm, depending on the generator setting, therefore the profile of the booster and horn are designed to amplify the amplitude and allow a range of ultrasonic amplitudes to be excited.

Figure 1: Ultrasonic transducer, booster and die horn

The die horn and booster are manufactured from titanium (Ti-6Al-4V) and were modelled fully with 3D quadratic elements in Abaqus. The FE model predicted that the longitudinal mode of the booster plus die horn occurs at 20.7 kHz. The modal frequency determined experimentally from EMA was found to be 20.8 kHz for the longitudinal mode. Fig. 2 shows a comparison of the FE results and those obtained using EMA. The amplification achieved by profiling the booster and die horn, i.e. the ratio of amplitudes at the output and input faces of the booster/horn system, was measured and calculated to be a gain of four.

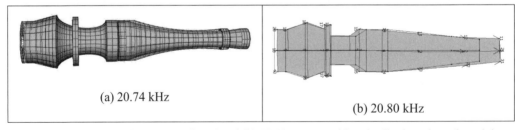

(a) 20.74 kHz

(b) 20.80 kHz

Figure 2: Comparison of (a) FE predicted and (b) EMA measured longitudinal mode and modal frequency.

Experimental Procedure

In these experiments, flat sheet metal specimens were compressed between a bowl-shaped female die and a round-nosed male die. The female die constituted the tip of the ultrasonic horn tuned to a longitudinal mode at 20.8 kHz. The die horn was excited by a piezoelectric transducer driven by an ultrasonic generator. The male die was connected to the cross-head of a Zwick-Roell test machine which provided a constant cross-head speed of 10 mm/min for these experiments. A Kistler force transducer, mounted between the male die and the cross-head as shown in Fig. 3, was used to measure the static-oscillatory force response during each test. The oscillatory displacement amplitude of the ultrasonic die was measured during each forming using a 3D laser Doppler vibrometer. The recorded signals were acquired using DataPhysics signal acquisition hardware and software for data processing.

Figure 3: Ultrasonic metal forming test setup.

Forming tests on four different metals were conducted to measure the effects of ultrasonic excitation of the forming die; aluminium A1050, aluminium alloy 7075, die cast magnesium AC50, and austenitic stainless steel grade 304. The specimens were all flat 3 mm thick plates. A series of static and ultrasonic forming tests were performed at a constant cross-head speed of 5 or 10 mm/min under dry surface conditions. In the experiments reported here, the specimen was compressed to approximately 1 mm displacement as measured by the machine cross-head at which point the ultrasonic excitation was applied to the ultrasonic die horn. The first set of force data was recorded from the load cell in the cross-head of the machine and two ultrasonic amplitudes of the die horn were excited; 12 μm and 20 μm. The second set of data was recorded for an ultrasonic amplitude of 20 μm and the force was measured using the piezoelectric force transducer mounted between the punch and the machine cross-head. For all tests reported here, the tests were stopped at a cross-head displacement of 3 mm. Table 1 shows the density and Young's modulus properties for the four metals.

Table 1: Materials used in the study.

Material	Density, ρ	Modulus of Elasticity, E
Aluminium A1050	2705 kg/m^3	70 GPa
Die cast magnesium AC50	1740 kg/m^3	44 GPa
Austenitic stainless steel 304	8030 kg/m^3	193 GPa
Aluminium alloy 7075 T73	2810 kg/m^3	73 GPa

Comparison of Static and Ultrasonic Forming Tests on Metals

Force measurement using the machine load cell. Fig. 4 shows the force-displacement results measured for static and static-ultrasonic forming tests on the four different metal specimens, measured by the load cell in the machine cross-head. During ultrasonic excitation of the female die the mean force is recorded by the test machine and clearly exhibits a reduction in the forming load in all tests.

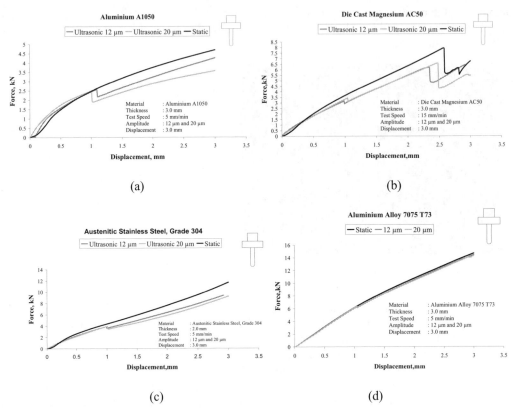

Figure 4: Die forming tests with and without ultrasonic excitation of the female die for (a) aluminium A1050 (b) die cast magnesium AC50 (c) austenitic stainless steel, grade 304 (d) aluminium alloy 7075 T73.

Table 2: Percentage reduction in the mean forming force.

Material	% Reduction in the forming force for 2 ultrasonic amplitudes	
	12 μm	20 μm
Aluminium A1050	16%	22%
Die cast magnesium AC50	7%	9%
Austenitic stainless steel, grade 304	7%	11%
Aluminium alloy 7075 T73	2%	3%

The results in terms of the percentage reduction in the forming force are summarised in Table 2. These results demonstrate that the effect of ultrasonic excitation of the die in metal forming is highly dependent on the material and its capacity to absorb ultrasonic energy. The soft aluminium specimens exhibited the largest reduction in the forming force which would indicate that the use of ultrasonic excitation is most beneficial in forming of this metal. The amplitude of the ultrasonic die horn is known to have a significant effect on the reduction in forming force [1] and this can be seen in Table 2. However, amplitude has little influence on force reduction in specimens where the reduction in forming force achieved is small.

If metals are heated to provide similar reductions in forming force, metal specimens generally benefit from improved ductility. However, the ultrasonic forming tests exhibit a reduction in the break force for the one specimen (die cast magnesium) that failed within the 3 mm cross-head displacement range of the tests. This result is consistent with previous findings that ultrasonic excitation leads to a reduction in the forming force and in the break force which can be explained by effective material softening due to the acoustoplastic effect [5] and dependency on factors such as acoustic impedance, internal friction, lattice imperfections, flow resistance, grain boundaries and impurities [6,7].

Force measurement using the piezoelectric force transducer. Measurement of the forming force without reference to the oscillatory force behaviour does not provide very meaningful interpretations of the effects of ultrasonic excitation because it relies on a reduction in mean load as being a direct measure of a beneficial effect. The measurement data presented in Fig. 5 superimposes the force measured from the piezoelectric force transducer on the force measured by the machine load cell. For all the tests shown in the figure, the ultrasonic amplitude of the die horn was 20 μm and the cross-head speed was 10 mm/min. The tests were carried out for a 3 mm displacement of the machine cross-head as before.

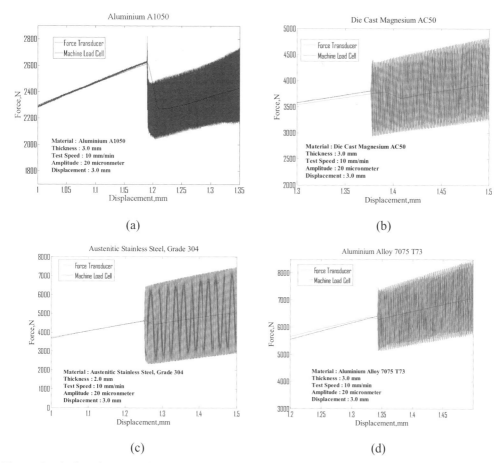

Figure 5: Die forming tests showing the measured oscillatory force for (a) aluminium A1050, (b) die cast magnesium AC50, (c) austenitic stainless steel grade 304, (d) aluminium alloy 7075 T73.

For each measurement, the peak-peak amplitude of the oscillatory force during ultrasonic excitation of the die horn is summarised in Table 3. The results for aluminium 1050 exhibit a path of maximum oscillatory force which is lower than the static load. These results are in good agreement with experimental results from cylindrical specimen compression tests reported by Daud [4]. For the other three metals, the maximum oscillatory force is larger than the static force even though, in all cases, there is a clear reduction in the mean load from the static load during the interval of ultrasonic excitation. Under the very high compressive loads required for compressive forming processes, it becomes difficult to achieve significant forming force reduction benefits from ultrasonic excitation of the forming tools for hard metal materials. For softer materials the benefits are significant.

Table 3: Amplitude of oscillatory force (pk-pk).

Material	Force amplitude (vibration amplitude 20 μm)
Aluminium A1050	500 N
Die cast magnesium AC50	1300 N
Austenitic stainless steel, grade 304	4000 N
Aluminium alloy 7075 T73	1800 N

Summary

The mean and oscillatory forces were measured during a simple ultrasonically assisted compressive forming test for four different metal specimens. The results agreed with previous studies of metal forming processes showing that the forming load is reduced by ultrasonic excitation of the forming tool. However, it was shown from measurements of the oscillatory force that a reduction in the mean forming force is not directly indicative of a benefit of ultrasonic excitation since, in many cases, the maximum oscillatory forming force can exceed the static forming force.

References

[1] Y. Daud, M. Lucas, Z. Huang: J. Mat. Proc. Tech., 186 (2007) 179-190.

[2] G.R. Dawson, C.E. Winsper, D.H. Sansome: Metal Forming, (1970), 234-238.

[3] S.A.A. Akbari Mousavi, H. Feizi, R. Madoliat: J. Mat. Proc. Tech., 187 (2007) 657-661.

[4] S.L. Peshkovsky, A.S. Peshkovsky: Ultrasonics Sonochemistry, 14 (2007) 314-322.

[5] M. Lucas, A. Gachagan, A. Cardoni: Proc. IMechE Pt.C JMES, 223 (2009) 2949-2965.

[6] O. Izumi, K. Oyama, Y. Suzuki: Trans. Japanese Institute of Metals, 7 (1966), 162-167.

[7] J.C. Hung, Y.C. Tsai, C. Hung: Ultrasonics, 46 (2007) 277-284.

Applied Mechanics and Materials Vols. 24-25 (2010) pp 317-322
© (2010) Trans Tech Publications, Switzerland
doi:10.4028/www.scientific.net/AMM.24-25.317

Modeling of Coating Stress of Plasma-sprayed Thermal Barrier Coatings

M. Arai[1, a]

[1]2-6-1 Nagasaka, Yokosuka-shi, Kanagawa-ken 240-0196 JAPAN

[a]marai@criepi.denken.or.jp

Keywords: Gas Turbine, Thermal Barrier Coatings, Coating Stress, Shear-lag Theory

Abstract. The surfaces of gas turbine components are coated with thermal barrier coatings (TBCs) using a plasma spraying technique. A lot of effort has been expended examining the TBC interfacial strength, however studies examining how residual stress is formed after the process and how the coating stress changes with temperature are limited. In this report, the residual stress prediction model is proposed based on the splat deposition process. A simplified model including the plasma sprayed process is developed based on shear-lag theory. The simplification is given in continuous particle deposition process. That is, continuous particle deposited coating is modeled as a single layer, which is called by "deposition layer". This deposition layer is assumed to impact directly onto the substrate. The binding layer is also introduced to express multiple cracks caused by quenching stress in splats and sliding deformation at splat boundary. It is shown that the numerical analysis has good agreement with the associated experiments.

Introduction

Ceramic thermal barrier coatings (TBCs), which play an important role in insulating components such as blades of gas turbines from high heat flux and/or high-temperature environments, are generally deposited using plasma-spraying technology. Residual stress in TBCs, which is generated during the spraying process and in service, is a considerable mechanical factor in the performance and lifetime of the coated components. The residual stress could lead to cracking within the ceramic coating, and to spallation of ceramic coating brought by interfacial crack propagation. Thus, it is necessary to fully understand the residual stress in TBCs, in order to assess precisely the lifetime up to the occurrence of TBC failure.

A plasma spraying process is characterized by the impingement of molten particles onto a substrate and rapid solidification of the particles spread along the substrate. Here, we focus on single splat particle movement in plasma spraying process. Spherical-shaped molten particles are compressed into a high-temperature plasma flow and spread onto the surface of substrate. Shrinkage of the flattened particle (splat) is constrained by the underlying substrate, and a tensile microstress (quenching stress) is generated in the splat during cooling up to the point at which it reaches the substrate temperature (T_p). The quenching stress can bring about cracking normal to the substrate surface forming a lamella structure which is a typical microstructure in TBCs. Another stress source, thermal stress occurs during cooling from the substrate temperature to room temperature (T_r) after finishing a complete particle flattening and solidification process. Due to differences in thermal expansion, thermal stress develops in the dual-layered structure. This continuous deposition process gives the residual stress value in TBCs, and consequently it needs to be considered in a model for predicting the residual stress.

In this study, we develop a comprehensive numerical model to estimate residual stress in TBCs generated during the plasma spraying process, based on the assumption of layer build-up structure modeled as the continuous splat deposition process. The model is developed upon a shear lag theory for taking into account sliding deformation at splat boundaries and stress relaxation caused by cracking in splat. A strain gage method is employed to measure the residual stress. We compare the results predicted by our numerical model with experimental results in order to verify the proposed model.

Modeling of Thermal Spray Deposition Process for TBC.
Layer Build-up Model. In this study, a continuous molten particle deposition process is idealized by impingement of a single layer (called the "deposition layer".) onto a substrate. This idealized model representation is shown in Fig. 1. Residual stress comprises both quenching stress, which is caused by contraction of the deposition layer on the underlying substrate, and thermal stress, which is generated by a thermal expansion mismatch between the deposition layer and the substrate.

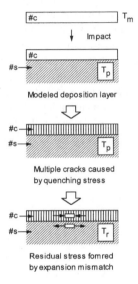

Fig.1 Schematic illustration for explanation of the model taking account of plasma -spraying process

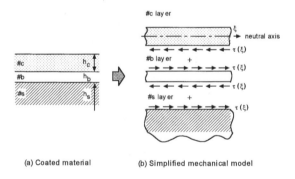

(a) Coated material   (b) Simplified mechanical model

Fig.2 The model used for formulation of coating stress

The deposition layer formed into a high-temperature plasma flow with spray temperature T_m impinges onto the substrate which is at a pre-heated temperature (substrate temperature) T_p through the drifting force of the plasma gas flow. The deposition layer is in melting/liquid state if the temperature in the deposition layer is also the same as the spray temperature. This deposition layer deposits and spreads on the surface of the substrate when impacting onto the substrate. The temperature in the deposition layer is cooled quickly up to the point of reaching the substrate temperature, and a contraction of the deposition layer is generated on the underlying substrate. This contraction brings about quenching stress in the deposition layer. The quenching stress will be developed by a temperature drop from the solidification temperature T_s to the substrate temperature T_p, because the effective temperature for causing thermal stress is limited in the region lower than the solidification temperature. The quenching stress generated during cooling will also cause cracking normal to the substrate surface as well as cracking phenomena observed in a single splat. After the temperature in the deposition layer reaches the substrate temperature, thermal stress arises in an overall dual-layered structure attached the deposition layer with a temperature drop from the substrate temperature to room temperature. Consequently, residual stress in TBC can be derived by summing the quenching stress and the thermal stress.

Formulation using Shear-lag Theory. The model utilized for formulation of coating stresses generated in the deposition layer during the plasma spraying process is shown in Fig. 2. Here, it should be noted that coating stress includes both quenching stress and thermal stress. In this study, a shear-lag theory is applied to obtain the coating stress, taking into account the fact that stress relaxation occurred in the deposition layer due to cracking.

Consequently, the coating stress generated in the deposition layer can be given by [1]:

$$\sigma_c = \frac{N_c}{h_c} = \frac{\sinh \beta(\xi - l) - \sinh \beta \xi + \sinh \beta l}{\sinh \beta l} \frac{E_c E_s h_s}{E_c h_c + E_s h_s} \int_{T_k}^{T} (\alpha_s - \alpha_c) dT \tag{1}$$

Numerical Analysis Procedure.

This section gives an explanation of the numerical analysis procedure to predict residual stress based on a layer build-up model.

Quenching stress analysis procedure:

(Step 1) Powder materials, solidification temperature and substrate temperature are set prior to the numerical analysis.

(Step 2) Deposition layer length is set to l and coating stress is set to zero. Initial temperature in the deposition layer is set to solidification temperature T_s.

(Step 3) Incremental coating stress $\Delta\sigma_c$ for associated temperature change ΔT is calculated using the following equation:

$$\Delta\sigma_c = \frac{\sinh \beta\left(\xi - \dfrac{l}{2^{i-1}}\right) - \sinh \beta\xi + \sinh \beta\dfrac{l}{2^{i-1}}}{\sinh \beta\dfrac{l}{2^{i-1}}} \frac{E_c E_s h_s}{E_c h_c + E_s h_s} (-\alpha_c(T_i))\Delta T \tag{2}$$

where i indicates the number of incremental calculations ($i=1, 2, ...$) and $T_i=T_{i-1}+\Delta T$. The term ($\dfrac{l}{2^{i-1}}$) means length divided by cracking generated in the deposition layer.

(Step 4) The coating stress at step i can be obtained from $\sigma_c^i=\sigma_c^{i-1}+\Delta\sigma_c$. If this coating stress does not reach the fracture strength σ_f of the ceramic material or the temperature T_i is larger than the substrate temperature T_p, the incremental calculation again returns to step (3).

(Step 5) If the coating stress exceeds the fracture strength, step n is progressed to $n+1$ by cracking initiation in the deposition layer. The coating stress at step $i=n+1$ is calculated using:

$$\sigma_c^i = \frac{\sinh \beta\left(\xi - \dfrac{l}{2^n}\right) - \sinh \beta\xi + \sinh \beta\dfrac{l}{2^n}}{\sinh \beta\dfrac{l}{2^n}} \frac{E_c E_s h_s}{E_c h_c + E_s h_s} \int_{T_k}^{T_i} (-\alpha_c) dT \tag{3}$$

and then the subsequent calculation returns to step (3) with Eq. (3).

Thermal stress analysis procedure:

(Step 6) The incremental coating stress $\Delta\sigma_c$ for the associated incremental temperature ΔT can be obtained by,

$$\Delta\sigma_c = \frac{\sinh\beta\left(\xi - \dfrac{l}{2^N}\right) - \sinh\beta\xi + \sinh\beta\dfrac{l}{2^N}}{\sinh\beta\dfrac{l}{2^N}} \frac{E_c E_s h_s}{E_c h_c + E_s h_s}(\alpha_c(T_i) - \alpha_c(T_i))\Delta T \qquad (4)$$

where N-1 is final number of incremental calculations after the quenching stress analysis.

(Step 7) The coating stress at step i is obtained from $\sigma_c^i = \sigma_c^{i-1} + \Delta\sigma_c$. If the temperature T_i is higher than room temperature T_r, the calculation returns to step (6). If the temperature T_i reaches room temperature, all calculations are finished. The coating stress at the final step consequently gives the residual stress in TBC.

Material Constants for Numerical Analysis.

Material constants are summarized as following;

(1) Substrate (Type 304 stainless steel)

$$\alpha_s(T) = 14.644 + 9.1534\times10^{-3}T - 5.149\times10^{-6}T^2 + 5.6839\times10^{-10}T^3 \qquad \left[\times10^{-6}1/K\right]$$
$$E_s(T) = 200 \qquad [GPa]$$

(2) Deposition layer (ceramic coating)

$$\alpha_c(T) = 3.8256 + 1.6493\times10^{-3}T - 1.2514\times10^{-5}T^2 + 3.1118\times10^{-3}T^3 \qquad \left[10^{-6}1/K\right], \quad T<1000[K]$$
$$= 10.916 \qquad \left[10^{-6}1/K\right], \quad T\geq1000[K]$$
$$E_c = 200 \qquad [GPa]$$
$$\sigma_f = 6159\times\exp(-0.0039503T) \qquad [MPa]$$

(3) Binding layer

$$h_b = h_c/5$$
$$\mu_b(v_p) = 0.0078127v_p^2$$

Other analysis parameters include, initial length of deposition layer of 25[mm], the thickness of layer h_c equals 1[mm] and the substrate thickness h_s is 5[mm]. Poisson's ratio v_c for deposition layer takes 0.07, which is same as one of freestanding ceramic coating, and one of substrate v_s is 0.3. In this case, solidification temperature T_s was assumed to be 2000[K]. This assumption is also reasonable from particle temperature measurement and associated material handbook data [2].

Numerical Analysis Results

Typical result for variation of coating stress with temperature is indicated in Fig.3. These result corresponds to the following plasma spraying condition; T_p=600[K] and v_p=175[m/s]. The broken line in this graph indicates the fracture strength curve of the sintering ceramic material [3]. It was found that the coating stress in deposition layer increased with temperature reducing from solidification temperature. The coating stress reaches the fracture strength and then causes crack in this layer. This coating stress or quenching stress in the plasma-spraying process is as described previously. The cracking reduces the coating stress in the deposition layer. The coating stress increases again with reduction of temperature, and again reaches the fracture strength of the sintering ceramic, which causes cracking in the deposition layer. Increase of coating stress, initiation of crack and drop of coating stress cyclically rise up to reaching to approximately 1700[K]. Multiple cracks, which were caused by the coating stress, lead to decrease of elastic modulus of deposition layer. Thus, the coating stress varies with lower slope against decreasing temperature. After reaching the substrate temperature, the coating stress in the deposition layer shifts to compressive due to thermal expansion mismatch between deposition layer and substrate. Finally, we have residual stress formed in the deposition layer, that is, in TBC. This process below substrate temperature corresponds to thermal stress process. This figure provides that the proposed simplified model based on plasma-spraying process gives us nice result for coating stress simulation in TBC.

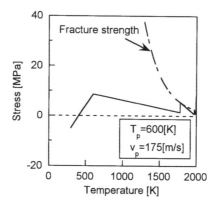

Fig.3 Simulation result of coating stress changing with temperature in case of $Tp=600[K]$ and $v_p=175[m/s]$

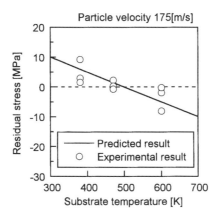

Fig.4 Comparison between numerical and experimental results in residual stress versus substrate temperature condition

   Relationship between residual stress and substrate temperature condition is indicated in Fig.4. In this figure, the circles indicate experimental results measured by the strain gage method. The solid line represents the numerical result from the simplified model. It is found that this comparison between experiment and numerical results gives good agreement over all substrate temperature condition. Relationship between residual stress and particle velocity condition is also shown in Fig.5. We can say that the experimental results almost correspond to the numerical one in this case, however some differences at higher particle velocities were observed.

Conclusion

In this report, the simplified model was developed in consideration of the spraying deposition process. The obtained results are summarized as following; the deposition process was modeled to simulate residual stress in TBC. Modeling is divided into two following processes; quenching stress process caused by contraction of splat when molten particle impacting onto the substrate, and thermal stress by thermal expansion mismatch. Simplification was achieved for continuous particle deposition by introducing a deposition layer. Another simplification was achieved by introduce a binding layer between the deposition layer and substrate, which models sliding deformation at splat boundary and multiple cracking in the coating caused by quenching stress. Numerical analysis was conducted for as-sprayed TBC. The numerical results showed good agreement with experimental one.

In this study, temperature distribution within the TBC was not taken into which would have an effect. Therefore we need to improve the model taking into account the effect of temperature distribution and associated stress distribution. However, we believe that the proposed model is quite simple and easy understood for engineers within the thermal spraying field, enabling them to choose the optimum particle energy and substrate temperature.

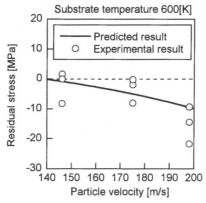

Fig.5 Comparison between numerical and experimental results in residual stress versus particle velocity condition

Acknowledgements

The author would like to acknowledgement for Mr. Ohno of Plasma Giken co. ltd. for conducting plasma-spraying process of TBC, and also the helpful comments for particle deposition process mechanism given by Dr. Fukanuma of the same company. Those experiments were conducted by Mr. Wada of Tokyo Institute of Technology. The author also would like to say thank you for him.

References

[1] M. Arai, E. Wada, K. Kishimoto: J. of Solid Mech. and Mat. Eng., Vol. 1, No. 10 (2007), pp.1251-1261.
[2] D. J. Green, R. H. J. Hnnink, M. V. Swain: "*Transformation Toughening of Ceramics*", CRC Press, Inc. (1989), for instance.
[3] R. Fujisawa, K. Matsusue, K. Kitahara: J. of the Society of Mat. Sci. Jpn, Vol. 35 (1986), pp.1112-1117.

Session 8A: Impact and Dynamics

Applied Mechanics and Materials Vols. 24-25 (2010) pp 325-330
© (2010) Trans Tech Publications, Switzerland
doi:10.4028/www.scientific.net/AMM.24-25.325

Tensile Properties of die-cast magnesium alloy AZ91D at high strain rates in the range between 300 s^{-1} and 1500 s^{-1}

I. R. Ahmad[a], SHU D. Wei[b]

[a,b] School of Mechanical and Aerospace Engineering, Nanyang Technological University,
Nanyang Ave., 639798 Singapore
[a] iram0002@ntu.edu.sg, [b] mdshu@ntu.edu.sg

Keywords: Hopkinson bar; magnesium alloys; AZ91D alloy; high strain rate; tensile properties; strain rate sensitivity

Abstract. Magnesium alloys have been increasingly used in the automobile, aerospace and communication industries due to their low density, high strength to weight ratio, good impact resistance and castability. Magnesium alloys, previously not used in load bearing components and structural parts are strongly being considered for use in such applications. Impact events in vehicles and airplanes as well as developments in weaponry and high speed metal working are all characterized by high rates of loading. Understanding of the dynamic behaviour of materials is critical for proper design and use in different applications. In the current study, a cast magnesium alloy AZ91D has been investigated at quasi-static and higher strain rates in the range between 300 s^{-1} and 1500 s^{-1}. The INSTRON machine was used to perform the quasi-static tests. High strain rate tests have been performed using the Split Hopkinson Tensile Bar (SHTB), a very useful and widely used tool to study the dynamic behaviour of variety of engineering materials. The results of a tensile testing indicate that the tensile properties including yield strength (YS), ultimate tensile strength (UTS) and the elongation at fracture (E_f) are affected by the strain rate variation. Higher stresses are associated with higher strain rates. The alloy AZ91D displays approximately 45% higher tensile stresses at an average strain rate of approximately 1215/s than at quasi-static strain rate. The dependence of the yield stress and tensile strength on the strain rate in the range of high strain rate above 1000 s^{-1} is larger than that at lower strain rates. The alloy AZ91D is observed to be more strain rate sensitive for strain rate higher than 1000 s^{-1}. A decrease in the strain rate sensitivity is also observed with the increasing strain in the specimen. It is observed that the hardening behaviour of the alloy is affected with increasing the strain rate. At high strain rates, the fracture of magnesium alloy AZ91D tends to transit from ductile to brittle.

INTRODUCTION

Light weight magnesium alloys due to their excellent strength to weight ratio and good impact strength are being investigated for use in the automotive, aerospace and electronics industries. AZ91D is being used extensively for manufacturing of gearboxes, steering column holders and brackets, cam covers, transmission housings, drive brackets, 4-wheel drive transfer cases, clutch and brake pedals and many other components. In recent years, AZ91D due to its suitable properties is also being used for information and communication technology (ICT) appliances such as mobile phones, cameras, laptops and compact and mini disc cases as well [1].

Magnesium alloys, which are rapidly becoming popular for fabricating structural and load bearing parts in the automobile, aerospace and electronic industries due to light weightiness, high specific strength and good impact resistance, have not been tested adequately at impact rates of loading. Lack of sufficient and conclusive data about their dynamic behaviour has steered the research to investigate these alloys at high strain rates. Aune et al [2] investigated the effects of strain rate on the dynamic properties of the die cast AZ91D, AM60B and AM50A. They found that the stress increases and percentage elongation is not affected by strain rates from 15 s^{-1} to 130 s^{-1}. No significant strain rate sensitivity variations were found for strain rate change from 15 s^{-1} to 130 s^{-1}. Ishikawa, Watanabe and Mukai [3] examined the compressive properties in the solution treated AZ91 at a strain rate of 10^3 s^{-1} and in the temperature range from 296 K to 723 K. At room temperature the maximum flow stress at a strain rate of 10^3 s^{-1} is about 15% higher than the stress at 10^{-3} s^{-1}. The dynamic stress-strain behavior of the AZ91 alloy at strain rates between 10^2 s^{-1} and 10^3 s^{-1} was investigated by Han, Xu and Liu [4]. The flow stress increased at first and then decreased with the strain rate increasing from 10^2 s^{-1} to 10^3 s^{-1}. The alloy exhibited both strain rate hardening and strain rate softening effects within the selected strain rate range. Shu, Zhou and Ma [5] found an increase in the tensile strength of magnesium alloys AM50 with increasing strain rate from 600 s^{-1} to 1350 s^{-1}. Abbott, Easton and Song [6] investigated the tensile properties of AZ91D in a strain rate range between 0.01 s^{-1} and 1 s^{-1} and found no significant effect of strain rate on AZ91D.

From the existing literature, it is observed that the micro-structural characteristics and tensile properties of magnesium alloys at quasi-static loading have been investigated by most of the authors but little is known about its mechanical properties at low to high strain rates between 10^2 s^{-1} and 10^3 s^{-1}. In the current study, AZ91D has been investigated under tensile loading using Split Hopkinson Tensile Bar (SHTB) in the range of strain rates between 300 s^{-1} and 1250 s^{-1}. The strain rate dependence of stress has been evaluated for the alloy.

EXPERIMENTAL SETUP AND DATA ANALYSIS

The Split Hopkinson Pressure Bar (SHPB) technique is a widely used technique for testing materials at high strain rates. As long as the equipment is calibrated accurately with proper preparations, this technique will offer a relatively high level of accuracy. The tensile Hopkinson bar apparatus consists of one input bar, one output bar, a striker tube, an anvil bar and an absorber bar at the end. Both the input and the output bars are fitted with a pair of strain gauges, each pair mounted 500 mm away from the specimen end. The strain gauges in a pair are fixed 180° apart on the bar to record the strain when impacted. The schematic diagram of the tensile Hopkinson bar is shown in Fig. 1

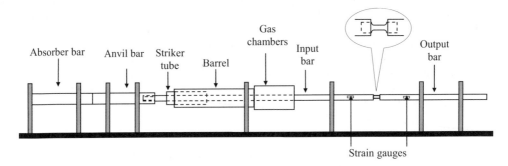

Fig. 1. A schematic diagram of Split Hopkinson Tensile Bar

The dog bone shaped specimen with threaded flanges is screwed between the input bar and the output bar. A gas gun drives the striking tube that is propelled from the barrel. The striking tube hits the anvil bar to initiate a compressive stress wave that propagates with speed c_o where $c_o = (E/\rho)^{1/2}$, being E the elastic modulus and ρ the bar material density, along the anvil bar into the absorbing bar until it reaches the free end of the absorbing bar where it reflects back as a tensile pulse. The tensile pulse will not be able to propagate back into the anvil bar and is absorbed in the absorber bar. A tensile wave also known as the incident wave propagates along the input bar towards the specimen/bar interface where it is measured by the strain gauges on the input bar. At this interface, the incident wave is partially reflected and partially transmitted through the specimen to the output bar as a transmitted wave. The reflected wave is again captured by the strain gauges on the input bar and the transmitted wave is measured by the strain gauges on the output bar. Measuring the amplitude of the incident, reflected and the transmitted pulses, ε_I, ε_R and ε_T respectively, shown in Fig. 2, it is possible to determine the stress-strain relationship of the specimen.

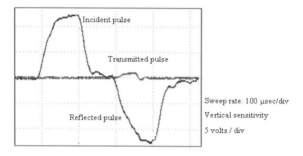

Fig. 2. Oscilloscope traces obtained from tensile Hopkinson bar on AZ91D alloy

When the specimen is deforming uniformly, the strain rate within the specimen is directly proportional to the amplitude of the reflected pulse. Likewise, the stress within the specimen is directly proportional to the amplitude of the transmitted pulse. The strain rate $\dot{\varepsilon} = \frac{d\varepsilon}{dt}$ in the specimen is calculated as

$$\frac{d\varepsilon_s}{dt} = -\frac{2c_o}{l_s}\varepsilon_R .$$ (1)

where l_s is the length of the specimen before impact. The strain is determined by integrating the strain rate from 0 to t, the total duration of the test.

$$\varepsilon_s(t) = -\frac{2c_o}{l_s}\int_0^t \varepsilon_R(t)dt .$$ (2)

Stress in the specimen can be calculated by using the following equation;

$$\sigma_{ave}(t) = E\frac{A_o}{A_s}\varepsilon_T(t) .$$ (3)

Where A_o and A_s are the cross sectional areas of the incident bar and the specimen.

Specimen

Fig. 3 shows the geometry of the specimens used in the present tests. The specimens were prepared from the same batches of AZ91D alloy to avoid any metallurgical effects.

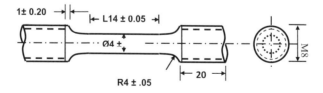

Fig. 3 Dimensions of the tensile specimen used

Test Results and Discussions

In order to study the tensile behavior of the magnesium alloy AZ91D, a series of impact tests were performed at various strain rates between 300 s^{-1} and 1250 s^{-1}. The strain rate variation with time is plotted in Fig. 4 for various tested strain rates. The values of strain rates shown in figure 4 are the average values. It is observed that the strain rate is relatively uniform at lower strain rates as compare to high strain rates.

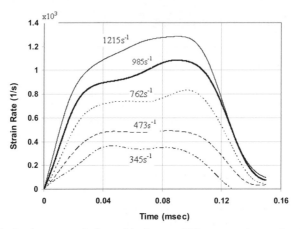

Fig. 4. Strain rate variation with time at different strain rate loadings

Effect of strain rate on dynamic properties of the alloy

Figure 5 depicts the resulting impact tensile stress-strain relations for the AZ91D alloy at various rates of strain. The alloy exhibit distinct yield points at high strain rates and the stress increases with increasing the strain rate monotonically. At a strain rate of 1215 s^{-1}, approximately 15 % and 45% high stresses are observed as compare to what is experienced at a strain rate of 345 s^{-1} at 1.5% strain and quasi-static test at 2% strain respectively. A similar trend is seen for other higher strain rates. The elongation to fracture of the specimen increases with increasing strain rate except the specimen tested at 473 s^{-1}. A slight work hardening during plastic deformation is also observed.

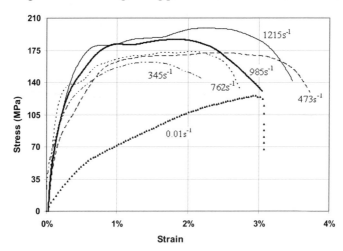

Fig. 5. Stress-strain relation for magnesium alloy AZ91D at different strain rates

Strain-hardening and strain rate sensitivity

The material deformation becomes increasingly difficult during plastic deformation due to increasing dislocation density. This process is called strain hardening and is depicted by the well known Hollomon equation;

$$\sigma = K\varepsilon^n .$$ (4)

K and n are the strength coefficient and the strain hardening exponent respectively. The hardening exponent can be found by determining the slope of a plot of $log\ (\sigma)$-vs-$log\ (\varepsilon)$. Figure 6, shows such a plot for the current study at various strain rate levels. The value of n calculated from this plot for three different strain rates 345, 762 and 1215 s^{-1} are 0.23, 0.196 and 0.185 indicating a slight decrease in the hardening exponent with increasing strain rate. In order to examine the effect of the strain rate on the tensile stress, the respective values of stress at different percentage strains are plotted in Fig.7 as a function of log of average strain rate. In order to evaluate the rate sensitivity of the stress the strain rate sensitivity m, is introduced in the following equation;

$$m = \left(\frac{\partial \sigma}{\partial \log \dot{\varepsilon}} \right)_{\varepsilon, T} \qquad (5)$$

Where T is the test temperature, which is room temperature in the present study

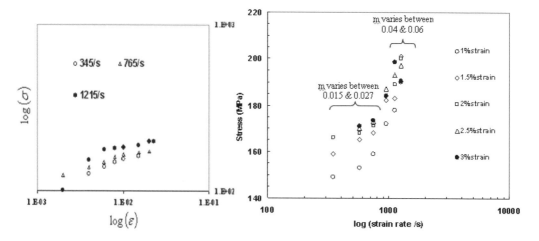

Fig. 6. log (σ)-vs-log (ε)   Fig. 7. Stress as a function of log strain rate.

The stress σ corresponds to different strain rates at some constant strain $\varepsilon = \varepsilon_o$. The value of m is plotted for different values of specimen strain. It is clear that the alloy AZ91D is a rate sensitive material. It is noted that the strain rate sensitivity of the AZ91D alloy varies between 0.015 and 0.027 for strain rate lower than 1000 s^{-1} and it is between 0.04 and 0.06 when the deformation rate is higher than 1000 s^{-1} indicating high rate sensitivity of the alloy at higher strain rates. The values of strain rate sensitivity calculated for the AZ91D alloy for strain rates up to 1000 s^{-1} are plotted as a function of the specimen strain in Fig. 8. As the strain increases, the strain rate sensitivity decreases monotonically. However, the decrease in strain rate sensitivity is minute for larger strains (> 2%) as compare to what is observed at lower strains. Similar behavior is observed for strain rate higher than 1000 s^{-1}.

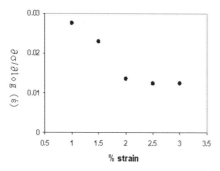

Fig. 8. Strain rate sensitivity m vs. specimen strain

CONCLUSION

A die cast magnesium alloy AZ91D has been investigated for its tensile properties at high strain rates. It is observed that the strain rate variation with time is relatively uniform when the specimen is deforming at lower strain rates as compare to high strain rates. A monotonic increase in the stress is observed with increasing strain rate from 345 s^{-1} to 1215 s^{-1}. At a strain rate of 1215 s^{-1}, approximately 15 % and 45% high stresses are observed as compare to what is experienced at a strain rate of 345 s^{-1} at 1.5% strain and quasi-static test at 2% strain respectively. The hardening exponent decreases with the increasing the strain rate. It is noted that the strain rate sensitivity for AZ91D alloy is lower for strain rate up to 10^3 s^{-1} and it increases sharply at strain rates higher than 10^3 s^{-1}. A decrease in the strain rate sensitivity is observed with increasing %strain in the specimen.

Future Recommendations

More tests with an extended range of strain rates up to 2000 and higher followed by numerical simulation using the Johnson-Cook materials mode, needed for better understanding of the dynamic behaviour of the alloys, are in progress.

Reference

1. Kainer, K., U. Magnesium-Alloys and Technologies. WILEY-VCH, Weinheim, 2003.
2. Aune, T. Kr., Darryl Albright, Hakon Westengen, Jojnsen Tor E., Andersson B., "Behavior of Die Cast Magnesium Alloys Subject to Rapid Deformation," 2000-01-1116, SAE International (2000)
3. Ishikawa, K., Watanabe, H. and Mukai, T., "High strain rate deformation behavior of an AZ91 magnesium alloy at elevated temperatures," Materials Letters (59), 1511-15 (2005)
4. Han, E. H., Xu, Y. B. and Liu, L., "Dynamic stress-strain behavior of AZ91 alloy at high strain rate," Materials Science Forum 546-549(1), 89-92 (2007)
5. SHU, D. W., Zhou, W. and MA, G. W., "Tensile Mechanical Properties of AM50 Alloy by Hopkinson Bar, Key Engineering Materials," (340-341), 247-254 (2007)
6. Song, W. Q., Beggs, P. and Easton, M., "Compressive strain-rate sensitivity of magnesium-aluminium die casting alloys," Materials and Design (30), 642-648 (2009)
7. Nicholas, T., "Tensile Testing of Materials at High Rates of Strain," Experimental Mechanics, (21),177-185 (1980)
8. Staab, G.H. and Gilat, A.,"A Direct-tension Split Hopkinson Bar for High Strain-rat Testing," Experimental Mechanics, (31), 232-235 (1991)

Applied Mechanics and Materials Vols. 24-25 (2010) pp 331-336
© (2010) Trans Tech Publications, Switzerland
doi:10.4028/www.scientific.net/AMM.24-25.331

The Dynamic Buckling of Stiffened Panels – A study using High Speed Digital Image Correlation

C. A. Featherston[1, a], J. Mortimer[1,b], M. Eaton[1,c], R. L. Burguete[2,d] and R. Johns[2,e]

[1]Cardiff School of Engineering, Cardiff University, Queens Buildings, The Parade, Cardiff, CF24 3AA, UK

[2] Airbus UK, New Filton House, Filton, Bristol, BS99 7AR, UK

[a]FeatherstonCA@cardiff.ac.uk, [b]MortimerJ@cardiff.ac.uk, [0]EatonM@cardiff.ac.uk, [d]Richard.Burguete@Airbus.com, [e]Rhiannon.Johns@cardiff.ac.uk

Keywords: Dynamic buckling, stiffened panel, digital image correlation, finite element analysis.

Abstract. For a structure subjected to an intermediate velocity impact in which the duration of loading is in the order of milliseconds and in excess of the period of it's first free vibration mode there is a relationship between impact duration and buckling load. Although this relationship results in higher buckling loads for shorter duration impacts, the precise nature of the correlation depends on a number of other factors, one of which is geometry. Since the design of many lightweight structures subject to dynamic loading in this intermediate range is based on the use of a static buckling load to which a load factor is then applied, it is essential that this factor accurately represents the relationship between the two and takes of account of any variations. Failure to do so will at least result in an over designed structure and at worst in catastrophic failure.

A series of finite element analyses (FEA) have been performed in order to determine the relationship between dynamic and static buckling loads for a range of stiffened panels with differing radii of curvature.

This paper describes preliminary tests performed to determine the feasibility of using high speed digital image correlation (DIC) to study such an impact and hence provide validation of the earlier FEA analyses. These are performed on a longitudinally stiffened panel subject to uniaxial compression, clamped within a rig designed to provide built-in end conditions and allow motion of one end in the direction of loading only. The specimen is tested using an accelerated drop test rig. Impact load is monitored throughout using a load cell. Full field displacement contours are obtained using a high speed DIC system. Results are presented which demonstrate deflection contours during and after impact enabling the path of the shock wave through the specimens to be determined. An initial comparison is then made the FEA results.

Introduction

Lightweight structures consisting of thin shells reinforced with uni-directional or bi-directional stiffeners are used extensively in a range of industries including the aerospace sector. Loading of these structures under combinations of in-plane compression, bending and shear creates the potential for failure by buckling. Extensive work has therefore been carried out in order to develop guidelines for the design of a range of different panel geometries under varying boundary conditions and load cases in order to prevent this mode of failure. These include design guides such as Bruhn [1] and data sheets including those published by ESDU [2]. In each case, however loading has been assumed to be applied slowly such that a static analysis can be performed.

In reality however, many of the loads applied to such structures for example those experienced during emergency braking and landing (Assler and Telgkamp [3])) are dynamic in nature. In such cases it has been shown that the relationship between the dynamic and static buckling loads can vary dramatically and that this variation is dependent on a number of factors. For example shorter duration impacts result in higher buckling loads (Weller et al [4]) and may be substantially in excess of the static buckling loads (Bisagni and Linde [5]) whilst longer duration impacts can result in much lower buckling loads . In order to meet future goals such as those set out by ACARE (The Advisory Council for Aeronautics Research in Europe) in their Strategic Research Agenda in Aeronautics 2020 [6] to reduce aircraft emissions by up to 80% it is essential that the former be taken into consideration to realize optimized structures and corresponding reductions in weight and therefore fuel consumption. The consequences of the latter eventuality could clearly be catastrophic.

One of the first researchers to investigate dynamic buckling was Zizicas [7] who developed a theoretical solution for the case of a simply supported rectangular plate under time-dependant inplane loads although a buckling criterion was not determined. Later Ari-Gur et al [8] showed that the dynamic buckling load of the plate was highly dependent on pulse duration and the existence of initial geometric imperfections based on experiments on a rectangular plate impacted by a mass moving in an in-plane direction. Weller et al [4] calculated the Dynamic Load Amplification Factor (DLF) for a series of beams and plates under differing pulse durations and geometric imperfection using the ADINA computer code. More recently work on stiffened panels has been performed by Bisagni and Zimmermann [9], Yaffe and Abramovich [10], Zhang et al [11] and Bisagni [12].

This paper presents research being carried out to extend the work described by examining the effects of geometry on the relationship between the static and dynamic buckling loads of a series of stiffened panels. An FEA model has been developed which has been used to study panels with differing radii of curvature. An experimental programme is now underway in order to validate this model and investigate this phenomena further. Results are presented here for a flat panel.

Finite Element Analysis

Finite element analyses have been carried out on a series of models representative of stiffened panels with radii of curvature of 400mm, 800mm and infinity (ie a flat plate) (detailed information regarding the geometry of the flat plate is given in the description of the test specimen below) in order to determine their dynamic buckling loads and relate these to their static behaviour. These were performed using the explicit solver (in which the solution is advanced kinematically from one increment to the next) available in the FEA software ABAQUS. Panels were modeled using quadrilateral S4RS elements (Shell, 4 nodes, Reduced integration, Small-strain) which are suitable for dynamic analysis of shell structures undergoing large-scale buckling behaviour with large rotations but relatively small amounts of membrane stretching and compression. In terms of boundary conditions movement in all five degrees of freedom (3 translations plus out-of-plane rotations) was prevented along the bottom of the panel, the sides remaining free and the top was prevented from translating in the x and z directions (Fig.3), and from rotating about the x and y axes. In each case both the plate and the stiffeners were constrained. Finally a distributed load was applied across the top edge of each of the panel, loading both plate and stiffeners according to the time profile illustrated in Fig. 1.

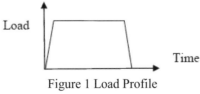

Figure 1 Load Profile

In addition to providing specific buckling loads, results demonstrated a relationship between dynamic and static buckling loads which varied depending upon the duration of impact and the radius of curvature of the panels (Fig. 2). These indicated that for short time duration loading the dynamic buckling load is increased in comparison to its static equivalent. As load duration increases however, the critical dynamic load is seen to fall to a value considerably lower than the static one. In terms of geometry the results demonstrate an increased sensitivity to load duration in the case of a flat panel when compared to a curved one.

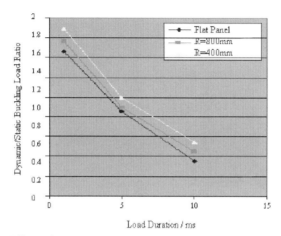

Figure 2 The Effect of Geometry on the Ratio of Dynamic to Static Buckling Load

Experimental Set-Up

Specimen. The panel tested comprised a flat aircraft grade duraluminium (BS1470 6082 – T6) plate 400mm long by 400mm wide and 0.5mm thick to which five longitudinal blade stiffeners each 50mm deep and 0.5mm thick were attached using Loctite Multibond 330 and rivets shown in Fig. 3.

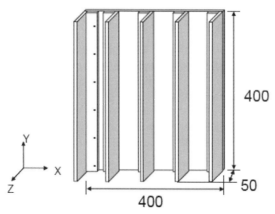

Figure 3 Panel Geometry

Test Rig. The panel was tested using the rig illustrated in Fig. 4. The loaded ends of the panel were mounted using an epoxy resin in slots machined in top and bottom plates designed to provide builtin boundary conditions. Guide rods located at the corners of the plates provided vertical alignment whilst enabling motion of the top plate in the direction of loading by means of a series of phosphor bronze bushes mounted in the upper plate. Impact loading was applied to the top plate using an

Instron Dynatup 9250HV with extended width base, capable of providing impact energies of up to 1.6kJ at velocities of up to 20 m/s using an accelerated drop weight system. Tests were performed using a mass of 5.7244 kg dropped from 0.04, 0.05 and 0.06m. This generated the load profiles presented in Fig. 5, which are representative of the load duration modelled and the anticipated dynamic buckling load.

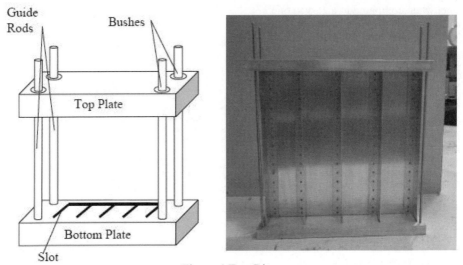

Figure 4 Test Rig.

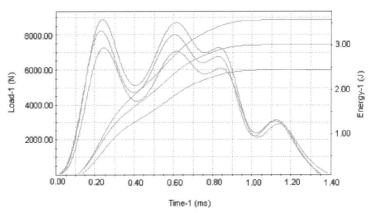

Time-1 (ms)

Figure 5 Load Profiles.

Digital Image Correlation. Full field displacement data was obtained for the plate only using 3D digital image correlation. This was achieved using the LIMESS system with AOS Technologies MOTIONeer high speed cameras synchronised using Correlated Solution's Vic Snap software to capture data from the panel at a resolution of 512 x 640 pixels and a rate of 1000 fps. This data was then post processed using Vic3D software. In addition a single camera was used to capture video footage at 10000 fps in order to obtain qualitative data on the deformation of the stiffeners.

Results

The results obtained in terms of deformation of the panel during impact are presented in Fig. 6. Out of plane deflections can be observed near the top of each stiffener 0.1ms after impact in both the high speed camera image (Fig 6a) and the FEA analysis (Fig 6b). These deflections are seen to move downwards through the stiffeners prior to the deformation moving into the panel (Fig 6c - 0.2, 0.3 and 1ms after impact). Due to the limitations of the high speed cameras used for the DIC which could operate at a maximum of 1000 fps for 512 x 640 pixel resolution it was not possible to obtain detailed experimental data regarding the behaviour of the back plate during impact since only one frame could be captured during the impact event. It was however possible to monitor the panel vibrations following the impact for comparison with results of the FEA. These results are provided for comparison in Fig 7. Again good correlation can be seen.

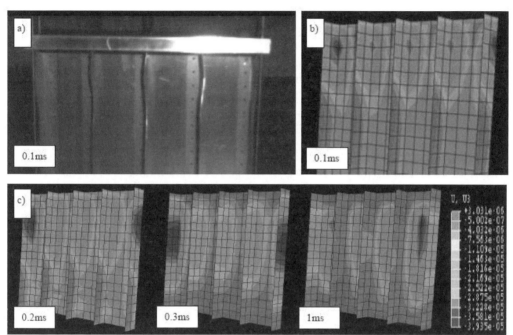

Figure 6 Out-of-Plane Displacements in the Stiffeners a), b) and Traveling Through the Panel in c).

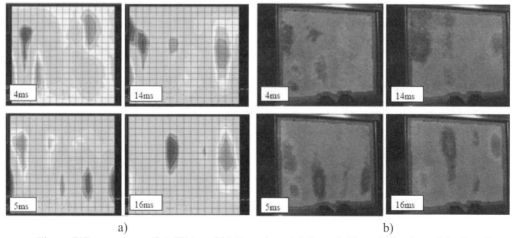

a)　　　　　　　　　　　　　　　　　　　　　　b)

Figure 7 Comparison of a) FEA and b) Experimental Out-of- Plane Vibration of the Panel Following Impact.

Conclusions

The feasibility of using high speed photography and digital image correlation to examine the dynamic behaviour of a stiffened panel has been demonstrated. The results obtained appear to validate those previously derived using explicit finite element analyses. Further work is required to obtain more qualitative data. This will be achieved using higher speed cameras with automated synchronization to enable faster data processing.

References

[1]   E.F. Bruhn: Analysis and design of flight vehicle structures. Jacobs Publishing Inc. (1973).

[2]   Engineering Sciences Data Unit (ESDU): Flat Panels in Shear. Buckling of long panels with transverse stiffeners. ESDU 02.03.02 (1971).

[3]   H. Assler, and J. Telgkamp: Design of aircraft structures under special consideration of NDT. 9th European Conference on NDT. Berlin (2006).

[4]   T. Weller, H. Abramovich and R. Yaffe: Dynamics of beams and plates subjected to axial impact. Computers and Structures, Vol. 32(3-4) (1981), p. 835-851.

[5]   C. Bisagni and P. Linde: Numerical simulation of the structural behaviour of orthotropically stiffened aircraft panels under short time duration loading. International Congress of Aeronautical Sciences. Hamburg (2006).

[6]   Advisory Council for Aeronautics Research in Europe. Strategic Research Agenda in Aeronautics 2020 (2004).

[7]   G.A. Zizicas: Dynamic buckling of thin plates. Trans. ASME, 74(7) (1952), p. 1257 1268.

[8]   J. Ari-Gur, J. Singer and T. Weller: Dynamic buckling of plates under longitudinal impact. Israel J Techno, Vol. 19 (1981), p. 57-64.

[9]   C. Bisagni and R. Zimmerman, Buckling of axially compressed fiber composite cylindrical shells due to impulsive loading. Proceedings of the European Conference on Spacecraft Structures, Materials and Mechanical Testing (1998), p. 557-562.

[10]   R. Yaffe and H. Abromovich: Dynamic buckling of cylindrical stringer stiffened shells. Computers and Structures, Vol. 81 (2003), p. 1031-1039.

[11]   T. Zhang, T. Liu, Y. Zhao and J. Luo: Nonlinear dynamic buckling of stiffened plates under inplane impact loading. Journal of Zhejiang University of Science, Vol. 5(5) (2004), p. 609-617.

[12]   C. Bisagni, Dynamic buckling test of cylindrical shells in composite materials. 24th International Congress of the Aeronautical Sciences, Yokahama (2004).

Applied Mechanics and Materials Vols. 24-25 (2010) pp 337-342
© *(2010) Trans Tech Publications, Switzerland*
doi:10.4028/www.scientific.net/AMM.24-25.337

Finite Element Model Updating of a Thin Wall Enclosure under Impact Excitation

O. S. David-west[1, a], J. Wang[2,b] and R. Cooper[1,c]

[1]School of Mechanical and Aerospace Engineering, Ashby Building, Queen's University, Belfast, Northern Ireland, United Kingdom, BT9 5AH

[2]Faculty of Engineering, Kingston University, Friars Avenue, Roehampton Vale London, United Kingdom, SW15 3DW

[a]o.david-west@qub.ac.uk, [b]J.Wang@kingston.ac.uk, [c]r.cooper@qub.ac.uk

Keywords: Impact excitation, Model Updating, Frequency Response Function.

Abstract

Simulation result of a structural dynamics problem is dependent on the techniques used in the finite element model and the major task in model updating is determination of the changes to be made to the numerical model so that dynamic properties are comparable to the experimental result. In this paper, the dynamic analysis of a thin wall structure (approx. 1.5±0.1 mm thick) was realized using the Lanczos tool to extract the modes between 0 and 200 Hz, but the interest was to achieve a good aggreement between the first ten natural frequencies. A shell element with mid size nodes was used to improve the finite element result and the model was tunned using the damping constant, material properties and discretization. The correlation of the results from the impact excitation response test and the finite element was significantly improved. A correlation coefficient of 0.99 was achieved after tunning the model.

Introduction

When numerical or analytical predictions for the dynamic response of a structure are compared with experimental results it is often noticed that there is some degree of mismatch. Due to advances in measurement and analysis techniques experimental measurements with acceptable error is believed to offer more confidence than the finite element model. An acceptable method to resolve this issue is to update the numerical model.

The most important task in model updating is the determination of the changes to be made to the numerical model and hence, maximise the correlation between the numerical and experimental data. D J Ewins [1] has presented some report on this subject and an extensive literature survey on this subject has been reported by Mottershead J E and Friswell M I [2]. J L Zapico et al [3] used a geometric scale of 1:50 to model the response of a bridge under transverse seismic loading; while C Mares et al [4] updated a test structure using the properties of the mode shapes and sensitivity parameters. Michael I Friswell et al [5] have mentioned the use of physical, geometric and generic element parameters and regularization techniques of singular decomposition, L-curves and cross validation for updating. N G Nalitolela et al [6, 7] adopted the use of mass and/or stiffness addition to update structural parameters.

In most situations there are variables from the test results of similar structures, to get a balance between these uncertainties; C Mares et al [8, 9] and Stefan Reh et al [10] have discussed updating using the stochastic tool of Monte-Carlo analysis. In this report, the numerical result have been improved by the use of discretization, mass and stiffness matrices and the inclusion of average damping constant obtained from experiment. The advances in technology has made enclosures used to reduced noise transferred to the environment in machines such as a generator more complex and hence the need to understand the behaviour precisely. Further studies on this research are tailored towards the development of an updating process for generator canopy using finite element tool.

Concept of Dynamic Analysis

The general dynamic analysis will solve the equation of motion which gives the time dependent response of every node point in the structure by including inertial forces and damping forces in the equation.

$$[M]\{\ddot{x}\} + [C]\{\dot{x}\} + [K]\{x\} = \{F\} \qquad\qquad [1]$$

where $[M]$ represents the structural mass matrix, $\{\ddot{x}\}$ the nodal acceleration vector, $[C]$ the structural damping matrix, $\{\dot{x}\}$ the node velocity vector, $[K]$ the structure stiffness matrix, $\{x\}$ the node displacement vector and $\{F\}$ is the applied time varying load.

In many engineering systems, the natural frequencies and mode shapes of vibration are of interest. However, for vibration modal analysis, the damping is generally ignored and considering a free vibration multi-degree of freedom system, the dynamic equation becomes.

$$[M]\{\ddot{x}\} + [K]\{x\} = \{0\} \qquad\qquad [2]$$

If the displacement vector $\{x\}$, has the form $\{x\} = \{X\}\sin\omega t$, then the acceleration vector is $\{\ddot{x}\} = -\{X\}\omega^2 \sin\omega t$ and substituting into equation 2, yields the eigenvalue equation.

$$([K] - \omega^2[M])\{X\} = \{0\} \qquad\qquad [3]$$

Each eigenvalue has a corresponding eigenvector and the eigenvectors cannot be null vectors.

$$\left|[K] - \omega^2[M]\right| = 0 \qquad\qquad [4]$$

Equation [3], represent an eigenvalue problem, where ω^2 is the eigenvalue and $\{X\}$ the eigenvector (or the mode shape). The eigenvalue is the square of the natural frequency of the system.

Structure Description and Experimental Procedure

The thin wall structure used for this study is a box made of steel measuring 900 x 600 x 750 (all in mm). The wall thickness is 1.5 mm (± 0.1 mm). Fig. 1, is a photograph of the structure and a schematic of the test set-up is shown in Fig. 2. The response vibration measurement of the box under the impact load were obtained under the fixed base boundary condition of the structure. The set up includes a digital Fourier Transfer analysis system having four input channels. The box was excited with an instrumented hammer, as recommended in BS 6897 [11] and the vibration data of the excited structure were acquired via a portable digital laser vibrometer (PDV 100) focused on strategic points on the structure.

Fig. 1. Photograph of the structure.

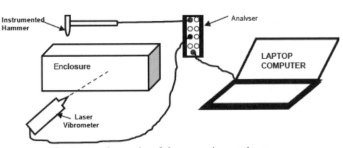

Fig. 2. A schematic of the experimental set-up.

The load and response signals sampled by a Bruel and Kjaer transient analyzer and the frequency response function were visualized via the computer. The test was repeated several times changing the point of input energy; the aim of which is to capture all the modes from the response of the structure.

Experimental Results

Fig. 3, shows the overlay of frequency response functions (FRF) of one set of the test and Fig. 4, is the sum of the FRFs.

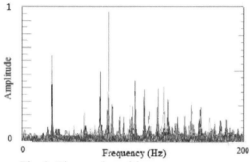

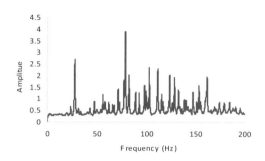

Fig. 3: The overlay of frequency response functions

Fig. 4: Sum frequency response function of a test

The dynamic properties of the structure were estimated from the sum frequency response function of every test by the so called peak-picking method. Peak picking is a frequency domain-based technique and a discussion of method is available in reference [12]

Updating Procedure

After an initial finite element model using ANSYS 11 commercial finite element tool, a sensitivity study was conducted to select the parameters for tuning the model. Fig. 5 is the flow chart illustrating the updating process. In this study we have used the element type, discretization, elastic modulus, average damping constant and material density for updating. The element mass and stiffness matrices is generally presented in the form shown in equations 5 and 6 [13].

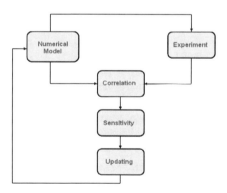

Fig. 5: Schematic of the process of updating

$$[m] = \int\limits_{-1}^{1}\int\limits_{-1}^{1}\int\limits_{-1}^{1} N^T \rho N \det(J) d\xi_1 d\xi_2 \xi_3 \qquad\qquad [5]$$

$$[k] = \int\limits_{-1}^{1}\int\limits_{-1}^{1}\int\limits_{-1}^{1} B^T DB \det(J) d\xi_1 d\xi_2 d\xi_3 \qquad\qquad [6]$$

Where ρ is the density, **D** the elasticity matrix, **N** the matrix of shape functions, **B** matrix of shape function derivatives, **J** the Jacobian matrix and (ξ_1,ξ_2,ξ_3) the local co-ordinates. As shown in the above relation, the material density, Young's modulus of the material and the element shape functions affects both the sensitivity and eigendata of a finite element model. The test structure has a thickness of 1.5 mm ($\pm$ 0.1 mm) and the measurement points grits of 50 x 50 (in mm) were marked on the panels.

Table 1: Natural frequencies

Experimental Result (Hz)	Result before Updating (Hz)	Result after Updating (Hz)
16.99	15.83	17.27
19.45	17.5	19.1
23.47	21.36	23.31
26.07	23.37	25.5
27.48	26.29	28.69
31.63	31.01	33.83
33.97	32.0	34.91
40.25	37.01	40.38
42.55	40.17	43.83
46.34	42.51	46.38

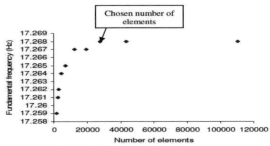

Fig. 6: Identification of convergence.

The initial model was coarse having 6900 elements with element thickness at the lower limit of 1.5 mm. In Table 1 is shown the natural frequencies before and after updating compared to the experimental result. The difference between the frequencies before updating is very high for instance, the fundamental frequency was 15.83 Hz before updating giving a difference of 1.16 Hz (ie 0.07 %)in comparison with the experimental result. With improvements obtained the first natural frequencies were 17.27 Hz from the numerical analysis and 16.99 Hz from the experiment. This being as expected as with finite element software an ideal fixed boundary condition can be generated. The mass and stiffness matrices were used to improve the finite element result using a shell thickness of 1.6 mm and elastic modulus of 210 GPa in the stead of 205 GPa used in the initial model. The structure was manufactured with steel procured from Yorkshire steeel company and the elastic modulus as obtained from the site varies from 195 GPa to 225 GPa.

The size of the total mass and stiffness matrices of the structure depends on the number of unconstrained nodal degrees of freedom in the model [13]. As the wall thickness of structure is very small it was modelled using a shell element, having mid side nodes in addition to corner nodes, since such elements give better accuracy of the result compared to shell elements with only the corner nodes. The threshold of convergence was as used for the numerical analysis taking cognizance of computational time and accuracy. As shown in Fig. 6, the result converging at the plateau. This being in aggrement with the report of E J Ewins that its very difficult to update coarse model [1]. The finite element model has 27600 elements.

Table 2: Details of updated model.

Element Type	Number of Elements	Shell thickness (mm)	Modulus of Elasticity (N/m²)	Mass Density (kg/m³)	Damping Constant	Material Type
Shell element with mid-side nodes	27600	1.6	2.1E+11	7850	0.02	Steel

Fig. 7: Finite element discretisation for the enclosure model.

The damping constant or ratio were obtained from the experiment at the resonance peaks of the response function, the value varies from 1.2% to 2.6% and an approximate average value of 2% was used for updating (Table 2) as the numerical tool gives the provision for a single value. In a dynamic analysis stiffness distribution is important as well as the mass distribution. The mass matrix [M] can be formulated to be 'consistent' or 'lumped'. The consistent formulation was adopted as it represents better the continuous nature of real structures whereas the lumped formulation is an approximation and save computational time. Fig. 7, shows the finite element discretisation for the box.

Finite Element Results

The dynamic analysis was realized using the Lanczos tool to extract the modes of the thin walled enclosure between 0 and 200 Hz.

Table 3: Finite element modes of the structure.

Modes number	Mode shapes	Description of Mode Shape
1,3,5		Breathing mode
2		Smaller side panels stationary and breathing at the top (double breathing) and longer side panels
4,7,8		Top and longer side panels, double breathing; smaller side panels single breathing.
6		Double breathing on all sides panels, but faster on the larger one.

The fundamental mode is a breathing mode at the natural frequency of 17.27 Hz and a description of the first eight modes are given in Table 3.

Correlation of Results

The data for the correlation analysis are the natural frequencies of the box obtained from the excitation response test and the finite element analysis. The relationship is presented in Fig. 8.

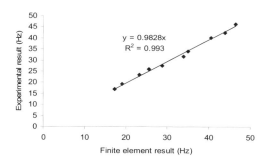

Fig. 8: Correlation of results.

Correlation coefficients is a measure that describe the strength of the relationship between the two quantities. The R^2 value of 0.99 implies a very good aggrement between the updated numerical result and the experiment.

Concluding Remarks

Impact excitation response tests were performed on the thin wall structure. The vibration was generated by the energy input into the structure using an instrumented hammer and the vibration data picked by the use of a laser vibrometer. The dynamic properties of the thin wall structure were estimated from the sum frequency response function of every test by the peak-picking method. The damping constants were obtained from the experiment at the resonance peaks of the response function, the about average value of 2% was used for updating.

To obtain a numerical solution the structure was discretized with a shell element having mid side nodes and the number of elements in the model was 27600. The results were further updated average damping constant, elastic material modulus and the material density. A satisfactory correletion of approximately one was obtained between the results from experiment and finite element simulation. The fundamental mode of vibration of the structure was breathing.

Enclosure produced by manufacturers for the reduction of noise to the environment in machines such as a generator is becoming more complex using present day technology and hence the need to understand the characteristics of such structures. This study is progressing towards the achievement of an updating tool for generator canopy.

References

[1] Ewins, D. J. Modal Testing : Theory and Practice Research Studies Press Ltd., 1984.
[2] Mottershead J E and Friswell M I, 'Model Updating in Structural Dynamics: A Survey', Journal of Sound and Vibration (1993) 167(2), 347 – 375.
[3] Zapico, J L, Gonzalez, M P, Friswell, M I, Taylor, C A and Crewe, A J, 'Finite Element Model Updating of a Small Scale Bridge', Journal of Sound and Vibration, 268 (2003) pp 993 – 1012.
[4] Mares C, Mottershead, J E and Friswell M I 'Results Obtained by Minimising Natural Frequency Errors and Using Physical Reasoning' Mechanical Systems and Signal Processing (2003) 17(1) 39 – 46.
[5] Friswell, M I; Mottershead J E and Ahmadian, H; 'Finite-element Model Updating using Experimental Test Data: Parametrization and Regularization', Phil. Trans. Royal Society of London A (2001), pp 169 – 186.
[6] Nalitolela, N; Penny, J E T and Friswell, M I; 'Updating Model Parameters by Adding an Imagined Stiffness to the Structure'; Mechanical Systems and Signal Processing (1993) 7(2), pp 161 – 172.
[7] Nalitolela, N G; Penny, J E T and Friswell M I; 'Mass or Stiffness addition Technique for Structural Parameter Updating', International Journal of Analytical and Experimental Modal Analysis', vol. 7 no 3, pp 157 – 168, Jul 1992.
[8] Mares, C; Mottershead, J E and Friswell, M I; 'Stochastic Model Updating: Part 1: Theory and Simulated Example', Mechanical Systems and Signal Processing 20 (2006) pp 1674 – 1695.
[9] Mottershead J E, Mares C, James S and Friswell M I; 'Stochastic Model Updating: Part 2: Application to a Set of Physical Structures', Mechanical Systems and Signal Processing 20 (2006) pp 2171 – 2185.
[10] Stefan Reh, Jean-Daniel Beley, Siddhartha Mukherjee and Eng Hui Khor; 'Probabilistic Finite Element Analysis using ANSYS', Structural Safety 28 (2006) pp 17 – 43.
[11] B S 6897 Part 5, 1995, 'Experimental Determination of Mechanical Mobility'.
[12] Jimin He and Zhi-Fang Fu: Modal Analysis, 1st ed: Butterworth-Heinemann 2001.
[13] M I Friswell and J E Mottershead 'Finite Element Model Updating in Structural Dynamics' Kluwer Academic Publishers, 1995.

Acknowledgement: *Thank   to   'Technology   Strategy   Board'   (project   number, TP/8/ADM/6/I/Q2017G) and 'EPSRC' (project number DT/F006829/1) for funding this project*

Applied Mechanics and Materials Vols. 24-25 (2010) pp 343-348
© (2010) Trans Tech Publications, Switzerland
doi:10.4028/www.scientific.net/AMM.24-25.343

What We Can Learn About Stick-Slip Dynamics

F. di Liberto[1, a], E. Balzano[1,b], M. Serpico[1,c] and F. Peruggi[1,d]

[1] Dipartimento di Scienze Fisiche, SPIN-CNR, INFN, Università di Napoli Federico II, via Cintia, I-80126 Napoli, Italy

[a]diliberto@na.infn.it, [b]balzano@na.infn.it, [c]serpicom@gmail.com, [d]peruggi@na.infn.it

Keywords: friction, crossover, low- high speed.

Abstract.
Stick-Slip motion is the basis for the description of a great variety of phenomena characterized by the presence of sliding friction between bodies with elastic features. In this article a simple experimental equipment for the analysis of this kind of dynamics is described. A wide set of possible experimental observations and measures are presented. This equipment has been tested at the university of Napoli Federico II in courses for undergraduate students and in the teacher training school for secondary education.

Introduction

A large number of natural phenomena characterized by very distinct qualitative and quantitative aspects, share common alternating dynamics in specific circumstances. These dynamics are characterized by alternating phases: *static* phases where the system accumulates potential energy and *dynamic* phases where this energy is transformed in kinetic energy. These phenomena are similar, in a more or less abstract way, to the problem of the sliding motion with friction of a body with elastic properties and are usually referred to as stick-slip processes [1-5]. Many other processes with two intermittent phases can be traced back to stick-slip processes: landslides, the motion of a windscreen wiper on a dry glass, the sound generated when a fingertip moves along the edge of a glass, the sound emission mechanism of a violin, are common examples of stick-slip processes. Many other examples may be found in a variety of mechanic processes, both artificial (braking of cars and trains) and natural (earthquake generation [10] and avalanche dynamics [9]).

The most simple physical system that exhibits a stick-slip dynamics consists of a body having mass M, that slides on a plane surface with friction. The body is pulled by a spring with elastic constant k, in such a way that the free end of the spring moves at a constant velocity v. For the sake of simplicity, a physical system equivalent to the former one can be considered, where the body is connected to a wall by means of a spring, while a rough plane slides at constant speed under the body. The acronym CS will be used to denote this Constrained Spring system (Fig. 1), and the acronym FS (Free Spring) to denote the previously mentioned system. The two systems are related to each other by means of a simple coordinate change.

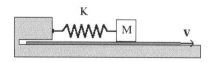

Figure 1 – Strained Spring System

Equation of motion

Let $x = x(t)$ be the position of the body in the SS system. Assume that the origin of the reference frame is located at the position of the body at the time $t = 0$, and that at $t = 0$ the spring is relaxed. At $t = 0$ the plane starts to move with speed v under the body. The equation of motion is given by

$$M\ddot{x} = F_s - F_{el} \tag{1}$$

where F_s is the static friction and F_{el} is the elastic force generated by the spring. At each instant following $t = 0$, the static friction equals the elastic force and the body remains united with the plane (stick phase). Therefore its motion is described by the equation

$$x(t) = vt \tag{2}$$

At a fixed time t_1, the elastic force of the spring

$$F_{el} = kx(t) = kvt \tag{3}$$

equals the critical value of the static friction force

$$F_{max} = \mu_s N = \mu_s Mg \tag{4}$$

where $N=Mg$ is the constraint reaction of the plane and μ_s is the static friction coefficient, which is assumed to be independent of the contact surface area [6,7]. Starting from t_1, the body slides back along the plane (slip phase), subject to the elastic force and dynamic friction force:

$$F_d = \mu_d N = \mu_d Mg \tag{5}$$

where μ_d is the dynamic friction coefficient.

It is assumed that the dynamic friction coefficient does not depend on the speed v, the mass M, and the contact surface area [6,7]. Thus, the equation of motion for the slip phase is given by

$$M\ddot{x} = F_d - kx(t) \tag{6}$$

with initial conditions

$$\begin{cases} x(t_1) = \dfrac{\mu_s N}{k} \\ \dot{x}(t_1) = v \end{cases} \tag{7}$$

If a new coordinate $y = y(t)$ for the position of the body is introduced

$$y(t) = x(t) - \frac{F_d}{k} \tag{8}$$

the equation of motion becomes

$$M\ddot{y} = -ky(t)$$

with initial conditions

$$\begin{cases} y(t_1) = \dfrac{(\mu_s - \mu_d)N}{k} \\ \dot{y}(t_1) = v \end{cases} \tag{9}$$

This equation admits a solution of the type

$$y(t) = A\cos(\omega t + \varphi) = A\cos\omega t\cos\varphi - A\sin\omega t\sin\varphi \tag{10}$$

where A and φ depend on the initial conditions, and $\omega^2 = k/M$.
Chosen $t_1 = 0$ as initial time, with the initial conditions (9), the solution can be written in the form

$$y(t) = \dfrac{(\mu_s - \mu_d)N}{k}\cos\omega t + \dfrac{v}{\omega}\sin\omega t \tag{11}$$

from which $x = x(t)$ in the slip phase is given by

$$x(t) = \dfrac{\mu_d N}{k} + \dfrac{(\mu_s - \mu_d)N}{k}\cos\omega t + \dfrac{v}{\omega}\sin\omega t \tag{12}$$

A low speed range for this equation can be considered, when the following inequality holds

$$\dfrac{v}{\omega} = v\sqrt{\dfrac{M}{k}} << \dfrac{(\mu_s - \mu_d)N}{k} \tag{13}$$

In such a case the term containing $\sin\omega t$ in equation (12) can be neglected and the motion of the block in the slip phase is a harmonic oscillation whose amplitude $(\mu_s - \mu_d)N/k$, for fixed values of M and k, is completely defined by the difference between static and dynamic friction coefficients (to get an idea of the orders of magnitude such that this occurs, observe that for $M = 1$ kg, $k = 1$ N/m, and $(\mu_s - \mu_d)N = 1$ N, equation (13) means that $v << 1$ m/s). The slip phase actually ends after half an oscillation, when the body acquires again the speed v, i.e. when it is again united with the plane. Subsequently, stick and slip phases alternate periodically. A combination of the time law for the stick phase and the time law for the slip phase gives the time law plotted in Fig. 2.

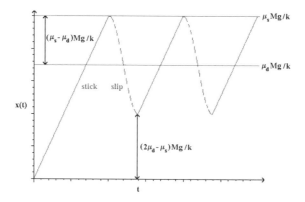

Figure 2 - Graph of $x(t)$

Dependence of the motion on the speed of the sliding plane

It's interesting to study the characteristic features of the motion when the speed of the sliding plane changes, forcing all the other parameters to stay constant. What happens is that, when v increases, the stick phase becomes shorter and shorter, until it disappears in the infinite speed limit. Let T_{stick} be the duration of the stick phase

$$T_{stick} = \frac{2(\mu_s - \mu_d)N}{kv} \approx \frac{const.}{v}. \tag{14}$$

The duration of the slip phases can be obtained differentiating relation (12) and requiring that

$$T_{slip} = 2\pi / \omega = T \tag{15}$$

This solution clearly represents the limit condition where no stick phase occurs. The other solutions of equation (8) are shown in Fig. 3 together with the values of T_{stick}. For small speeds, the stick time diverges while the slip time tends to $T/2$, i.e. just the value that has been used in the low speed approximation. When the speed increases, T_{stick} exhibits a fast decrease and asymptotically tends to zero, while T_{slip} tends to its asymptotic value T. In the neighborhood of a certain critical value v_{crit} of the sliding speed (in the case shown in Figure 7, $v_{crit} = 0.42$ m/s) there is a cross-over phenomenon that quickly brings the system from configurations where $T_{stick} \gg T_{slip}$ to configurations where $T_{stick} \ll T_{slip}$. For speeds much greater than this critical speed, the stick phase is practically non-existent and the motion can be approximated as a simple harmonic oscillation.

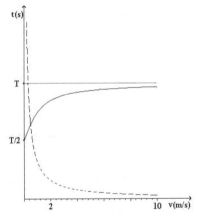

Figure 3 - Graphs of T_{stick} (red) and T_{slip} (blue) for the following values of the parameters: $M = 1$ kg, $k = 25$ N/m, $\mu_s = 1$, $\mu_d = 0.5$. The period T of the harmonic oscillations is about 1.3 s

The experimental device

The block is a wooden parallelepiped whose weight is approximately 650 g (inclusive of the weight of the position sensor mounted on top of it) with a side covered by a rubber layer. This side of the block is put on a strip of cloth whose sliding motion can be driven by means of a handle connected to a cylinder. It is not difficult to move the handle of the cylinder in such a way that the cloth slides on the workbench at an approximately constant speed. The block is connected by means of a spring to a force sensor that measures the tension of the spring. The force sensor is inserted in a flat screen, which is orthogonally locked to the wooden table around which the strip slides. On top of the block there is a position sensor (sonar) that measures the position of the block relative to the flat screen. Both the sensors are interfaced with a PC (by means of the LabPro software package). The software generates and displays directly the graphs of the observables (spring tension, position, speed, and acceleration of the block, and all the physical functions that can be evaluated from them) as functions of time. In Fig. 4-6 the graphs of some of the measured quantities are shown (For M = 0.650 kg, k = 22 N/M, and v = 0.06 m/s).

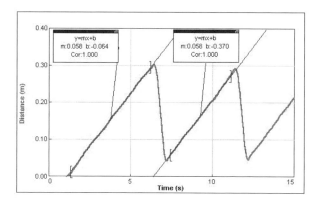

Figure 4 – Measured position of the block

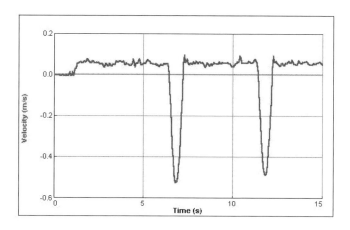

Figure 5 – Measured velocity of the block

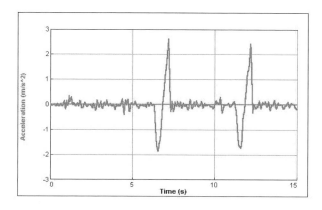

Figure 6 – Measured acceleration of the block

Conclusions

Some of the didactic opportunities given by a simple stick-slip device designed and implemented at the University of Naples are presented in this paper. What seems to be peculiar in this experiment is that one has, on the one hand, the opportunity to perform very simple observations related to the dynamics of the block (time law, elastic force, friction) and to measure all the physical quantities related to it; on the other hand, one has the opportunity to go deeper inside the subject with more complex observations. Think of the possibility to explicit the dependence of the stick-slip dynamics on the sliding velocity or on the initial conditions, to the exploration of different types of dynamics and their connection with the model, to the possibility of displaying real-time the phase space trajectory and the mechanic energy trends. The use of computer aided sensors allows also to introduce the problems connected with transduction, sampling and numerical calculus, whose implications can also be observed real-time. During the last years, these didactic materials have been used in lab-sessions for both undergraduate students and participants to the teacher training school for secondary education at the University of Naples "Federico II".

References

[1] F.P. Bowden and D. Tabor: *Friction and Lubrification* (Oxford University Press 1954).

[2] E. Rabinowicz *The intrinsic Variables sffecting the Stick-Slip Process* Proc. Phys. Soc. **71** (1958) 668-675, and *Stick and slip* Sci. Am. **194** (1956),109-118.

[3] F.J. Elmer *Nonlinear dynamics of dry friction* J. Phys. A: Math. Gen. **30** (1997),6057-6063.

[4] B.N.J. Persson and V.L. Popov *On the origin of the transition from the slip to stick* Solid State Communications **114** (2000) 261-266.

[5] A.J. McMillan *A non-linear model for self-exicited vibrations* Journal of Sound and Vibration **205** (2000) 323-335.

[6] M. Denny *Stick-slip motion : an important example of self-excited oscillation.* Eur. J. Phys **25** (2004) 311-322.

[7] R. Cross *Increase in friction force with sliding speed,* Am. J. Phys. **73** (2005) 812-816.

[8] F.-J. Elmer *The Friction* Lab. http://monet.physik.unibas.ch/~elmer/flab/index.html.

[9] S.R. Nagel I*nstabilities in a sandpile.* Rev. of Mod. Phys. **64** (1992), 32. L.A.N: Amaral and K.B. Lauritsen *Energy avalanches in rice-pile model*, Physica.A **231** (1996), 608-614. J. Rajchenbach *Dynamic of grain avalanches*, Phys.Rev.Lett. **88** (2002),14301.

[10] M.A. Moser, (1986) *The response of stick-slip systems to random seismic excitation* Technical Report: CaltechEERL:1986.EERL-86-03,   California   Institute   of   Technology http://caltecheerl.library.caltech.edu/167/00/8603.pdf; J. Galeano, P. Español and M.A. Rubio *Experimental and theoretical results of stress relaxations in a model of earthquake dynamics*, Europhys. Lett., **49** (2000) 410.

[11] L. Gratton, S. Defrancesco, *A simple measurement of the sliding friction coefficient*, Physics Education, vol 41, 232 (2006).

Applied Mechanics and Materials Vols. 24-25 (2010) pp 349-355
© *(2010) Trans Tech Publications, Switzerland*
doi:10.4028/www.scientific.net/AMM.24-25.349

Determination of High Strain-Rate Compressive Stress-Strain Loops of Selected Polymers

T. Yokoyama[1, a] and K. Nakai[1, b]

[1] Department of Mechanical Engineering, Okayama University of Science, Okayama 700-0005, Japan

[a] yokoyama@mech.ous.ac.jp, [b] nakai@mech.ous.ac.jp

Keywords: Dynamic energy absorption, Hopkinson bar, Polymer, Strain rate, Stress-strain loop

Abstract. Compressive stress-strain loops of selected polymers at strain rates up to nearly 800/s are determined in a strain range of nearly 8% on the standard split Hopkinson pressure bar. Four different commercially available extruded polymers are tested at room temperature. The compressive stress-strain loops at low and intermediate strain rates are measured on an Instron testing machine. The effects of strain rate on the Young's modulus, flow stress and dissipation energy are discussed. It is shown that the area included within the stress-strain loop increases with increasing strain rate as well as a given strain, that is, all four extruded polymers tested exhibit intrinsic strain-rate dependent viscoelastic behavior and a high elastic aftereffect following complete unloading.

Introduction

Polymeric materials with low mechanical impedance have widely been used in automotive, aerospace and portable electronics applications for shock and vibration absorption. In order to ensure the structural integrity of these applications from the product design stage, it is needed to have a precise knowledge of the stress-strain and energy dissipation behavior of these materials under impact loading. So far, the dynamic compressive [1-3], tensile [4, 5] and torsional [6] stress-strain responses of various polymers have often been determined on the conventional [7] or modified split Hopkinson pressure bar (SHPB). However, except for Ref. [8], most of the previous studies focused only on the dynamic stress-strain behavior during loading, and did not consider that during unloading process. Consequently, the dynamic energy absorption characteristics of polymeric materials have not been rigorously understood as yet.

The purpose of the present paper is to obtain valid compressive stress-strain loops of selected polymers at strain rates up to about 800/s using the conventional SHPB. Four different commercially available extruded polymers or ABS (Acrylonitrile Butadiene Styrene), HDPE (High Density Polyethylene), PP (Polypropylene) and PVC (Polyvinylchloride) were tested at room temperature. Cylindrical specimens with a slenderness ratio l/d (= length/diameter) of 0.5 were used in the SHPB tests, and those with $l/d = 2.0$ as recommended in the ASTM Designation were used in the low and intermediate strain rate tests. The compressive stress-strain loops at low and intermediate strain rates were measured on an Instron 5500R testing machine. The influences of strain rate on the Young's modulus, flow stress at a fixed strain of 5% and dissipation energy were examined. It is shown that the area enclosed by the stress-strain loop increases with increasing strain rate as well as strain, that is, all four polymers tested display inherent strain-rate dependent viscoelastic behavior and delayed reversible deformation following complete unloading. It is expected that the present results will lead to a better understanding of the dynamic mechanical response of polymers.

Experimental Procedure

Test Polymers and Specimen Preparation. Four different common polymers, i.e., two amorphous polymers (ABS, PVC) and two semi-crystalline polymers (HDPE, PP) were chosen for compression tests at room temperature (see, Fig.1) Cylindrical specimens were machined out of commercial extruded rods with a diameter of nearly 10 mm into short cylinders with a diameter of 9 mm. The static specimen's length was determined to be $l/d = 2.0$ in accordance with the ASTM Designation E9-89a [9] (see, Table 1). The impact specimen's length was determined so that the slenderness ratio of l/d ($d = 9$ mm) value is equal to 0.5, which corresponds to the optimum specimen geometry for metallic materials suggested by Davies and Hunter [1] (see, Table 2). All specimens of the different four polymers were tested in the as-received state.

Table 1 Geometry of static compression specimen

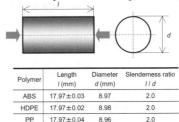

Polymer	Length l (mm)	Diameter d (mm)	Slenderness ratio l/d
ABS	17.97±0.03	8.97	2.0
HDPE	17.97±0.02	8.98	2.0
PP	17.97±0.04	8.96	2.0
PVC	17.94±0.04	8.96	2.0

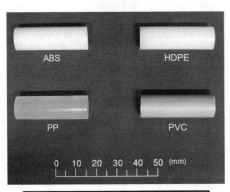

Abbreviation	Full name for polymers
ABS	Acrylonitrile-butadiene-styrene
HDPE	High Density Polyethylene
PP	Polypropylene
PVC	Polyvinylchloride

Fig.1 Picture of four different polymers tested

Table 2 Geometry of impact compression specimen

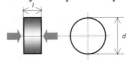

Polymer	Length l (mm)	Diameter d (mm)	Slenderness ratio l/d
ABS	4.44±0.01	8.98	0.5
HDPE	4.43±0.03	8.98	0.5
PP	4.50±0.07	8.97	0.5
PVC	4.44±0.03	8.97	0.5

High Strain-Rate Compression Testing. The general arrangement of the SHPB set-up is given in Fig. 2. The SHPB set-up consists of two 2024-T4 Al alloy bars of 2000 mm long and 10.1 mm diameter, which remain elastic during the tests. A striker bar is a rod of the same material and diameter, having a length of 350 mm. The mechanical properties of the bars are as follows: Young's modulus $E = 73$ GPa; longitudinal elastic wave velocity $c_0 = 5130$ m/s; mechanical impedance $Z = \rho c_0 = 14.2 \times 10^6$ kg m^{-2}s^{-1}; yield strength $\sigma_Y = 325$ MPa. The 2024-T4 Al alloy bars with low mechanical impedance are used to reduce a drastic impedance mismatch between the polymer specimen ($Z \doteq 1 \sim 2 \times 10^6$ kg m^{-2}s^{-1}) and the conventional steel bars ($Z \doteq 40 \times 10^6$ kg m^{-2}s^{-1}), which results in a transmitted strain signal with a very low signal-to-noise ratio. The specimen is held in place between the input and output bars by applying a very small pre-compression load with turning of the head of a support block. As in the static tests, lubricant (petroleum jelly) is applied to the bar/specimen interfaces to minimize the frictional effects. A pulse shaping technique [8] is used to generate well-defined compressive strain pulses without inherent high-frequency noise in the input bar. Namely, a thin 1050 aluminum disk of nearly 10 mm diameter and 0.2 mm thick is attached onto the impact end of the input bar using a thin layer of petroleum jelly. When the input bar is impacted with the striker bar launched through the gun barrel, a compressive strain pulse (ε_i) is generated in the input bar and travels towards the specimen. At the bar/specimen interface, because of the impedance mismatch, part of the strain pulse is reflected back into the input bar (ε_r) and the remaining part is transmitted through the specimen into the output bar (ε_t). Figure 3 gives a Lagrangian x-t diagram illustrating the details of the strain pulse propagation in the Hopkinson bars. Note that in the impact testing of the polymers, the duration of the reflected and transmitted strain pulses commonly becomes much longer than that of the incident strain pulse (ε_i). This is entirely due to a very long retardation time [10] of the polymers themselves. The incident, reflected and transmitted strain pulses are then recorded with two pairs of strain gages (gage length=1 mm) mounted on the input and output bars. The output signals from the strain gages are fed through a bridge circuit into a 10-bit digital storage oscilloscope, where the signals are digitized and stored at a sampling time of 1 µs/word. The digitized data are then transferred to a 32-bit personal computer for data processing.

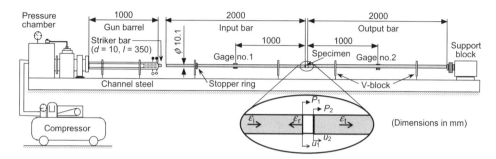

Fig.2 Schematic diagram of conventional split Hopkinson pressure bar set-up
(recording system not shown)

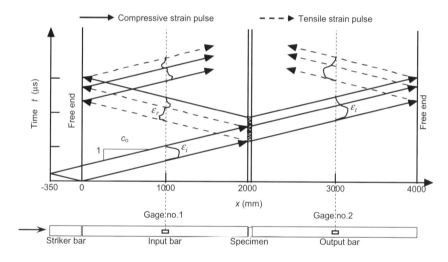

Fig.3 Lagrangian diagram for conventional split Hopkinson pressure bar

By applying the elementary one-dimensional theory of elastic wave propagation, we can determine the nominal strain $\varepsilon(t)$, strain rate $\dot{\varepsilon}(t)$ and stress $\sigma(t)$ in the specimen from the SHPB test records as [11]

$$\varepsilon(t) = \frac{u_1(t) - u_2(t)}{l} = \frac{2c_o}{l} \int_0^t \{\varepsilon_i(t') - \varepsilon_t(t')\} dt' \tag{1}$$

$$\dot{\varepsilon}(t) = \frac{\dot{u}_1(t) - \dot{u}_2(t)}{l} = \frac{2c_o}{l} \{\varepsilon_i(t) - \varepsilon_t(t)\} \tag{2}$$

$$\sigma(t) = \frac{P_2(t)}{A_S} = \frac{AE}{A_S} \varepsilon_t(t) \tag{3}$$

Here u and P are, respectively, the displacement and the axial force on both ends of the specimen (see the inset in Fig. 2). Equations (1) to (3) are derived under the assumption of dynamic force equilibrium ($P_1(t) = P_2(t)$) across the specimen, which can be expressed as

$$\sigma_1(t) = \sigma_2(t) \quad \text{or} \quad \varepsilon_i(t) + \varepsilon_r(t) = \varepsilon_t(t) \tag{4}$$

where

$$\sigma_1(t) = \frac{P_1(t)}{A_S} = \frac{AE}{A_S}[\varepsilon_i(t) + \varepsilon_r(t)], \quad \sigma(t) = \frac{P_2(t)}{A_S} = \frac{AE}{A_S}\varepsilon_t(t)$$

Here A and E denote the cross-sectional area and Young's modulus of the Hopkinson bars; l and A_S denote the length (or thickness) and cross-sectional area of the specimen. Eliminating time t through Eqs. (1) to (3) yields the nominal (or engineering) compressive stress-strain and strain rate-strain relations. The compressive stress and strain are assumed to be positive in this work.

Results and Discussion

A number of the SHPB tests were conducted on the four different polymers at room temperature. Figure 4 indicates typical oscilloscope records from the SHPB test on ABS. The top trace gives the incident and reflected strain pulses (ε_i and ε_r), and the bottom trace gives the transmitted strain pulse (ε_t). The recorded signal data are neither smoothed nor averaged electronically. It is very important to notice that the duration ($\fallingdotseq 380$ μs) of the reflected and transmitted strain pulses is much longer than that ($\fallingdotseq 270$ μs) of the incident strain pulse. This is because it takes much time for the specimen to recover to zero stress during unloading, causing often the overlapping between the transmitted strain pulse and its strain pulse reflected from the free (or right) end of the output bar. This is graphically represented in the Lagrangian x-t diagram shown in Fig. 3. In the present tests, the overlapping of the two strain pulses is successfully avoided by the use of the long output bar. Figure 5 shows the dynamic compressive stress and strain time histories for ABS. Note that the strain increases more slowly than the stress, and then decreases asymptotically to about 3.5% strain until the stress very gradually reduces to zero at nearly 380 μs. Figure 6 presents the resulting dynamic stress-strain loop and strain rate-strain relations in compression. The strain rate does not remain constant during loading as well as unloading, and hence the strain rate $\dot{\varepsilon} = 680$/s given denotes the average one during loading process, which is calculated by dividing the area under the strain rate-strain curve up to the maximum strain (= 8%) by the value of its strain. As in the low and intermediate strain rate tests, the dynamic stress-strain loop is not closed, and, consequently, a residual strain of about 3.5% is not completely recovered to zero in time.

Figures 7(a) to 7(d) show the compressive stress-strain loops of the four different polymers at three different strain rates. The intermediate strain rate stress-strain loops were measured on the Instron 5500R testing machine at a crosshead speed of 100 mm/min. It is observed that the initial slope (or Young's modulus E) and the area within the loop increase greatly with increasing strain rate. In an effort to evaluate the effects of strain rate on the overall compressive properties, the measured values for the Young's modulus, flow stress at a given strain of 5% and dissipation energy are plotted in Figs. 8 to 10, respectively, as functions of the average strain rate $\dot{\varepsilon}$ during loading process. Note that in the plots of the Young's modulus in Fig. 8, the strain rate in the initial elastic portion of the dynamic stress-strain loops does not exactly correspond to the average strain rate during entire loading process and, hence, the strain rate is re-evaluated by dividing the area under the initial strain rate-strain curve up to a strain at the proportional limit by the value of its strain. Thus, the average strain rate in the initial elastic region for each polymer is used in Fig.8. In Fig. 10, the dissipation energy U_d is obtained as the area enclosed by the loop, which can be calculated by subtracting the area under stress-strain curve during unloading from that during entire loading.

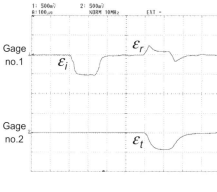

Sweep rate: 100 μs/div
Vertical sensitivity:
　Upper trace: 500mV/div (1266 με/div)
　Lower trace: 500mV/div (1274 με/div)

Fig. 4 Oscilloscope traces from SHPB test on ABS
$(V_S = 13.4 \text{ m/s})$

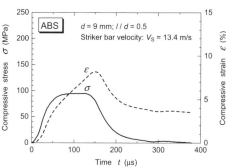

Fig. 5 Dynamic stress and strain time histories
for ABS

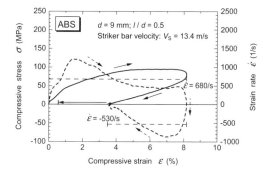

Fig. 6 Dynamic stress-strain and strain
rate-strain loops in compression for ABS

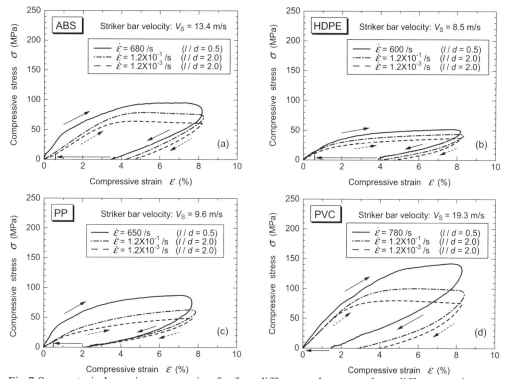

Fig.7 Stress-strain loops in compression for four different polymers at three different strain rates

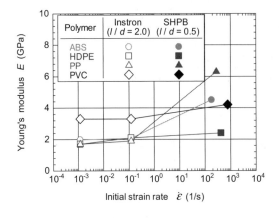

Fig.8 Young's modulus versus initial strain rate
for four different polymers

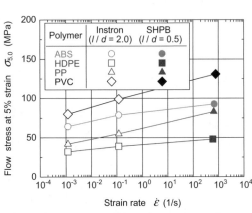

Fig.9 Flow stress at 5% strain versus strain rate
for four different polymers

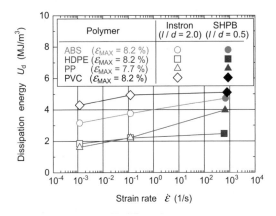

Fig.10 Dissipation energy versus strain rate for
four different polymers. Note: strain values in
parentheses denote the maximum compressive
strain achieved in each SHPB test

As can be seen from Figs. 8 to 10, the Young's modulus, flow stress at the fixed strain of 5% and dissipation energy increase significantly with increasing strain rate. Amorphous polymers (ABS and PVC) exhibit a higher flow stress and a larger dissipation energy than semi-crystalline polymers (HDPE and PP) at any strain rate. The dissipation energy is mostly converted to heat during high rate deformation, which causes the adiabatic temperature rise within the specimen. The temperature rise ΔT_d can be evaluated based on the dynamic dissipation energy U_d as (see Ref. [12]):

$$\Delta T_d = \frac{\beta_C}{\rho c_p}\oint \sigma(\varepsilon)\,\mathrm{d}\varepsilon = \frac{\beta_C}{\rho c_p}U_d \tag{5}$$

Here the numerical integration is performed over $[0, \varepsilon_{max}]$; β_C $(= 0.9)$ denotes the conversion factor; ρ and c_p are the mass density and the specific heat listed in Table 3; $\sigma(\varepsilon)$ is the dynamic stress-strain loop. The temperature rise in each polymer calculated from Eq.(5) is given also in Table 3. Of the four polymers, the highest temperature rise is clearly observed in PVC during high rate deformation up to nearly the same strain of 8%.

Table 3 Temperature rise in four polymers tested during high rate deformation

	ABS (●)*	HDPE (■)	PP (▲)	PVC (◆)
ρ (kg/m³)	1040	960	900	1420
c_p (kJ/kg·K)	(1.53)**	(2.30)	(1.93)	(1.01)
U_d (MJ/m³)	4.72	2.47	3.97	5.10
ΔT_d (K)	2.67	1.01	2.05	3.20

* Note: symbols in parentheses correspond to those given in Fig. 10
** Note: values in parentheses are provided by manufacturers

Conclusions

From the present experimental work, we can draw the following conclusions:

(1) All four polymers exhibit intrinsic strain-rate dependent viscoelastic behavior and a high elastic aftereffect following complete unloading.

(2) The Young's modulus, flow stress at the fixed strain of 5% and dissipation energy for all four polymers increase greatly with increasing strain rate.

(3) The highest temperature rise is found in PVC during high strain-rate deformation up to nearly 8% strain.

(4) Amorphous polymers (ABS and PVC) have higher flow stress and larger dissipation energy than semi-crystalline polymers (HDPE and PP) at any strain rate.

References

[1] E.D.H. Davies and S.C. Hunter: J. Mech. Phys. Solids, Vol. 11 (1963), p. 155.

[2] S.S. Chiu and V.H. Neubert: J. Mech. Phys. Solids, Vol. 15 (1967), p. 177.

[3] E.W. Billington and C. Brissenden: Int. J. Mech. Sci., Vol. 13 (1971), p. 531.

[4] W. Chen, F. Lu and M. Cheng: Polymer Testing, Vol. 21 (2002), p. 113.

[5] V.P.W. Shim, J. Yuan and S.-H. Lee: Exp. Mech., Vol. 41 (2001), p. 122.

[6] N.A. Fleck, W.J. Stronge and J.H. Liu: Proc. Roy. Soc. Lon., Vol. A429 (1990), p. 459.

[7] H. Kolsky: Proc. Phys. Soc., Vol. B62 (1949), p. 676.

[8] B. Song and W. Chen: Exp. Mech., Vol. 44 (2004), p. 622.

[9] ASTM E9-89a: *Annual Book of ASTM Standards*. Vol. 03.01, p. 98, American Society for Testing and Materials, Philadelphia (1995).

[10] I.M. Ward and J. Sweeney: *An Introduction to the Mechanical Properties of Solid Polymers*, 2nd Edition, p. 57, John Wiley & Sons, Chichester (2004).

[11] U.S. Lindholm: J. Mech. Phys. Solids, Vol. 12 (1964), p. 317.

[12] M.A.Meyers: *Dynamic Behavior of Materials*, p. 377, John Wiley & Sons, London (1994).

Session 8B: Identification

Applied Mechanics and Materials Vols. 24-25 (2010) pp 359-364
© (2010) Trans Tech Publications, Switzerland
doi:10.4028/www.scientific.net/AMM.24-25.359

Identification of the shape of curvilinear beams and fibers

M. L. M. François[1,a], B. Semin [1,b], H. Auradou [1,c]

[1] Laboratoire FAST, Université Paris Sud 11,
Bâtiment 502, 91405 Orsay Cedex, France

[1,a] marc.francois@u-psud.fr
[1,b] benoit.semin@fast.u-psud.fr
[1,c] harold.auradou@fast.u-psud.fr

Keywords: analytical shape, curve detection, virtual image correlation, fiber, filament, loop

Abstract. This work concerns the shape identification of curvilinear objects, for example bent beams or wires in mechanics. The beam's digital picture is analyzed with the introduced Virtual Image Correlation method. This one consists in finding the optimal correlation between the beam's image and a virtual beam, whose curvature field is described by a truncated series. The gray level and amplitude of the virtual beam does not need to reproduce exactly the ones of the physical beam image. The analytical form of the optimal shape allows one to derive mechanical properties: the identification of the Young's modulus of a bar is given as an example. We will also show the robustness of the method with regards to the quality of the image.

Introduction

Full-field strain measurements recently took a major place in experimental mechanics [1, 2]. However, these methods do not apply to uni-dimensional objects, such as wires or beams images, whose width can be of the order of one pixel. Furthermore such objects may show large deformations, even including the formation of loops.

Methods issued from the image processing field, such as skeletonizing, level set methods, ridge following methods [3] or Radon transform based methods provide numerical results consisting in a set of pixels. This kind of discrete result is unadapted for beam mechanics that requires the computation of the curvature field (the second derivative of the shape is very noisy in this case).

The proposed method consists in finding the best correlation between the physical beam image and a virtual beam. From an approximate initial shape consisting of a set of straight segments, the virtual beam is gradually "deformed" until it perfectly matches (with respect to a quadratic distance) the physical beam image.

Unidimensional sketch of the method

This section explains the method in a simplified 1D case in which x is the abscissa. The goal is to recover the symmetry axis of a physical image f_0 from the digitalized image f. In the present example, f_0 has a symmetry axis at $x_s = 1.4$, ranges from 0 (black) at the border to 1 (white) at its center (Fig. 1a). The available information is the digital image f (Fig. 1b), where the gray level of each pixel is computed as the mean value of f_0 along the pixel (the pixel size corresponds here to an unit). Due to the choice of x_s in between two pixels, the digital image, 4 pixel width, has no symmetry axis. The proposed method for the recovery of x_s from f consists in finding the abscissa x_0 for which the quadratic distance between f and the virtual image g is minimum. This one, defined for any $x \in [-R, R]$ is expressed as:

$$g(x) = \frac{1}{2}\left(1 + \cos\left(\frac{\pi x}{R}\right)\right). \tag{1}$$

It has a symmetrical bell shape (Fig. 1c, shifted by a trial value $x_0 = 5$) and its width and shape are *a priori* different from the one of f_0. In order to maximize the precision of the method, g is discretized over a mesh (not represented in figure 1) thinner than the pixel size of f. The computation of the quadratic distance (Eq. 2) requires the computation of f^*, the (cubic) interpolation of f over the fine mesh of g (Fig. 1d). Generally, f^* is not similar to f_0.

$$\phi(x_0) = \int_{x_0-R}^{x_0+R} (f^*(x) - g(x - x_0))^2 \mathrm{d}x \qquad (2)$$

For the sake of simplicity, calculi over f^* and g are expressed here as continuous integrals but are computed numerically (over the fine mesh of g).

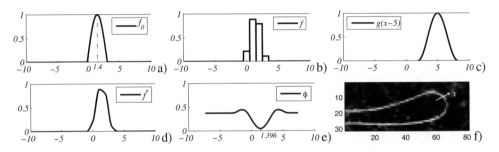

Figure 1: 1D sketch of the method: a) physical image, b) digital image, c) virtual image, d) interpolated digital image, e) quadratic distance, f) 1D mean line detection

The figure 1e shows the variation of the quadratic distance ϕ as a function of the shift x_0: it is minimum at $x_0 = 1.396$ pixels. Compared to the prescribed $x_s = 1.4$ pixels, this result shows that the proposed method reaches a subpixel precision for this example. Extra tests with various values x_s and square or triangle functions f_0 show quite similar accuracy as soon as f_0 is symmetric. Sub-pixel performance is usually observed with such algorithms used in the Digital Image Correlation field; mathematical explanations can be found in [2].

This simplified 1D analysis can be seen as a beam's mean line detection process along some axis x in a 2D image (Fig. 1f); function f corresponding to a cross section of the 2D image.

The virtual beam

We consider an image f of a physical beam (Fig. 1f, 3, 5). Compared to the 1D case (previous section), the researched mean line of the beam is no more a position but a curve. This curve represents the mean line of the virtual beam g, defined from its curvatures $\gamma(s)$, where s is the curvilinear abscissa. They are described by a truncated series:

$$\gamma(s) = \sum_{n=0}^{N} A_n \tilde{\gamma}_n(\tilde{s}), \qquad (3)$$

where $\tilde{s} \in [0, 1] = s/L$ and L is the (fixed) overall length of the beam. The functions $\tilde{\gamma}_n$ are the basis functions of the series description (Lagrange or Fourier in the present case) and the A_n are their respective weights. The angles $\theta(s)$ and positions $\mathbf{x}(s)$ of the virtual beam mean line are obtained by successive integration; introducing the integration constants $\theta_0 = \theta(0)$ and $\mathbf{x}_0 = \mathbf{x}(0)$, the problem consists in determining the $(N+4)$ terms $\{\mathbf{x}_0, \theta_0, A_0, ..., A_N\}$ for which the virtual beam is the closest to the physical one. The gray level of the virtual beam $g(\mathbf{X})$ is defined, similarly to Eq. (1), from the distance $r \in [-R, R]$ from the current point $\mathbf{X}$ to the virtual beam mean line then g is a sole function of r:

$$g(r) = \frac{1}{2}\left(1 + \cos\left(\frac{\pi r}{R}\right)\right). \tag{4}$$

The virtual beam g is not defined out of its definition domain D_g ($r \in [0, R]$, $s \in [0, L]$).

Figure 2: Example of virtual beam. Dashed line: mean line.

Optimization process

The searched parameters are the coefficients A_k of the series and the two integration constants θ_0 and the ordinate $x_{0,2}$ of $\mathbf{x}_0$ (abscissa $x_{0,1}$ is fixed because if not the problem is underdefined). These terms are joined together in a pseudo vector $V_k = \{x_{0,2}, \theta_0, A_k\}$. The optimization consists in finding the minimum, over V_k of the function Φ defined, similarly to Eq. 2, as:

$$\Phi = \iint_{D_g} (f - g)^2 \, \mathrm{d}S, \tag{5}$$

that is the quadratic distance between the virtual (g) and physical (f) beam images, where D_g is the definition domain of g and $\mathrm{d}S$ the surface element. For the sake of clarity this section uses continuous integrals and does not distinguish f from f^*. Writing the condition of minimum $\partial\Phi/\partial V_k = 0$ (and considering $g(\mathbf{X})$, where $\mathbf{X}(V_k)$ is a current point of the virtual beam) gives:

$$\oint_{\partial D_g} \left((f - g)^2 \, \mathbf{n} \cdot \frac{\partial \mathbf{X}}{\partial V_k} \mathrm{d}V_k\right) \mathrm{d}l - 2 \iint_{D_g} (f - g)\left(\mathbf{grad}(g) \cdot \frac{\partial \mathbf{X}}{\partial V_k} \mathrm{d}V_k\right) \mathrm{d}S = 0. \tag{6}$$

The first term, a line integral over the contour ∂D_g of D_g ($\mathbf{n}$ is the outer normal vector) is neglected on the two (small) ends of the beam at $s = 0$ and $s = L$. Its value over the lateral borders ($r = \pm R$) can be also neglected if one supposes that f, at these points (in the background of the physical image), is a constant value $f|_{\partial D_g}$: since $g(\pm R) = 0$ (see Eq. 4), using the divergence theorem leads to

$$\oint_{\partial D_g} (f - g)^2 \, \mathbf{n} \cdot \frac{\partial \mathbf{X}}{\partial V_k} \mathrm{d}V_k \quad \mathrm{d}l = (f|_{\partial D_g})^2 \frac{\partial}{\partial V_k}\left(\iint_{D_g} \mathrm{div}(\mathbf{X}) \mathrm{d}S\right) \mathrm{d}V_k. \tag{7}$$

As $\mathrm{div}(\mathbf{X})$ is constant (equals to 2), the surface integral equals $2S$ where S is the surface of the virtual beam. As S does not change with respect to the shape parameters V_k (a kinematic property of the beams of constant length), the right hand term equals to zero and the optimization condition (Eq. 6) reduces to the one proposed by Hild and Roux [2] for the DIC method:

$$\iint_{D_g} (f - g)\left(\mathbf{grad}(g) \cdot \frac{\partial \mathbf{X}}{\partial V_k}\right) \mathrm{d}S = 0. \tag{8}$$

The analytical definition of g (Eq. 3, 4) allows a fast computation of $\mathbf{grad}(g)$ and $\partial \mathbf{X}/\partial V_k$ (not shown here). The problem is discretized in order to use a Newton iterative process:

$$g(V_k + \Delta V_k) = g(V_k) + \mathbf{grad}(g) \cdot \frac{\partial \mathbf{X}}{\partial V_p} \Delta V_p. \tag{9}$$

This, with Eq. (8), gives:

$$\Delta V_p \iint_{D_g} \left(\mathbf{grad}(g).\frac{\partial \mathbf{X}}{\partial V_k} \right) \left(\mathbf{grad}(g).\frac{\partial \mathbf{X}}{\partial V_p} \right) \mathrm{d}S = \iint_{D_g} \left(\mathbf{grad}(g).\frac{\partial \mathbf{X}}{\partial V_k} \right) (f - g)\mathrm{d}S, \quad (10)$$

which is a matrix equation, that can be rewritten as $M_{kp}\Delta V_p = L_k$. Its solution ΔV_p defines the updated shape of the virtual beam. The iterative process stops with respect to a speed convergence criterion.

From a numerical point of view, Eq. (10) is computed with discrete integrals. The mesh of g is obtained from a discretization of s and r; it is more refined than the pixel size (about 3 times in the examples). Computations require the value of f on the mesh of g: as presented in the first section, this is done with a cubic interpolation (the result was referred as f^* in the first section). As the shape of the virtual beam changes at each step, this interpolation is reevaluated at each step.

Preliminary identification of the beam shape

Fig. (1e) shows that, even in 1D, the quadratic distance is convex only close to the solution $x_0 = x_s$. In this condition, the Newton algorithm requires the initial step to be close to the solution, in other words that D_g contains the beam image in f since the first iteration.

For this reason, a preliminary step by step algorithm identifies the beam shape segment by segment. A segment correspond to a small straight virtual beam (of typical length $4R$) whose gray level is defined by Eq. 4. Each new segment starts at the middle of the previous. Its angular orientation is defined from the smallest quadratic distance (Eq. 5) between it and the physical image f (the computation of Φ uses again the interpolation of f on the fine mesh of g).

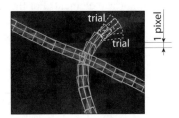

Figure 3: Segment by segment preliminary identification.

This method has proven to be robust and able to deal with loops. Fig 3 illustrates such preliminary identification. The image is extracted from a film of a thin fiber transported by a fluid flow in a transparent fracture [4]. No preliminary image processing was done. Despite the low definition of the image, the identification is already smooth and accurate. The half-width of the virtual beam was set at $R = 1$ pixel and the computing time was a few seconds on a current computer.

Technical aspects

Once the preliminary identification is made, the user can choose the type (Fourier or Legendre for instance) and the order N of the series. The initial values of the parameters V_k are obtained from this preliminary identification. The process starts and the shape of the virtual beam is gradually deformed until it fits the physical beam shape.

Fig. 4 shows the result of the identification, using a Fourier series et the order $N = 80$. The virtual beam perfectly matches the physical one, even in the detail view. Due to the analytical

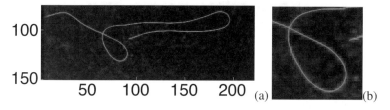

Figure 4: Shape identification. (a) global view. (b) loop detail.

definition of the virtual beam, this shape remains smooth (and infinitely differentiable). The precision of the identification depends upon the retained order of the series. Orders too low (here $N < 8$) do not allow the virtual beam to match the physical shape. The quadratic distance Φ decreases monotonically as N increases and asymptotically reaches a plateau whose value, generally nonzero, depends only upon the similarity between physical and virtual beams (but is not correlated to the precision of the identification).

Validation of the method: identification of the mechanical properties of a cantilever beam

A 2017-T4 aluminium straight bar (diameter 4.95 mm, length 2459 mm) has been fixed in the horizontal chuck of a milling machine in front of a black curtain. It bends under its own weight. The proposed method provides the shape, angles and the curvatures of the bar. This last one has

Figure 5: Aluminium bar bending under its own weight.

a strong mechanical meaning in the Timoshenko's beam theory as it is related to the bending moment. Considering that the external actions only reduce to the gravity, the beam theory writes as:

$$M(s) = \int_s^L \rho g \left(x_1(\xi) - x_1(s) \right) d\xi, \tag{11}$$

$$M(s) = E \frac{\pi R^4}{4} \gamma(s), \tag{12}$$

in which M is the flexural moment, ρ the linear mass density, g the standard gravity, E the Young's modulus and x_1 denotes the horizontal axis. In the large transformation framework, this problem accepts a numerical solution. This solution is used to validate the precision of our technique. The criterion retained for the comparison is the least square distance between the ordinates x_2. The best fit was obtained for $E = 72$ GPa, that corresponds to the value found in the literature. Furthermore, this value was confirmed by a three point bending test that gave $E = 72.6$ GPa.

The mean discrepancy between the ordinates of the analytical shape and the one measured by the present method was only 2 pixels (1.2 mm). Compared to the beam size, this gives a relative error of 5.10^{-4}. This identification was done with a Legendre series at an order $N = 8$.

Figure 6: The aluminium bar seen in the unwrapped frame of the virtual beam

Another tests, in which the chuck was turned of an half turn, shown that this was comparable to the imperfect straightness of the beam. The optical aberrations were not compensated in the present study, taking them into account would increase the precision of the method. Another strong visual information is given by the image of the physical beam in the straightened frame of the virtual beam g (Fig. 6): a correct identification theoretically leads to a symmetrical image, whatever the beam is curved.

Conclusions

The Virtual Image Correlation method presented here provides an analytical identification of the shape of any elongated object. It does not require any previous image processing and the full process only takes a few minutes for an high resolution image and a standard computer. Various test have shown the accuracy of the result, even on ill-defined images. The method uses only the pixels that belong to the beam image and its close vicinity, reducing the computing time and inducing insensitivity to far field artifacts (such as stones in Fig. 5). Furthermore, all the pixels of the beam are taken into account by the correlation (and not only the brightest ones); this contributes to the precision of the method.

This method can be applied in various fields, for example medicine (intestine shape identification, for example), biology (bacterium shape), engineering (beam measurement) and have already be successfully used in fluid mechanics for the shape identification of thermal plumes.

Using some relevant procedure (for example [5]), it will be possible to extract automatically the flexural properties of a fiber or beam from their images under various loadings. Further developments will concern image sequences, in which the result of an image will be used as initial values for the next one. This should allow the identification of the loop formation.

References

[1] M. Grédiac, E. Toussaint, and F. Pierron. Special virtual fields for the direct determination of material parameters with the virtual fields method. 1-Principle and definition. *International Journal of Solids and Structures*, 39(10):2691–2705, 2002.

[2] F. Hild and S. Roux. Digital image correlation: from displacement measurement to identification of elastic properties - a review. *Strain*, 42(2):69–80, 2006.

[3] S. R. Aylward and E. Bulitt. Initialisation, noise, singularities, and scale in height ridge traversal for tubular object. *IEEE Transactions on Medical Imaging*, 21(2):61–75, 2002.

[4] M. V. D'Angelo, B. Semin, G. Picard, M. Poitzsch, J.-P. Hulin, and H. Auradou. Single fiber transport in a fracture slit: influence of the wall roughness and of the fiber flexibility. *Transport in porous media, Published on line, DOI 10.1007/s11242-009-9507-x*, 2009.

[5] E. M. Meacham and J. F. Doyle. Implementation of a versatile two-stage inverse solution method using non-linear image data. In *Proceedings of the SEM 2007 conference*, 2010.

Applied Mechanics and Materials Vols. 24-25 (2010) pp 365-370
© (2010) Trans Tech Publications, Switzerland
doi:10.4028/www.scientific.net/AMM.24-25.365

Construction of shape features for the representation of full-field displacement/strain data

Weizhuo Wang[1,a], John E Mottershead[1,b], Amol Patki[2,c] and Eann A Patterson[2,d]

[1] Department of Engineering, University of Liverpool, Liverpool, UK, L69 3GH
[2] Department of Mechanical Engineering, Michigan State University, East Lansing, MI, USA
[a] wangweizhuo@gmail.com [b] j.e.mottershead@liverpool.ac.uk
[c] patkiamo@egr.msu.edu [d] eann@egr.msu.edu

Keywords: full-field strain pattern, shape features, Zernike moment, model updating

Abstract

The achievement of high levels of confidence in finite element models involves their validation using measured responses such as static strains or vibration mode shapes. A huge amount of data with a high level of information redundancy is usually obtained in both the detailed finite element prediction and the full-field measurements so that achieving a meaningful validation becomes a challenging problem. In order to extract useful shape features from such data, image processing and pattern recognition techniques may be used. One of the most commonly adopted shape feature extraction procedures is the Fourier transform in which the original data may be expressed as a set of coefficients (coordinates) of the decomposition kernels (bases) in the feature space. Localised effects can be detected by the wavelet transform. The acquired shape features are succinct and therefore simplify the model validation, based on the full-field data, allowing it to be achieved in a more effective and efficient way. In this paper, full-field finite element strain patterns of a plate with a centred circular hole are considered. A special set of orthonormal shape decomposition kernels based on the circular Zernike polynomials are constructed by the Gram-Schmidt orthonormalization process. It is found that the strain patterns can suitably be represented by only a very small number of shape features from the derived kernels.

Introduction

Numerical model updating [1] with the availability of full-field measurements [2] involves handling a huge amount of data with a high level of redundancy. It is necessary to extract useful information from the full-field responses, e.g. stress/strain, mode shapes etc, in a succinct form. Shape feature extraction [3,4] is one of the feasible ways to tackle such problems.

Simple shape features such as perimeter of boundaries, area, diameter, circularity, orientation and eccentricity describe the shape in a general way. More substantial shape features may be obtained by functional transformations [4]. In this case, the original shape may be expressed as a set of coefficients (coordinates) of the decomposition kernels (bases) in the feature space. The choice of transformation bases is problem dependent. For instance, periodic features can easily be represented by the Fourier transform; Local significance and global approximation can be detected by the wavelet transform and the Zernike moment is especially powerful in the discrimination of circular shapes.

Proper selection of the kernel functions usually results in a small number of significant and effective shape features by the nature of orthonormal kernels. However, the orthonormality holds only if the domain of the shape matches the definition domain of the orthonormal kernels. For example, the circular Zernike decomposition kernels, known as the Zernike polynomials, are orthonormal over the unit circle [6]. When applying the Zernike moment to non-circular structures, modification of the kernels has to be carried out to satisfy the orthonormality. Mahajan [5] employed the Gram-Schmidt orthonormalisation (GSO) process to produce a set of orthonormal Zernike annular polynomials. The GSO process is theoretically applicable to any structures.

In this paper, full field strain patterns of a square specimen with a central circular hole are considered. Sets of orthonormal kernels based on the modified Zernike polynomials defined over the specimen are generated by the GSO process. Results show that the full-field strain pattern with thousands pixels can efficiently be described by only 10 shape features by the constructed Zernike kernels.

Construction of shape features – modified Zernike moments

The general form of transform-based shape features may be expressed as,

$$\mathfrak{D}_i = \int_\Omega \mathfrak{K}_i^*(x,y) S(x,y) dxdy \tag{1}$$

where $S(x,y)$ denotes the continuous shape pattern, Ω denotes the domain of definition, $\mathfrak{K}_i(x,y)_{i=1,2,...}$ are the set of transformation kernels and $*$ denotes the complex conjugate. For instance, Fourier features can be obtained by assigning complex sinusoids as the kernel functions.

When the orthogonal Zernike polynomials $\{V_{i,\ i=1,2,...}\}$ defined over a unit circle are adopted as the kernel functions, equation (1) is expressed as

$$\mathfrak{D}_{Z_i} = \int_{x^2+y^2\leq1} V_i^*(x,y) S(x,y) dxdy \tag{2}$$

where the Zernike polynomials

$$V_i \equiv V_{n,m}(x,y) \equiv V_{n,m}(\rho,\vartheta) = R_{n,m}(\rho)e^{im\vartheta} \tag{3}$$

where $R_{n,m}(\rho)$ represents the radial polynomials [7]. Thus, $\mathfrak{D}_{Z_i}$ may now be called the Zernike moment descriptor (ZMD).

Gram-Schmidt orthonormalization of the Zernike polynomial over a non-circular domain

A set of orthonormal Zernike polynomials defined over an arbitrary domain, e.g. a rectangular plate with a circular hole as shown in figure 1, may be determined by Gram-Schmidt orthonormalization (GSO) as

$$P_1' = V_1 \quad \text{and} \quad P_\ell' = V_\ell - \sum_{k=1}^{\ell-1} \text{proj}_{P_k'}(V_\ell) , \quad \text{for } \ell, k \in \mathbb{N} \text{ and } \ell \geq 2 \tag{4}, (5)$$

where $\mathbb{N}$ is the set of natural numbers, $\text{proj}_{P_k'}(V_\ell)$ denotes the projection of V_ℓ onto P_k' with respect to the definition domain $H(x,y)$ which is a binary (zero or one valued) function that defines the shape of the structure, V_ℓ is the circular Zernike polynomial; $\{P_1', P_2', ..., P_N'\}$ are the Gram-Schmidt orthogonalized arbitrary Zernike polynomials which can further be normalised by $P_i = P_i'/\|P_i'\|$. Thus, $\{P_1, P_2, ..., P_N\}$ are the orthonormal arbitrary Zernike polynomials. For example, the function $H_{RC}(x,y)$ for a rectangular plate with a circular hole as shown in Fig 1 can be defined as

$$H_{RC}(x,y) = \begin{cases} 1, & |x| \leq \dfrac{a}{2}; \ |y| \leq \dfrac{b}{2} \text{ and } \sqrt{x^2+y^2} \geq \varepsilon \\ 0, & otherwise \end{cases} \tag{6}$$

Thus, shape features, i.e. the Zernike moment defined over the non-circular structure, can be obtained by substituting the Gram-Schmidt orthonormalized kernels $P_k(x, y)$ into equation (1) as

$$z_k = \iint_{H_{RC}} S(x, y) P_k^*(x, y) dx dy \tag{7}$$

where z_k is called the kth Zernike moment of the strain patterns $S(x, y)$.

By definition, the number of decomposition kernels $\{P_{k, \ k=1,2,...}\}$ is infinity and so is the number of shape feature descriptors. It is essential to reduce the number of the descriptors without significant loss of information. Appropriate selection of the kernels according to the problem in hand may result in only a small number of significant descriptors. However, certain criteria must be met to ensure the retained small number of significant shape features is sufficient to represent the original pattern. Comparing the norm of the reconstructed pattern by retaining the significant terms with the original is a feasible criterion [7].

Full-field strain pattern decomposition using the modified Zernike polynomials

A thin square plate with a circular hole in the centre is modelled. The dimension of the plate is 0.1m ×0.1m and the radius of the hole is 0.01m. The specimen is modelled by plane stress elements as shown in Fig 2. Elastic-plastic material with isotropic hardening is considered. Tensile pressure ramp loading is applied steadily from time 0.0 to 3.0 seconds to the two side edges. Four particular points in the ramp loading history are shown in Table 1. Thus, the responses at 1.0, 2.0 and 3.0 seconds are considered in this example.

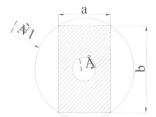

Fig 1. Rectangular test specimen with a circular hole normalised inside an unit circle - constrains:
$$a^2 + b^2 = 4; \varepsilon < \min\left(\frac{a}{2}, \frac{b}{2}\right)$$

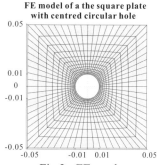

FE model of a the square plate with centred circular hole

Fig 2 : FE mesh

Table 1: Ramp Loading (Pressure)

		Step 1	Step 2	Step 3
Time (Sec)	0.0	1.0	2.0	3.0
Pressure (MPa)	0.0	80.0	160.0	240.0

The shape patterns of the maximum principal strain for the three loading stages are shown in the first row of Fig 3. Full-field shape pattern descriptors are not necessarily concentrated on the regions of high strain (such as around the circumference of the central hole) and special measures are therefore necessary to ensure that the shape can be represented using an acceptably small number of descriptors. The continuous strain pattern of the whole structure can be interpolated from the nodal strains using element shape functions. Then the important shape features of these full-field data may be extracted by the chosen decomposition kernels. An intuitive selection is the Zernike polynomial because the maximum and minimum strains appear in the regions arranged symmetrically around the circumference of the central hole. For example, the 14th circular Zernike polynomials, as shown in Fig 4 (a), exhibit patterns that are arranged similarly to the strain distributions. A set of orthonormal kernels defined over the domain of the specimen can be obtained by applying the *Gram-Schmidt orthonormalization* (GSO) based on the circular Zernike polynomials. The 14th kernel of the GSO-processed Zernike polynomials, denoted as the GSO_Z1 kernel, is shown in Fig 4 (b). It is obvious that the high-strain locations in the GSO_Z1 kernel is not close enough to the centre, when compared to the strain patterns. This problem may be overcome by transforming the radial coordinates of the circular Zernike polynomial before carrying out the GSO. For instance, the radial coordinate ρ in the radial function $R_{n,m}(\rho)$ of equation (3) may be replaced with $\rho' \equiv \rho'(\rho, v) = \rho^v$ where $s \in \mathbb{R}$ and $0 < v < 1$. Thus, the transformed radial function is expressed as

$$R'_{n,m}(\rho) \equiv R_{n,m}(\rho') = R_{n,m}(\rho^v) \tag{8}$$

Taking $\rho' = \rho^{0.25}$ for example, another set of orthonormal kernels, denoted as GSO_Z2, can be obtained by applying the GSO process to the coordinate-transformed Zernike polynomials. The 14th GSO_Z2 as illustrated in Fig 4 (c) is closer to the strain patterns than GSO_Z1 for the region near the central hole. However, large values remain in the regions around the outside edges of the GSO_Z2 kernels, which are unrepresentative of the strain distributions. A decaying weight function $w(\rho, t)$ may be applied to the radial functions of the coordinate-transformed Zernike polynomials to reduce this effect,

$$R^w_{n,m}(\rho, t) \equiv w(\rho, t)R'_{n,m}(\rho) = \rho^t R'_{n,m}(\rho), \qquad t \in \mathbb{R}^- \tag{9}$$

Similarly, a new set of kernels, denoted as GSO_Z3, can be created by the GSO processing on the weighted and coordinate transformed Zernike polynomials. The GSO_Z3 version of the 14th kernel is shown in Fig 4 (d) when assigning $t = -3$ to the weight function in equation (9). As seen from Fig 4, the GSO_Z3 kernel is now very similar to the strain patterns. Therefore, it is practical to employ the GSO_Z3 kernels to extract shape features from the strain patterns because they are more capable of extracting the essential shape information close to the central hole. Thus, the Zernike moment descriptors of the strain patterns for the loading step 3 as shown in Fig 3 are determined by substituting the GSO_Z3 kernels to equation (7) and its contribution are plotted in Fig 5. Also, the individual kernels of the 2nd to 5th largest terms sorted decreasingly for the strain pattern at 3rd step are shown in Fig 6. It can be seen from Fig 5 that only a very small number of terms are significant. The largest ZMD term is z_1 which represents the mean value of the shape pattern. The second most significant ZMD is z_6 which corresponds to the 6th GSO_Z3 kernel as illustrated in Fig 6, representing very closely the maximum and minimum strain patterns around the central hole. Further shape features corresponding to the 27th, 13th and 14th kernels etc also contain significant shape information.

In order to examine the development of these significant ZMDs in the loading history, more stages within the 3 steps are considered. A total of 13 stages is shown on the horizontal axis of Fig 7. The growth of the 7 largest ZMDs is shown in Fig 7. The ZMD z_1, representing the average strain, grows rapidly after 2 seconds because the whole specimen begins to yield. The next most dominant shape feature is ZMD z_6, indicating the distribution of the maximum and minimum strains around the hole , as shown in Fig 7. Several other important shape features $\{z_{27}, z_{13}, z_{14}, z_{15}, z_5\}$ also increase considerably as the stain field develops with increasing load.

Since only a small number of ZMD terms are significant, the number of shape features needed to recover the original pattern may be determined by comparing the norm of the original shape and reconstructed pattern by the retained ZMD terms [7]. The reconstruction from the 10 largest terms is shown in Fig 3. It is seen that the recovered images are very similar to the original patterns. The ratios between the norms of the original and reconstructed strain patterns by 10 terms is 97.2% meaning that the retained 10 terms are sufficient to recover more than 97% of the strain pattern.

Therefore, the retrained shape feature terms are succinct and may be assembled as a shape feature vector (SFV) to represent the original shape. Furthermore, the comparison of the full-field responses between analytical predictions and experimental measurements can be transferred to the distance measurement between the SFVs [4]. For example, cosine distance may be applied to measure the angular distance of two SFVs.

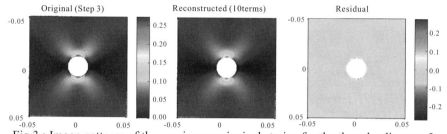

Fig 3 : Image patterns of the maximum principal strains for the three loading step3

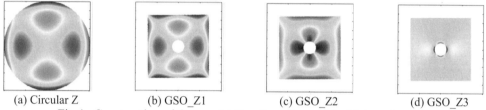

(a) Circular Z   (b) GSO_Z1   (c) GSO_Z2   (d) GSO_Z3

Fig 4 : Construction of the special Zernike polynomials (#14, n=4, m=2)

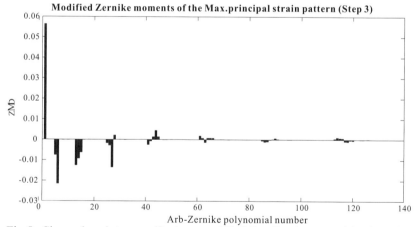

Fig 5 : Shape-descriptor amplitudes - modified Zernike decomposition kernels.

Fig 6 : Individual decomposition kernels (modified Zernike bases) of the strain pattern of step 3 – sorted in the decreasing order by the magnitude of the Zernike moment descriptor (2nd to 5th).

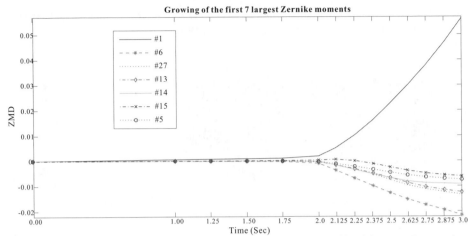

Fig 7 : Development of the 7 largest Zernike moments with loading history – 3 seconds

Summary

Image processing and pattern recognition techniques provide effective and efficient ways to describe the full-field structural responses such as vibration mode shapes and static stress/strain patterns. Shape decomposition based on transformation is one of the most common procedures. Appropriate selection of shape decomposition kernels usually result in a very small number of significant shape features. It is also feasible to construct satisfactory kernels to suit the special problem in hand. Sensitivities of the shape features with respect to the structural parameters were determined. An example problem was presented for demonstration. A modified Zernike moment descriptor was constructed to describe the full-field strain patterns of a square plate with a central circular hole and succinct shape features are obtained. The development of plasticity of the specimen could be reflected by the retained significant shape features.

Acknowledgements

The authors wish to acknowledge the support of EC project ADVISE (Advanced Dynamic Validations using Integrated Simulation and Experimentation), Grant number 218595.

References

[1]   Mottershead J., and Friswell M., 1993, "Model updating in structural dynamics: a survey," Journal of sound and vibration, 167(2), p. 347–375.

[2]   Whelan M. P., Albrecht D., Hack E., and Patterson E. A., 2008, "Calibration of a speckle interferometry full-field strain measurement system," Strain, 44, pp. 180-190.

[3]   Wang W., Mottershead J. E., and Mares C., 2009, "Mode-shape recognition and finite element model updating using the Zernike moment descriptor," Mechanical Systems and Signal Processing, 23, pp. 2088-2112.

[4]   Wang W., Mottershead J. E., and Mares C., 2009, "Vibration mode shape recognition using image processing," Journal of Sound and Vibration, 326, pp. 909-938.

[5]   Mahajan V. N. , 1981, "Zernike annular polynomials for imaging systems with annular pupils," J. Opt. Soc. Am., 71, pp. 75-85.

[6]   Wang W., Mottershead J. E., Patterson E. A., and Patki A., "Shape features and finite element model updating from full field strain data," International Journal of Solids and Structures, submitted.

[7]   Zernike F., "Translated: diffraction theory of the cut procedure and its improved form, the phase contrast method," Physica, 1.

Applied Mechanics and Materials Vols. 24-25 (2010) pp 371-377
© (2010) Trans Tech Publications, Switzerland
doi:10.4028/www.scientific.net/AMM.24-25.371

An Innovative Own-Weight Cantilever Method for Measuring Young's Modulus in Flexible Thin Materials Based on Large Deflections

Atsumi OHTSUKI [1, a]

[1] Meijo University, Department of Mechanical Engineering,

1-501, Shiogamaguchi, Tempaku-ku, Nagoya, 468-8502 JAPAN.

[a] email: Hohtsuki@ccmfs.meijo-u.ac.jp

Keywords: Material Testing, Cantilever, Young's Modulus, Mechanical Property, Own-Weight Deformation, Large Deflection, Thin Material

Abstract. This report deals with an innovative method (*Own-Weight Cantilever Method*) to measure Young's modulus of flexible thin materials. A newly developed method is based on the large deformation theory considering large deformation behaviors due to own-weight in flexible thin materials. Analytical solutions are derived by using Bessel Functions. By means of measuring the horizontal displacement or the vertical displacement at a free end of a cantilever, Young's modulus can be easily obtained for various flexible thin and long materials. Measurements were carried out on a piano wire. The results confirm that the new method is suitable for flexible thin wires.

Introduction

In recent years, flexible thin materials with very high performance are widely used. Therefore, large deformation analyses of these flexible materials have attracted attention considerably in the design of mechanical springs, fabrics and various thin-walled structures (aerospace structures, ship, car, etc.). An investigation of large deformation behaviors occurring in flexible materials is necessary for evaluation of mechanical properties such as Young's modulus. Here, a new Young's modulus measuring method (*Own-Weight Cantilever Method*) is proposed. Analytical solutions are derived by using Bessel Functions. By using this method, Young's modulus of thin and long flexible materials can be easily obtained by just measuring the horizontal displacement or the vertical displacement at the free end of the cantilever.

In order to assess the applicability of the proposed method, several experiments were carried out using a piano wire. As a result, it becomes clear that the new method is suitable for flexible thin wires. Besides the Own-Weight Cantilever Method studied here, the *Cantilever Method* [1], the *Circular Ring Method* [2,3] for a flexible single-layered material have already been reported, based on the nonlinear large deformation theory. Moreover, the *Cantilever Method* [4,5] for a flexible multi-layered material have been developed, based on the nonlinear large deformation theory.

Theory

For small deformations, there are some testing methods, e.g. three- or four-point bending test for evaluating mechanical properties of various materials. These conventional tests are commonly used to obtain Young's modulus because of their simplicity. However, since the conventional methods are based on the small deformation theory, these methods are inapplicable directly to large deformation problems. Therefore, as the deformations of specimen grow larger, a more exact analysis is required to obtain accurate results. From this point of view, a new testing method (*Own-Weight Cantilever Method*) is derived considering large deformation behaviors. The new method can be applied to various thin, long fiber materials (Glass fibers, Carbon fibers, Optical fibers, etc.) and thin sheet materials.

Basic Equations [6,7]

A typical illustration of deflections is given in Fig.1 for a cantilever subjected to own-weight w where w is the distributed load per unit length with a supporting angle θ_0. The horizontal displacement is denoted by x, the vertical displacement by y, and θ is the deflection angle. Furthermore, the arc length is denoted by s, the radius of curvature by R and the bending moment by M. The relationships between R, M, s, x, y and θ are given by:

$$1/R = -d\theta/ds, \quad M/EI = -d\theta/ds, \quad dx = ds \cdot \cos\theta, \quad dy = ds \cdot \sin\theta. \tag{1}$$

where E is Young's modulus and I, the second area moment.

Putting V_0 as the vertical reaction at the origin O, M_0 as the bending moment at the origin O and X as the integral of $\overline{X}$, the bending moment M applied at an arbitrary position Q (x, y) is

$$M = V_0 \cdot x - M_0 - \int_0^s w(x - \overline{x})d\overline{s} = wL \cdot x - M_0 - w[x\overline{s} - X]_0^s$$
$$= wL \cdot x - M_0 - w\{xs - X(s) + X(0)\} \tag{2}$$

The basic equation is derived from Eqs.1 and 2 in the form of:

$$EI \cdot d^2\theta/ds^2 + w(L-s)\cos\theta = 0. \tag{3}$$

Introducing the following non-dimensional variables and transforming the variables,

$$\xi = x/L, \ \eta = y/L, \ \zeta = s/L, \ \gamma = wL^3/(EI), \ \beta = ML/(EI) \quad . \tag{4}$$

equation 3 reduces to Eq.5

$$d^2\theta/d\zeta^2 + \gamma(1-\zeta)\cos\theta = 0. \tag{5}$$

Assuming the following relationships in Eq.5,

$$\psi = \theta - (\theta_A + \theta_0)/2 \qquad [-(\theta_A - \theta_0)/2 \leq \psi \leq (\theta_A - \theta_0)/2]. \tag{6}$$

finally, the nonlinear differential equation is obtained as follows.

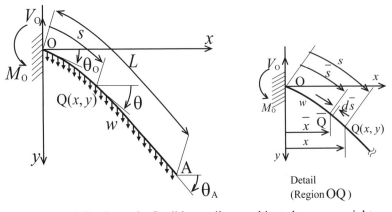

Fig.1 Large deflections of a flexible cantilever subjected to own-weight.

$$\frac{d^2\psi}{d\zeta^2} - \frac{4J_1\{(\theta_A-\theta_o)/2\}\cdot\sin\{(\theta_A+\theta_o)/2\}}{\theta_A-\theta_o}\cdot\gamma\cdot\psi\cdot(1-\zeta)$$

$$= -J_o\{(\theta_A-\theta_o)/2\}\cdot\cos\{(\theta_A+\theta_o)/2\}\cdot\gamma\cdot(1-\zeta) \tag{7}$$

where

$$\sin\psi \cong 4J_1\{(\theta_A-\theta_o)/2\}\cdot\{\psi/(\theta_A-\theta_o)\}. \tag{8}$$

$$\cos\psi \cong J_o\{(\theta_A-\theta_o)/2\}. \tag{9}$$

The functions $J_n(x)$ appearing in Eqs.7-9 are Bessel Functions.

Considering the conditions $d\psi/d\delta\big|_{\delta=0}=0$ $\left(d\theta/ds\big|_{\zeta=1}=0\right)$ and $\psi\big|_{\delta=0}=(\theta_A-\theta_o)/2$ $\left(\theta\big|_{\zeta=1}=\theta_A\right)$, the non-dimensional maximum horizontal displacement $\xi_A(=\xi_{max}=x_A/L)$ (x_A:horizontal displacement at the free end A) and the non-dimensional maximum vertical displacement $\eta_A(=\eta_{max}=y_A/L)$ (y_A: vertical displacement at the free end A) are obtained as follows.

$$\xi_A = \int_0^{\zeta_A} \cos\theta d\zeta$$

$$= \int_0^{\zeta_A} \cos\left[\left\{\frac{\theta_A-\theta_o}{2}-q\right\}\cdot\left\{1+\frac{m^2\cdot\delta^3}{2\cdot(4/3)}+\right.\right.$$

$$\left.\left.\frac{m^4\cdot\delta^6}{2\cdot4\cdot(8/3)\cdot(10/3)}+\cdots\right\}+q+\frac{\theta_A+\theta_o}{2}\right]d\zeta \tag{10}$$

$$\eta_A = \int_0^{\zeta_A} \sin\theta d\zeta$$

$$= \int_0^{\zeta_A} \sin\left[\left\{\frac{\theta_A-\theta_o}{2}-q\right\}\cdot\left\{1+\frac{m^2\cdot\delta^3}{2\cdot(4/3)}+\right.\right.$$

$$\left.\left.\frac{m^4\cdot\delta^6}{2\cdot4\cdot(8/3)\cdot(10/3)}+\cdots\right\}+q+\frac{\theta_A+\theta_o}{2}\right]d\zeta \tag{11}$$

where

$$\delta = \gamma\cdot(1-\zeta) \qquad [0\le\delta\le\gamma]. \tag{12}$$

$$q = \frac{(\theta_A-\theta_o)\cdot J_o\{(\theta_A-\theta_o)/2\}\cdot\cos\{(\theta_A+\theta_o)/2\}}{4J_1\{(\theta_A-\theta_o)/2\}\cdot\sin\{(\theta_A+\theta_o)/2\}}. \tag{13}$$

$$m^2 = \frac{16J_1\{(\theta_A-\theta_o)/2\}\cdot\cos\{(\theta_A+\theta_o)/2\}}{9\gamma^2\cdot(\theta_A-\theta_o)}. \tag{14}$$

and the non-dimensional distributed load γ is calculated from Eq.15.

$$-\frac{\theta_A - \theta_o}{2} = \left\{ \frac{\theta_A - \theta_o}{2} - q \right\} \times$$

$$\left\{ 1 + \frac{m^2 \cdot \gamma^3}{2 \cdot (4/3)} + \frac{m^4 \cdot \gamma^6}{2 \cdot 4 \cdot (8/3) \cdot (10/3)} + \cdots \right\} + q \quad (15)$$

Equations 10, 11 and 15 are the fundamental formulae to obtain the Young's modulus of the own-weight cantilever with an arbitrary supporting angle θ_o, based on the nonlinear large deformation theory. From the viewpoint of the experiment, however, it is very simple and convenient to support a horizontal cantilever as a support technique of beams. Therefore, considering the availability of the new method, Young's modulus measurement at the supporting angle of $\theta_o=0$ (that is, horizontal) is described in the following section as a special case of *"Own-Weight Cantilever Method"*

Measuring Techniques (Horizontal cantilever)

As known from Eqs.10, 11 and 15, the formulae for measuring Young's modulus, based on the nonlinear large deformation theory are complicated in general. Therefore, for the sake of simplicity, the usage of the chart is recommended here by the author.

Although there are various methods in order to measure Young's modulus, two representative methods are introduced in this paper. The original data is a horizontal displacement x_A or a vertical displacement y_A. Two charts (Nomographs) of γ - x_A relation (see Fig.2) and γ - y_A relation (see Fig.3) are presented, which were computed previously by using Eqs.10, 11 and 15. The calculation is repeated until the assumed angle θ_A at the free end coincides with the calculated angle θ_A based on Eq.(15). When the two values agree with each other, the deformed shape is determined ultimately. In Figs.2 and 3 (symbol •), various sketches represent deformed beams on the specific conditions x_A, y_A.

The following formula based on Eq.4 is utilized, when the chart is used for calculating Young's modulus.

$$E = wL^3 / (\gamma I). \quad (16)$$

Method 1: Measurement of x_A

The usage of this method is shown below. A chart (Nomograph) is given in Fig.2, illustrating the relationship of γ and x_A/L (ξ_A). Using this chart, Young's modulus E in a cantilever can be calculated from the relational expression given in Eq.16. As an example, Young's modulus E is obtained for a piano wire (SWP-A) with diameter: $d=0.9$ mm, distributed load per unit length: $w=49.3 \times 10^{-3}$ N/m.

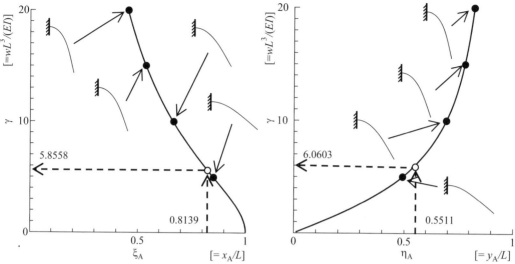

Fig.2 Non-dimensional chart for the finding the parameter γ when the horizontal displacement x_A is given.

Fig.3 Non-dimensional chart for the finding the parameter γ when the vertical displacement y_A is given.

Fig.4 Experimental set-up

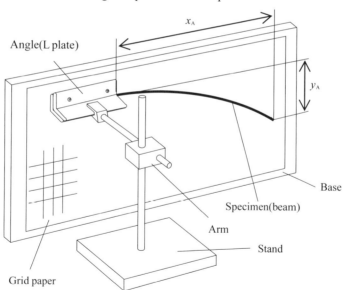

When L=900.0 mm, x_A =732.5 mm (i.e., ξ_A= x_A/L =0.8139) is measured and then γ is taken from Fig.2 (γ=5.8558). Therefore, Young's modulus E is calculated from Eq.16 as follows.

$$E=\frac{wL^3}{\gamma I}=\frac{49.3\times10^{-3}\times(0.9)^3}{5.8558\times3.221\times10^{-14}}\fallingdotseq190.3\times10^9\ \text{N}\big/\text{m}^2=190.3\,\text{GPa}.$$

Method 2: Measurement of y_A

A similar chart (Nomograph) is given in Fig.3, illustrating the relationship of γ and y_A/L (η_A). Using this chart, Young's modulus E can be calculated from Eq.16. As an example, Young's modulus E is obtained for a piano wire (SWP-A) mentioned above. When L=900.0 mm, y_A =496.0 mm (i.e., η_A= y_A/L =0.5511) is measured and then γ is taken from Fig.3 (γ=6.0603). Therefore, from Eq.16, Young's modulus E is calculated as follows.

$$E = \frac{wL^3}{\gamma I} = \frac{49.3 \times 10^{-3} \times (0.9)^3}{6.0603 \times 3.221 \times 10^{-14}} \fallingdotseq 183.9 \times 10^9 \ \mathrm{N/m^2} = 183.9 \ \mathrm{GPa}.$$

Experimental Investigation

In order to assess the applicability of the proposed method, several experiments were carried out using a thin piano wire (SWP-A) with diameter: d=0.38 mm, distributed load per unit length: w=8.78×10^{-3} N/m. An experimental set-up is shown in Fig.4. Young's modulus of SWP-A by applying the Method 1 and Method 2 are shown in Figs. 5 and 6, respectively. In the experiment, a horizontal displacement x_A and a vertical displacement y_A at the free end are measured by using a grid paper with 1mm spacing. Here, the influence of the length upon Young's modulus was examined for several lengths of a cantilever.

The measured values of Method 1 and 2 remain nearly constant for various lengths L in the range of 350.0-700.0 mm and the standard deviation (S.D) is small. That means Young's modulus shows a

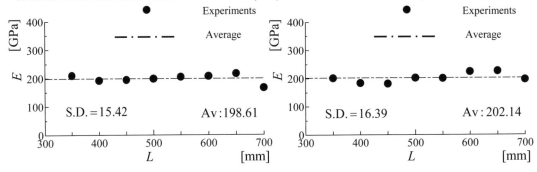

Fig.5 Young's moduli for ϕ=0.38mm piano wires (SWP-A)(from Data x_A).

Fig.6 Young's moduli for ϕ=0.38mm piano wires (SWP-A)(from Data y_A).

stable value considerably within the experimental range of the slenderness ratio L/d≈900-1800. As a whole, the mean Young's moduli determined by the two methods are reasonably in good agreement with each other although Method 1 or 2 has a little scatter in the values. Therefore, there is little reason to choose between Method 1and Method 2 to measure Young's modulus. As a reference, the value measured by using conventional three-point bending test based on the small deformation theory is E_0=205.8 GPa.

Conclusions

The Own-Weight Cantilever Method is developed as a new and simpler material testing method for measuring Young's modulus in a flexible thin material.

The principal conclusions are drawn as follows from the results of theoretical and experimental analyses.

(1) The new method is based on the nonlinear large deformation theory.

(2) Two representative charts are drawn on the basis of the proposed theory for the sake of simplicity.

(3) Based on the new idea, a set of testing devices was designed.

(4) A thin piano wire (SWP-A) was tested. Experimental results clarify that the new method is suitable for measuring Young's modulus in a flexible thin material.

(5) Based on the assessments, the proposed method is applicable widely to Young's modulus measurement in thin sheets and fiber materials (e.g., steel belts, glass fibers, carbon fibers, optical fibers, etc.).

References

[1] A.Ohtsuki: Proceedings of the 4th International Conference on Advances in Experimental Mechanics, 3-4, 53-58 (2005).

[2] A.Ohtsuki and H.Takada: Transactions of Japan Society for Spring Research, 47, 27-31 (2002).

[3] A.Ohtsuki: Proceedings of the 2005 SEM Annual Conference & Exposition on Experimental and Applied Mechanics, Section 72, 113(1)-113(8) (2005) [CD-ROM].

[4] A.Ohtsuki: Advances in Experimental Mechanics V(5th BSSM 2007), 195-200 (2007).

[5] A.Ohtsuki: Thin-Walled Structures (5th ICTWS 2008), 189-196 (2008).

[6] A.E.H.Love: A treatise on the mathematical theory of elasticity, Dover Pub., 399-412(1944).

[7] S.P.Timoshenko and J.M.Gere: Theory of elasticity, McGrawHill, 76-82(1961).

Applied Mechanics and Materials Vols. 24-25 (2010) pp 379-384
© *(2010) Trans Tech Publications, Switzerland*
doi:10.4028/www.scientific.net/AMM.24-25.379

Identification of the Mechanical Properties of Superconducting Windings Using the Virtual Fields Method

J.-H. Kim[1, a], F. Nunio[2, b], F. Pierron[1, c] and P. Vedrine[2, d]

[1]LMPF, Arts et Métiers ParisTech, Rue Saint-Dominique, BP 508,

51000 Châlons en Champagne, France

[2]CEA, CEN de Saclay, SM/IRFU/SIS/LCAP, Bat. 123, 91191 Gif-Sur-Yvette cedex, France

[a]jin.kim@chalons.ensam.fr, [b]francois.nunio@cea.fr, [c]fabrice.pierron@chalons.ensam.fr

Keywords: full-field measurement, mechanical properties, superconducting coils, virtual fields method

Abstract. Tensile tests were performed in order to identify the stiffness components of superconducting windings in the shape of rings (also called 'double pancakes'). The stereo image correlation technique was used for full-field displacement measurements. The strain components were then obtained from the measured displacement fields by numerical differentiation. Because differentiation is very sensitive to spatial noise, the displacement maps were fitted by polynomials before differentiation using a linear least-square method. Then, in the orthotropy basis, the four in-plane stiffnesses of the double pancake were determined using the Virtual Fields Method.

Introduction

Strong magnetic fields induced by superconducting windings result in significant deformation of the windings. Therefore, the manufacturing of large superconducting magnets makes it necessary to determine accurately their elastic properties. These windings have the shape of a flat rectangular section ring with two layers of winded superconducting cables. These are called 'double pancakes' in the text. Standard stiffness measurement techniques are based on homogeneous stress/strain fields in the specimens and local strain measurements through strain gauges. In the case of anisotropic materials such as superconducting windings, the number of parameters increases, so several tests need to be performed. Moreover, during the mechanical tests, such homogeneous fields are not easily obtained in superconducting windings due to their cylindrical specimen geometry. The present study aims at taking advantage of the availability of non-contact full-field measurements and inverse identification procedures in order to identify the windings equivalent homogeneous stiffnesses. A very detailed FE model of the double pancake indicated that the out-of-plane displacements are significant, showing a complex 3D behaviour. In this study, a stereo image correlation technique with back-to-back cameras was chosen to observe the 3D behaviour of the double pancake. Full-field heterogeneous displacement fields were measured through the stereo image correlation technique with back-to-back cameras and then strain components were obtained from the measured displacement fields by numerical differentiation after spatial smoothing. The virtual fields method (VFM) was used as an inverse procedure to process strains for the identification of the stiffness components. The VFM is based on the principle of virtual work, which describes the global equilibrium of the solid. A relevant use of the equilibrium equation leads to the identification of the constitutive parameters. In the orthotropy basis, four stiffness components were determined from a single tensile test of the double pancake using the VFM.

Methodology

Stereo image correlation The stereo image correlation setup is shown in Fig. 1(a). Calibration was performed using the standard procedure provided by the system manufacturer. The speckle pattern was applied by spraying a first layer of white paint over the specimen that had been polished first to provide a smooth surface. This was necessary because of the different layers of cables that exhibited a rather rough surface. Then, droplets of black paint were sprayed over the white background to produce the pattern shown in Fig. 1(b). The subset was 21 by 21 pixels with a shift of 5 pixels.

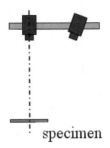

specimen

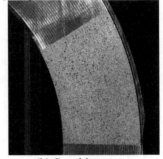

(a) Stereo image correlation set-up   (b) Speckle pattern

Fig 1. Stereo image correlation set-up and speckle pattern.

The Virtual Fields Method. The principle of virtual work, describing the global equilibrium of the specimen, can be written as (if body forces are neglected):

$$-\int_V \sigma_{ij}\varepsilon_{ij}^* dV + \int_{\partial V} T_i u_i^* dS = 0 \tag{1}$$

where V is the volume of the specimen, ∂V its boundary, σ the stress tensor, ε^* the virtual strain field, T the surface load density and u^* the virtual displacement field associated to ε^*. In cylindrical coordinate system and assuming a linear elastic orthotropic behaviour, the stiffness components to be identified relate the in-plane stress to the in-plane strain components as follows:

$$\begin{pmatrix} \sigma_\theta \\ \sigma_r \\ \sigma_s \end{pmatrix} = \begin{bmatrix} Q_{\theta\theta} & Q_{\theta r} & 0 \\ Q_{\theta r} & Q_{rr} & 0 \\ 0 & 0 & Q_{ss} \end{bmatrix} \begin{pmatrix} \varepsilon_\theta \\ \varepsilon_r \\ \varepsilon_s \end{pmatrix} \tag{2}$$

where σ is the stress tensor, ε the strain tensor and the Q_{ij}'s are the stiffnesses to be determined (expressed in the orthotropy basis).

The principle of the VFM is to substitute the stress information in the above equation from the constitutive equation. Eq. 1 then becomes:

$$\int_S Q_{\theta\theta}\varepsilon_\theta\varepsilon_\theta^* dS + \int_S Q_{rr}\varepsilon_r\varepsilon_r^* dS + \int_S Q_{\theta r}(\varepsilon_r\varepsilon_\theta^* + \varepsilon_\theta\varepsilon_r^*)dS + \int_S Q_{ss}\varepsilon_s\varepsilon_s^* dS = \frac{Fu^*(M)}{e} \tag{3}$$

where F is the applied tensile load, e the specimen thickness and u^* (M) the virtual displacement of the point were the load is applied. When the material is homogeneous, the stiffness components can be moved outside of the integration sign and the choice of a particular set of virtual fields will provide a linear system relating the unknown stiffnesses to the external forces (measured by the load cell) and weighted integrals of the actual strains that can be measured from full-field measurements. The key issue of the VFM is the choice of appropriate virtual fields among the infinite possibilities. Several studies have been performed using virtual fields defined empirically. Recently, with the development of the so-called special virtual fields [1] and the optimization of these special fields with respect to noise sensitivity [2], this problem has been solved efficiently. In the paper, two approaches for determining virtual fields will be employed: manually defined, as in [3], or using special optimized virtual fields as in [4]. These will be referred to as M for manually defined or S for special optimized.

When the identification is robust and the manually defined virtual fields well selected, the two approaches should give similar results, S being more accurate when noise is low. This is therefore used here as a qualitative check of identification quality.

Results and Discussion

Test configuration and area of interest. In this study, the double pancake is tested according to the load configuration in Fig. 2. This test configuration yields a heterogeneous strain field at the surface of the specimen, therefore, the four rigidities can be retrieved from this single tensile test. The inner radius (R_0) of the pancake is 238.5 mm, the outer radius (R_1) 342.5 mm, and its thickness is 13 mm. Owing to the specimen size and the camera pixel aspect ratio, only a 30° angular section of the pancake specimen was investigated. The best range of the region of interest was selected as 0-30° from a finite element (FE) simulation following the selection method used in [4], so this region will be used for the rest of the study.

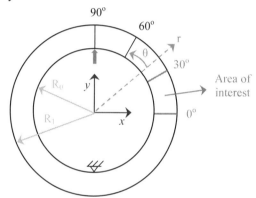

Fig 2. Geometry and test configuration of the double pancake.

Validation on FE simulated data. In order to investigate the structural behaviour of the double pancake and validate the proposed approach, a detailed 3D FE model was built up using the Cast3M FEA software developed by CEA. The 3D FE model is shown in Fig. 3.

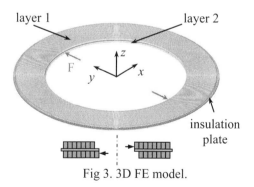

Fig 3. 3D FE model.

The material properties used as inputs in the finite element calculations are:
Conductor (isotropic): E = 113 GPa, ν = 0.33
Insulation (transverse isotropic: 1=warp, 2=fill, 3=normal):
E_1 = 24 GPa, ν_{12} = 0.14, ν_{21} = 0.14, G_{12} = 5 GPa
E_2 = 24 GPa, ν_{13} = 0.34, ν_{31} = 0.14, G_{13} = 5 GPa
E_3 = 10 GPa, ν_{23} = 0.34, ν_{32} = 0.34, G_{23} = 5 GPa

For the 1.5 mm insulation plate between two pancakes, the warp and fill directions are θ and r (radial) and the normal direction is z (axial). For the insulation tape between each turn of conductor, the warp and fill directions are θ and z and the normal direction is r. The applied force is 100 kN. The calculated displacement fields on the front and back surfaces of the area of interest are shown in Fig. 4 (a) and (b). As expected, the displacement is negative in the x-direction and positive in the y-direction. However, the out-of-plane displacement is significant. The magnitude of the out-of-plane displacements is of the same order of that of the in-plane displacements, and moreover the gradient of UZ is higher than those of UX and UY. It is considered that this peculiar behaviour of the double pancake is associated with interaction between the two layers due to the offset loading condition as shown in Fig. 3. To realistically obtain the strain fields, the differentiation process that will be used experimentally is applied. The displacement fields were smoothed using 4th order polynomial curve fitting. The strains were then obtained by analytical differentiation of these polynomials. The strain fields are presented in Fig. 4 (a) and (b). The x strain changes from negative (compression, inner area) to positive (tension, outer area), and the y strain changes from positive (tension, inner area) to negative (compression, outer area) showing typical strain fields in a curved beam in bending. However, the shear strain is not zero in the $0°$ area and significant difference was observed between front and back surfaces.

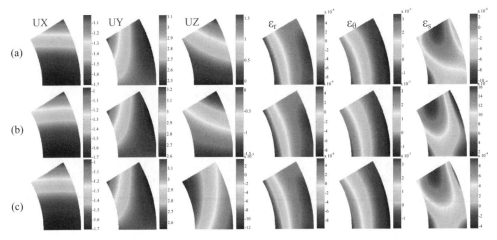

Fig 4. Displacement [mm] and strain fields from FEA (a) front face (b) back face (c) average.

The identified results of stiffnesses of the double pancake using the VFM are shown in Table 1.

Table 1. Identified stiffnesses for the double pancake (M: Manual VFM, S: Special VFM).

Rigidities		front		back		average	
		M	S	M	S	M	S
E_{rr}	GPa	108.3	111.1	73.1	58.5	69.4	65.6
$E_{\theta\theta}$	GPa	172.1	139.1	70.9	85.3	100.4	100.3
$\nu_{\theta r}$		0.29	0.31	0.31	0.29	0.31	0.31
G_{ss}	GPa	18.7	-15.8	26.6	9.12	23.8	25.2

In this case, unexpected results were observed in the identification results. The identified results are very inconsistent between the front and back surfaces. $E_{\theta\theta}$ is considered as the stable term to identify, but the results are significantly different even between the manual and special VFM and far form the FE input (113 GPa). The identified results are negative in some cases for G_{ss}. After further investigation of the shear strain fields of both faces, it was observed that there exist high bending strains on both sides due to the specific geometrical offset configuration of the double pancake. Therefore, the average value of both side displacements was calculated in order to get rid of the bending strain effects. The average displacement fields and the actual in-plane strain obtained from the average displacement fields are shown in Fig. 4 (c). The identification results using the actual in-plane strains are shown in Table 1. Interestingly, the identified results are very consistent between the manual and special VFM and the identified $E_{\theta\theta}$ is much closer to the FE input (113 GPa).

Experimental identification results. The initial step was to identify the material properties of the monolithic type conductor which is used for the double pancake superconducting wires. Full-field displacement fields were measured through digital image correlation (DIC). The strain fields were obtained from the displacements by numerical differentiation after spatial smoothing. Here, the diffuse approximation method [5] with r=20 was used for the smoothing. The cross section size of the specimen is 5.53 x 2.78 mm^2. The identification was performed using the manual VFM and the identified results were: E = 92.6 GPa, v = 0.34.

The next step was to identify the material properties of the double pancake. In the FE simulation, very complex 3D behaviour of the double pancake was observed. To get rid of the bending effects, it was decided to measure the displacement fields on both sides of the double pancake using back-to-back cameras. It should be noted here however that when the experiment was performed using a single camera for each surface, the bending effect was not cancelled out. This is due to the fact that the location of the inspected areas (front and back surfaces) is in a slightly offset position, resulting in torsional effect in the strain fields. Therefore, stereo image correlation technique using two cameras for each face was finally chosen to get rid of this effect. The displacement fields on both sides of the specimen were measured and the average displacement fields were obtained. Then, the final actual in-plane strain fields were processed to identify the stiffnesses of the double pancake. The experiment was performed using the stereo image correlation set-up shown in Fig. 5.

Fig 5. Stereo image correlation set-up.

The double pancake was mounted on a tensile test machine through specially designed grips and two 2048 x 2048 pixels CCD cameras with an incidence angle observe the specimen surface. The stereo configuration shown in Fig. 1 (a) was used. The displacement fields were computed from the obtained speckle pattern images using the Vic3D software. A pre-load of 1 kN was applied to avoid a settling effect before the main test. The applied load was 9.24 kN for measurement on the front and 9.25 kN for the back face. The measured displacement fields on the front and back surfaces are shown in Fig. 6 (a) and (b). The patterns of the displacement fields are very close to that from FE analysis, even though the boundary conditions are slightly different: point load for the model whereas in reality, the load is transferred through a certain angular sector. However, the measurement area being far away from the loading areas, Saint-Venant assumptions results in similar patterns. The patterns of out-of-plane displacements and the signs are very similar between the experimental measurements and the FEA as shown in Figs. 4 and 6, but the magnitude is different due to the different boundary conditions. Nevertheless, the values are significant enough to influence the results of UX and UY if a 2D technique was used. It was found that the order of magnitude of UZ is 10^{-2} mm, making the correct measurements of UZ very difficult. The displacement fields were smoothed using 4th order polynomial curve fitting to obtain strain fields. The strain maps are presented in Fig. 6. ε_θ is of the order of 10^{-4} and ε_r and ε_s are of the order of 10^{-5}.

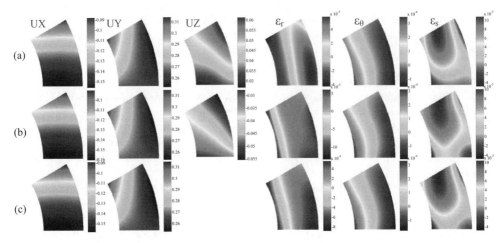

Fig 6. Displacement [mm] and strain fields from experiment (a) front (b) back (c) average.

In the same fashion, the average displacement fields were obtained from the front and back displacements and the actual in-plane strains were calculated to identify the global stiffnesses on the double pancake. The identified results are reported in Table 2. As can be seen in Table 2, the identified $E_{\theta\theta}$ is close to the identified longitudinal modulus of the conductor (92.6 GPa). In addition, it was observed that E_{rr} from experimental measurements is much lower than that of the simulated ones, which means the material properties of the insulation tape between the conductor wires used in the FEA is probably overestimated.

Table 2. Identified stiffnesses for the double pancake (M: Manual VFM, S: Special VFM).

Rigidities		Identified values					
		front		back		Back to back	
		M	S	M	S	M	S
E_{rr}	GPa	23.5	20.9	61.4	55.8	36.8	33.3
$E_{\theta\theta}$	GPa	66.4	65.8	109.7	103.8	82.9	80.6
$\nu_{\theta r}$		0.31	0.34	0.31	0.31	0.31	0.32
G_{ss}	GPa	19.6	17.2	22.7	23.6	20.9	19.6

References

[1] M. Grédiac, E. Toussaint and F. Pierron: Int. J. Solids Structures Vol. 39 (2002), p. 2691–2705

[2] S. Avril, M. Grédiac and F. Pierron: Comp. Mech. Vol. 34 (2004), p. 439-452

[3] F. Pierron, S. Zhavoronok and M. Grédiac: Int. J. Solids Structures Vol. 37 (2000), p. 4437–4453

[4] R. Moulart, S. Avril and F. Pierron: Composites: Part A, Vol. 37 (2006), p. 326–336

[5] S. Avril, P. Feissel, F. Pierron and P Villon: European J. Comp. Mech. Vol. 17 (2008), p. 857-868

Applied Mechanics and Materials Vols. 24-25 (2010) pp 385-390
© *(2010) Trans Tech Publications, Switzerland*
doi:10.4028/www.scientific.net/AMM.24-25.385

Measuring the Static Modulus of Nuclear Graphite from Four-Point Flexural Strength Tests and DIC

J. D. Lord[1, a], N. J. McCormick[1, b], J. M. Urquhart[1, c], G. M. Klimaytys[1, d]
and I. J. Lingham[2, e]

[1]National Physical Laboratory, Hampton Road, Teddington, Middlesex, TW110LW, UK
[2]National Nuclear Laboratory, Bristol Road, Stonehouse, Gloucestershire, GL10 3UT, UK

[a]jerry.lord@npl.co.uk,, [b]nick.mccormick@npl.co.uk, [c]jeannie.urquhart@npl.co.uk,
[d]grant.klimaytys@npl.co.uk, [e]ian.j.lingham@nnl.co.uk

Keywords: Digital Image Correlation, DIC, modulus, bend test, flexural strength, nuclear, graphite

Abstract

The paper describes a novel method based on Digital Image Correlation (DIC) for measuring the static modulus of active PGA graphite specimens during conventional four-point bend strength tests. DIC has been used in combination with finite element and numerical modelling to monitor the displacement fields developing in the specimen during testing, and calculate representative modulus values.

Details of the model and results are presented for two specimen geometries and a range of materials with different levels of exposure and density, tested in the perpendicular and parallel orientations. The calculated static modulus values from the DIC measurements confirm the trend between modulus, flexural strength and density. Comments and observations on the uncertainty in the measurement are also presented.

Introduction

Flexural strength, modulus and density are used as important parameters for gauging the condition of the graphite moderator in the UK nuclear reactors. Structural integrity assessments for the continued operation of the reactors are reliant upon property data obtained from material removed from the reactor cores. Currently the flexural strength and modulus are measured separately from four-point bend tests and ultrasonic dynamic Young's modulus (DYM) tests, but there are distinct advantages if this data could be obtained from a single test, and this was the purpose of this work.

Conventional contact displacement and strain measurement methods could not be used due to the nature and condition of the graphite material, so a method using DIC has been developed for the active graphite specimens. DIC is a full field non-contact strain measurement technique that was first developed in the 1980s [1,2]. It has seen significant development and uptake in recent years due to the availability of higher resolution cameras and increased computer power. DIC works by comparing images of a component or testpiece at different stages of deformation and tracking blocks of pixels to measure surface displacement and build up full-field 2D and 3D deformation vector fields and strain maps [3]. The position of the centre of the pixel blocks is determined to sub-pixel accuracy over the whole image using sophisticated correlation functions, from which the vector and strain components can be calculated. In this work DIC has been used in combination with inverse modelling to monitor the displacement field in the specimen during testing, calculating representative modulus values and stress-strain data from the tests, and offering the potential for additional supporting data for the graphite structural integrity assessment.

Experimental Setup

Initial test method development and validation was carried out at NPL, and tests on active material then carried out at the National Nuclear Laboratory (NNL), Sellafield using an identical test set up. The image acquisition system consisted of a PC and two 5-Megapixel cameras interfaced to a data acquisition unit for synchronizing the image capture with the analogue load signal from the Zwick test machine, Figure 1. The cameras were attached to the test machine baseplate at a fixed height and working distance to image the whole sample during testing, and illuminated with high intensity LED lighting. The specimen was imaged from both sides to improve the measurement statistics and capture any local variations in the structure or properties. The DIC data was calculated from each set of camera images and individual modulus values calculated, which were then averaged to present a single value of modulus for the specimen.

Most of the tests were carried out on rectangular beam-shaped specimens approximately 50 x 8 x 8 mm in dimension, in 4 point bending with a 13.3/40 mm roller separation; but some were made on shorter beams, approximately 40 x 8 x 8 mm in dimension, in four-point bending with a 10/30 mm roller separation, as shown in Figure 2. Often with DIC, specimens are prepared by spraying a speckle paint pattern on the surface, but this was not necessary in this case as the DIC analysis was possible using the natural texture and machined surface of the graphite. The typical grain structure of the material can be seen clearly in Figure 2. In all cases the tests were carried out according to ASTM C651[4], at a constant displacement rate of 0.12 mm/min, and images captured at 2Hz, so each frame increment corresponds to a crosshead displacement of 1 μm.

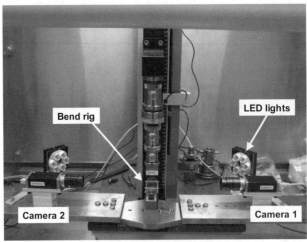

Figure 1: DIC system installed on the test machine at NNL

For each test between 100 and 200, 8-bit images were captured. Processing of the images was carried out by initial manual examination of the load-displacement data to identify the linear range over which the modulus would be calculated, followed by cropping of the images to isolate the specimen from the background and reduce the image size for calculation.

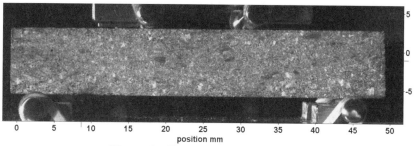

Figure 2: Close up of graphite beam

This gave a typical image size of 2100 x 600 pixels and a scaling of ~ 40 pixels/mm. The images were then processed using DIC to calculate the vertical displacement of the beam, Vy. The 2D DIC analysis was carried out using the *LaVision Strainmaster* software [5], using a "window" size of 32 x 32 pixels, with 50% overlap, giving an effective unit size of 16 x 16 pixels. This meant that approximately 2800 displacement vectors were calculated over each of the beam faces imaged. Figure 3 shows a typical load-displacement plot from the flexural test with the linear range highlighted (Frames 40-120 in this case), and the evolution of the Vy displacement field shown at different points in the test. All displacements were calculated relative to the start of the linear range - in this case selected as image 40 - but depending on the range chosen, between 4-10 profiles were generated for a particular specimen.

The displacement profiles (Vy) along the central length of the beam were then calculated and the load data and displacement profiles then input into the model for the calculation of modulus. Figure 4 shows the evolution of the beam displacement profiles during the test. The maximum deflection of the beam in the centre is 0.045 mm.

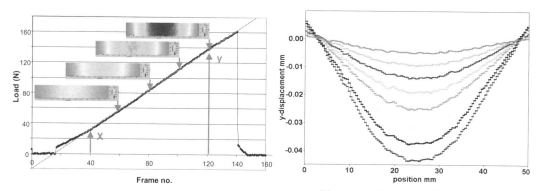

Figure 3: Typical load-displacement trace

Figure 4: Corresponding beam displacement profiles

Modelling

A FEA analysis was carried out using the same sample geometry and bend rig configuration used in the test set-up, using a Young's modulus of 3 GPa and Poisson's ratio 0.1. A 3D model was developed that was linear elastic and used 8 node cuboid elements in the *Abaqus* package. The size and shape of the elements chosen was not critical as only the elastic displacements in the central section of the beam were being considered, and it was not necessary to refine the mesh to model the stress concentrations under the rollers as this did not have a significant effect on the elastic behaviour and calculated centre-line displacements. Point loading for the lower rollers was achieved by using a line of nodes corresponding to the lower contacts, which were fixed in the vertical direction but free to move in the other directions. An area-loading model was used for contact for the top rollers. This system was used to calculate infinitesimal displacements in the vertical and horizontal directions for a plane of nodes at the mid plane of the model, assuming linear elastic behaviour. To enable an efficient analysis of the data a simple numerical model was used to fit to the FEA prediction of the y-displacement profile. It was found that using a function of the form:

$$f(x) = a + bx^2 + cx^4$$

gave a good fit (Figure 5), where a, b, c are fitting parameters.

The DIC displacement profile data was first normalised by dividing the displacement data by the load increment with respect to the "reference" frame to give the unit load-deflection for the specimen. These were then averaged to improve the signal-to-noise levels, generating a single reference curve for the sample, Figure 6.

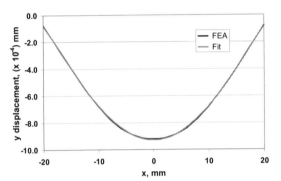

Figure 5: Comparison of FEA and analytical models   **Figure 6: Normalised unit displacement profile**

Results

Initial calibration tests were carried out on a series of Perspex (PMMA) specimens to check the test setup using the same test fixtures and procedures used for the graphite tests. Perspex was chosen as the "reference" material as it is homogeneous and of similar modulus to the low density irradiated material. Results were in excellent agreement with the published manufacturer's data, which gives a value for elastic modulus of 3.1 GPa.

The flexural strength data for the graphite specimens is plotted against density in Figure 7. Figure 8 shows the corresponding DIC-derived modulus data obtained from the same series of tests. In both cases the data has been normalised with respect to the properties of the virgin PGA graphite in the respective orientations. The density of the graphite beams was measured using mensuration with an assumed geometry for the volume.

Two separate batches of material, subjected to different levels of radiation damage were tested. Batch 1 samples were tested with the graphite grain structure/extrusion direction running parallel and perpendicular to the orientation of the beam; Batch 2 only in the parallel orientation.

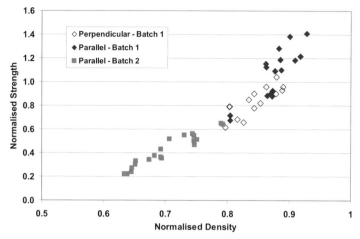

Figure 7: Flexural strength vs density

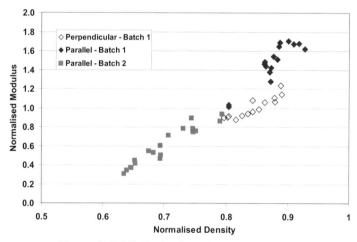

Figure 8: DIC- derived modulus data vs density

Both strength and modulus data show the same behaviour – the Batch 1 material is significantly stronger and stiffer than the Batch 2, and the material tested in the perpendicular orientation is generally less affected by the radiation (for the limited amount of samples tested). For the shorter beams from Batch 2 separate DYM measurements were made on samples cut from the end for comparison. The DIC-derived modulus data were systematically lower than the dynamic modulus values, in the ratio of ~0.92. This agrees well with the corrections used historically in the modelling and prediction of the static properties, when only dynamic modulus values were available.

The change in properties depends on the level of radiation received. In all cases there is a reduction in the density, but very different response depending on the cumulative radiation dose received. Initially the behaviour is dominated by irradiation hardening which leads to an increase in strength and modulus (Batch 1, low dose), but after further exposure radiolytic corrosion leads to weight loss and a reduction in properties (Batch 2, high dose). The two mechanisms operate simultaneously, but the effect of irradiation hardening soon saturates and the behaviour is then dominated by radiolytic corrosion.

Factors affecting the accuracy and uncertainty

Several assumptions were made in the modelling and measurement procedure, relating to the homogeneity of the specimen, specimen shape and dimensions, perfect alignment of the rig and the positioning and alignment of the specimen within the bend rig. A pragmatic approach was taken, because although there is some anisotropy present due to specimen orientation, the structure of the graphite itself is very complex and heterogeneous, with local 3D variations in constituent materials and density, particularly with the irradiated material, which would be significantly more complex to model. At a fine scale therefore the concept of a single, constant value for modulus is bound to be insufficient, but for the purposes of measuring the representative properties of the specimen then this assumption was required.

Due to the nature of the material being tested there is potential for variations in the properties along the length of the bar. Modelling results indicate that for small variations in modulus the specimen deflections are relatively unaffected, but the shape of the curve becomes slightly asymmetric. Indeed asymmetric displacement profiles were observed in some of the tests, but further work is necessary to isolate the precise cause at this stage as there may also be contributions from misalignment of the rollers and specimen within the test jig, machining damage and dimensional variations, and local heterogeneous microstructure.

Of all the factors considered the most likely source of measurement uncertainty is probably due to specimen misalignment within the jig. There is also likely to be a contribution from material inhomogeneity and the local variations in the graphite microstructure. Based on these observations and the modelling results it is predicted that the typical uncertainty in modulus obtained from the DIC/flexural strength tests is of the order 1-2%.

Summary

DIC has been used successfully for measuring the static modulus of active graphite samples during the four-point flexural strength test. The calculated static modulus values confirm the trend between modulus and density, and modulus and flexural strength, and the technique is able to distinguish between materials tested in different orientations. Typical uncertainties in modulus of 1-2% were achieved.

References

[1]   M A Sutton and W J Wolters, Determination of displacement using an improved digital
      image correlation method, *Image Vision Computing*, 1983, 1(3), 133–139.
[2]   T C Chu, W F Ranson, M A Sutton and W H Peters, Applications of digital image correlation
      techniques to experimental mechanics, *Experimental Mechanics*, 1985, 25(3), 232–244.
[3]   M.A. Sutton, J-J. Orteu, and H.W.Schreier. Image Correlation for Shape, Motion and
      Deformation Measurements. Springer, 2009. ISBN; 978-0-387-78746-6
[4]   ASTM C651 - 91(2005)e1 Standard Test Method for Flexural Strength of Manufactured
      Carbon and Graphite Articles Using Four-Point Loading at Room Temperature
[5]   LaVision Strainmaster Davis 7.0 User Manual, November 2004

Acknowledgements

The work described was carried out as part of a contract with Magnox, who are thanked for their support and assistance.

Session 9: Composite and Cellular Materials 2

Applied Mechanics and Materials Vols. 24-25 (2010) pp 393-400
© *(2010) Trans Tech Publications, Switzerland*
doi:10.4028/www.scientific.net/AMM.24-25.393

Experimental and numerical buckling analysis of delaminated hybrid composite beam structures

M.M. Nasr Esfahani[1, a], H. Ghasemnejad[2, b] and P.E. Barrington[3, c]

[1,2,3]Kingston University, Faculty of Engineering, London, SW15 3DW, UK

[a]M.M.N.Esfahani@gmail.com, [b]H.Ghasemnejad@kingston.ac.uk, [c]P.Barrington@kingston.ac.uk

Keywords: Buckling; Delamination; Hybrid; Composite; ANSYS

Abstract. In this paper the effect of delamination position on the critical buckling load and buckling mode of hybrid composite beams is investigated. Experimental and numerical studies are carried out to determine the buckling load of delaminated composite beams. The laminated composite beams with various laminate designs of $[G_{90}]_6$, $[C_{90}]_8$, $[C_0/G_0]_4$ and $[C_{90}/G_{90}]_4$ were manufactured and tested to find the critical buckling load. Three different defect positions were placed through the thickness to find three main buckling modes. It was found that delamination position and lay-up can affect the buckling mode and also the critical buckling load. By approaching the delamination position to the outer surface of the specimen the buckling load decreases. The buckling process of hybrid and non-hybrid composite beams was also simulated by finite element software ANSYS and the critical buckling loads were verified with the relevant experimental results.

Introduction

Fibre-reinforced polymer (FRP) composite materials are widely used because of their high strength-to-weight and stiffness-to-weight ratios as compared with many traditional materials [1,2]. Delamination is one of the most serious failure modes in laminated composite materials. It can occur in several ways during manufacturing process, maintenance and impacts of foreign objects [2,3]. Laminated composite materials containing one or more delamination can buckle at a lower level of compressive load than laminated composites without delamination, and this level depends on the size, position and shape of the delamination [4].

Buckling and postbuckling behaviour in delaminated composite structures have been studied to investigate the buckling resistance loads [5, 6, 7, 8, 9]. Most of these previous works are about buckling response of non-hybrid laminated composite structures. Hybrid laminated composites are an effective method to improve the mechanical properties of laminated composite structures under various loading conditions.

In the present study, the effect of delamination position on the critical buckling load of non-hybrid and hybrid composite laminates of CFRP and GFRP with single delamination position is studied. The laminate designs of $[G_{90}]_6$, $[C_{90}]_8$, $[C_0/G_0]_4$ and $[C_{90}/G_{90}]_4$ were chosen to compare the buckling resistance of hybrid and non-hybrid composite beams. The delaminated composite beams were manufactured and tested to find the critical buckling load. The buckling behaviour of hybrid and non-hybrid composite beams was also simulated using ANSYS software to compare numerical and experimental buckling modes and the numerical critical buckling loads were verified with relevant experimental results.

2. Experimental details

2.1. Manufacturing

The delaminated composite beams of $[G_{90}]_6$, $[C_{90}]_8$, $[C_0/G_0]_4$ and $[C_{90}/G_{90}]_4$ were manufactured from the unidirectional GFRP and CFRP composites. The thickness of all laminated beams was 2 mm. Three defect positions of $H/t = 0.5, 0.25$ and 0.125 were chosen and Teflon film of 13 µm was placed at these particular positions to model the delaminated composite beams. The geometry of a composite beam with a single delamination is shown in Fig. 1.

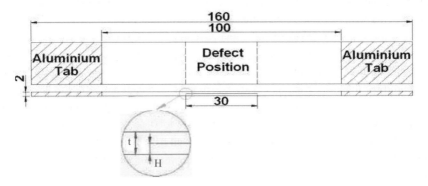

Figure 1. Geometry of delaminated composite plate with various delamination positions.

2.2. Mechanical properties

The mechanical characteristics of carbon/epoxy and glass/epoxy were obtained in accordance with the relevant standards [10-13]. These tests were tensile, shear, fibre volume fraction and coefficient of friction. All specimens were manufactured from carbon fibre reinforced plastic (CFRP) and glass fibre reinforced plastic (GFRP) materials of density 1.8 g/cm^3 and 1.6 g/cm^3 with epoxy resin. A summary of the findings for tensile, shear and fibre volume fraction are summarised in Table 1.

Table 1. Material properties of the unidirectional CFRP and GFRP composites.

CFRP

E_1 (GPa)	E_2 (GPa)	G_{12} (GPa)	υ_{12}	σ_u 0° MPa	σ_u 90° MPa	τ_s MPa	V_f (%)
138±12	10.5±1	6.3±0.4	0.1	330±14	32±4	147±14	42

GFRP

E_1 (GPa)	E_2 (GPa)	G_{12} (GPa)	υ_{12}	σ_u 0° MPa	σ_u 90° MPa	τ_s MPa	V_f (%)
35.1±4	9.6±1	4±0.5	0.32	807±10	21.3±2	97.9±4	40.3

2.3. Buckling test

Each specimen was tested at the rate of 0.2mm/min using a Universal Testing Machine with 50 kN load cell. For each test configuration three specimens were tested. The force-displacement diagrams were recorded automatically for each test (see Fig. 2).

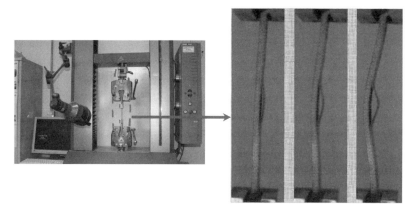

Figure 2. Composite beam in Universal Testing Machine.

All experimental force-displacement results for various laminate designs at different delamination positions are presented in Fig 3.

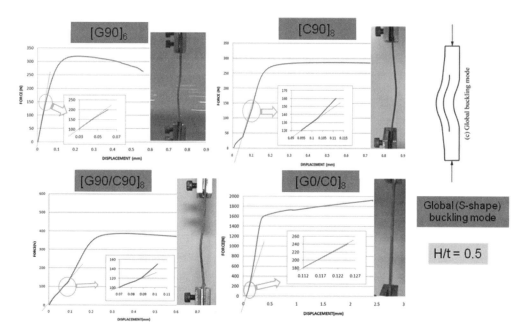

Figure. 3.a Experimental force-displacement results for global buckling mode at delamination position of H/t = 0.5.

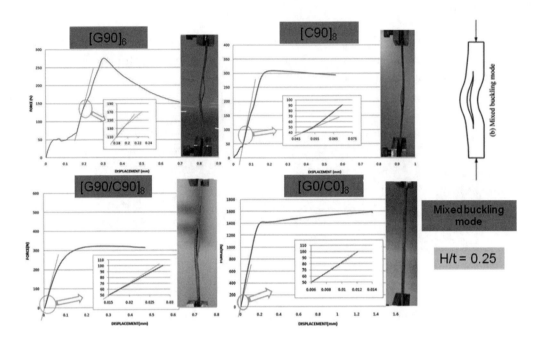

Figure. 3.b Experimental force-displacement results for mixed buckling mode at delamination position of H/t = 0.25.

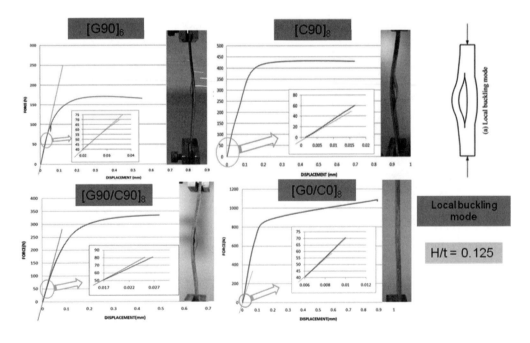

Figure. 3.c Experimental force-displacement results for local buckling mode at delamination position of H/t = 0.125.

3. Numerical study

In the present work, Eigen-buckling analysis was performed to simulate the buckling behaviour of delaminated composite beams. The element type of SOLID46 layered element with six degree of freedom was chosen to create mesh for laminated composite beam. To simulate the boundary conditions all degrees of freedom at one end of composite beam were set equal to zero.

The delaminated composite beam was designed with 6 volumes to model the single delamination (see Fig. 4). To simulate the delamination area at the first interface, top and bottom area of volumes were not glued in the interfacial areas. Thus, double nodes occur at the same coordinates of the interfacial areas. During the buckling simulation the double nodes separate from one another. In this case, other areas which were not glued represent delamination in the laminated composite beam.

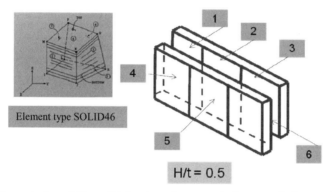

Figure. 4. Numerical modeling of delaminated composite beam at defect position of H/t = 0.5.

The numerical results and their comparison with experimental results are presented in Fig. 5.

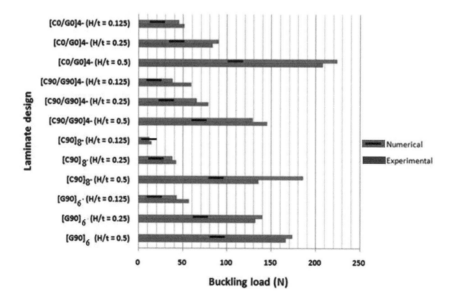

Figure 5. Comparison of experimental and numerical critical buckling loads.

4. Results and discussion

The experimental and numerical critical buckling loads of various lay-ups are presented in Figs. 6 and 7. The experimental results are the average of three tests for each specimen. The critical buckling load at different defect positions were compared together and good agreement were found between experimental and FE results. It is noteworthy that this type of simulation is suitable for linear buckling behaviour of composite beam. However in this study post-buckling and delamination failure are not considered. Our results indicate that hybrid composite beams are able to show more resistance against buckling load in comparison with non-hybrid composite beams. By decreasing the H/t ratio the critical buckling resistance load decreases for all non-hybrid and hybrid laminated composite beams.

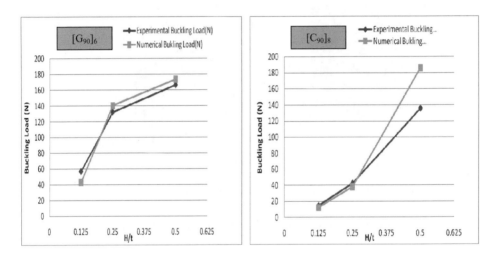

Figure 6. Comparison of buckling load in three defect positions for non-hybrid composite beams with laminate designs of $[G_{90}]_6$ and $[C_{90}]_8$.

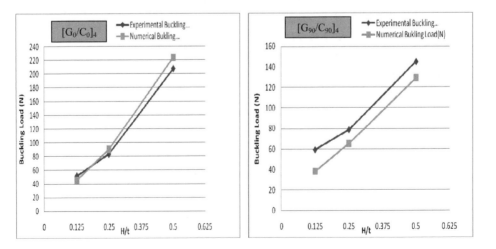

Figure 7. Comparison of buckling load in three defect positions for hybrid composite beams with laminate designs of $[G_{90}/C_{90}]_4$ and $[G_0/C_0]_4$.

The buckling modes of global, local and mixed are compared for three defect positions in laminated composite beams. Good agreement was found between numerical and experimental buckling modes in all delaminated composite beams. This simulation method could be a new method to predict the critical buckling load of delaminated composite materials. The comparison between numerical and experimental results for three laminate designs of $[G_{90}]_6$, $[C_{90}]_8$, $[C_{90}/G_{90}]_4$ indicates that global buckling mode occurs for the defect position of $H/t = 0.5$, as well as mixed buckling mode for defect position of $H/t = 0.25$, and local buckling mode for defect position of $H/t = 0.125$ (see Fig. 8).

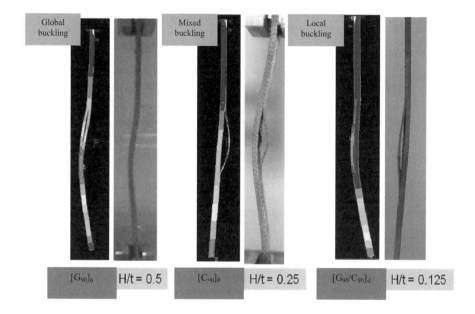

Figure 8. Three distinct buckling modes in delaminated composite beams.

References

[1] H. Ghasemnejad, H. Hadavinia, and A. Aboutorabi, in: Effect of Delamination Failure in Crashworthiness Analysis of Hybrid Composite Box Structures, Materials and Design, 31(3),pp.1105-1116,(2010).

[2] H. Ghasemnejad, BRK Blackman, H Hadavinia & B Sudall, in: Experimental studies on fracture characterisation and energy absorption of GFRP composite box structure, Composite Structures, 88(2),pp.253-261,(2008).

[3] Z. Aslan, and M. Şahin, in: Buckling behaviour and compressive failure of composite laminates containing multiple large delaminations. Composite Structures, 89(3), pp. 382-390 (2009).

[4] M. Zor, F. Sen, ME. Toygar, in: An investigation of square delamination effects on the buckling behavior of laminated composite plates with a square hole by using three-dimensional FEM analysis. J Reinf Plast Compos, 24(11),pp.19–30 (2005).

[5] YC. Wee, CG. Boay, in: Analytical and numerical studies on the buckling of delaminated composite beams. Composite Structures, 80(2), pp.307-319 (2007).

[6] F. Cappello, D. Tumino, in: Numerical analysis of composite plates with multiple delaminations subjected to uniaxial buckling load. Composites Science and Technology, 66,pp. 264–272,(2006)

[7] G. Li, SS. Pang, Y. Zhao, SI. Ibekwe, in: Local buckling analysis of composite laminate with large delaminations induced by low velocity impact. Polymer Composites, 20(5),pp. 634-642,(2009).

[8] S. Moradi, F. Taheri, in: Delamination buckling analysis of general laminated composite beams by differential quadrature method. Composites Part B: Engineering, 30(5),pp.503-511,(2009).

[9] WM. Kyoung, CG. Kim, CS. Hong, SM. Jun, in: Modelling of composite laminates with multiple delaminations under compressive loading. Journal of Composite Materials, 10(32),pp.951-968,(2009)

[10] BS EN ISO 2747, Glass Fibre Reinforced Plastics-Tensile Test, British Standard Institute, London,1998.

[11] BS EN ISO 14129, Fibre Reinforced Plastics Composite-Determination of the In-plane Shear Stress/Shear Strain Response, Including the In-plane Shear Modulus and Strength by the ±45 Tension Test Method, British Standard Institute, London,1998.

[12] ASTM D 3171-99, Standard test method for constituent content of composite materials. Annual book of ASTM standards, West Conshohocken, PA, 2002.

[13] BS EN ISO 15024:2001, Fibre-reinforced plastic composites. Determination of mode I interlaminar fracture toughness, G_{IC}, for unidirectional reinforced materials. BSI, 2002.

Applied Mechanics and Materials Vols. 24-25 (2010) pp 401-406
© *(2010) Trans Tech Publications, Switzerland*
doi:10.4028/www.scientific.net/AMM.24-25.401

Effects of Bonded Splice Joints on the Flexural Response of Pultruded Fibre Reinforced Polymer Beams

G. J. Turvey[1, a]

[1]Engineering Department, Lancaster University, Bailrigg, Lancaster, LA1 4YR, UK

[a]g.turvey@lancaster.ac.uk

Keywords: Bonded joints, beam bending, PFRP material.

Abstract. Three Pultruded Fibre Reinforced Polymer (PFRP) 152 x 152 x 6.4mm Wide Flange (WF) beams were fabricated with a central two-plate splice joint. The 6.4mm thick PFRP splice plates were 210, 410 and 610mm long. Each beam was tested in symmetric four-point bending about its major and minor-axis and deflections, rotations and surface strains were recorded. Beam transverse stiffnesses, support rotations and splice rotational stiffnesses were quantified and compared with theoretical predictions. Predicted deflections were 3.5% to 18.5% larger and support rotations were 10% smaller to 14.2% larger than the experimental values. Splice end rotations were generally poorly predicted.

Introduction

The use of PFRP profiles in structural frameworks has been increasing over the past decade and a half. There are a number of reasons for the increasing structural use of PFRPs. The first is greater awareness amongst the civil and structural engineering design community of the potential of PFRPs, especially their positive attributes of low self-weight and high corrosion resistance. Another important factor is the increasing volume of data reported on the structural performance of PFRPs. Thus, even though there are no statutory design codes for PFRPs, sufficient data/guidance exists on the load – deformation response of structural elements and bolted joints to enable PFRP frames to be designed by adapting the Simple Design Method developed for steel frames.

Despite the existence of useful information on bolted tension and beam-to-column joints [1,2] and bonded tension joints [3,4], improved understanding of the behaviour of PFRP joints is still required. This is mainly because PFRPs are orthotropic elastic brittle materials with low through-thickness strength. Moreover, future research should also take account of the difference between the respective dominant design criteria for steel and PFRP structures, namely strength and stiffness.

Much of the bonded joint research has been focused on ultimate strength, but joint stiffness may be equally important for PFRP frames. Furthermore, as is clear from [3,4], most research on bonded PFRP joints has been concerned with their behaviour in tension. Flexural behaviour of bonded PFRP joints remains to be addressed. Such joints, known as splice joints, are used to join PFRP profiles end-to-end to produce continuous beams or to repair damaged beams.

The objective of this paper is to describe an investigation of splice joints in PFRP beams. Details are given of bending tests on three bonded splice joints connecting the ends of PFRP WF profiles. Deformation data (deflections and rotations) are presented which enable the transverse stiffnesses of the beams and the rotational stiffnesses of the splice joints to be quantified. Formulae are given for the calculation of beam deformations and their accuracy is quantified by comparison with the experimental deformations.

Material Properties

An EXTREN® 500 series 152 x 152 x 6.4mm WF section (Note: Reference to a trade name is solely for the purposes of factual accuracy.) was selected for the beams. The same series 6.4mm thick plate section was used for the splices.

The *minimum* longitudinal elastic moduli of the WF and plate sections are given in [5] as $17.2 kN/mm^2$ and $12.4 kN/mm^2$ respectively. Elastic moduli determined from tension tests on coupons cut longitudinally out of the web and flanges of WF and plate sections are often as much as 20% higher than the minimum values.

Splice Joint Fabrication Details

Three nominally identical 152 x 152 x 6.4mm WF profiles were each cut to a length of 3.1m. Each beam was then cut in half. Six rectangular splice plates - two for each beam - were cut out of the 6.4mm thick plate material such that their longer sides were parallel to the pultrusion direction. The plates were nominally 152mm wide, i.e. equal to the width of the flanges of the WF profiles. The lengths of the pairs of splice plates were 210mm, 410mm and 610mm.

The beam splice joints were fabricated in stages. The whole of one surface of each splice plate was abraded with sandpaper to remove the surface veil. The outer surfaces of the top and bottom flanges at each end of the beam halves were similarly abraded over a length slightly greater than the half-length of the splice plate. Resin dust was then removed from the abraded surfaces. Adhesive tape was applied to the edges of the splice plates, the edges and ends of the abraded flanges and also across the width of the flanges at the ends of the splices. The purpose of the tape was to define the bond areas and to facilitate removal of adhesive spew after bonding. The beam halves were then aligned lengthwise (flanges upright) on two trestle tables so that their ends were 10mm apart and bond areas were accessible between the ends of the tables. A two-part adhesive (Araldite® 2015) was then applied to the abraded surfaces of the splice plates and the beam flanges. Several 1mm diameter wire spacers were placed in the adhesive applied to the splice plates to ensure uniform bond thickness. Thereafter, the splice plates were brought into contact with the flanges, adjusted to ensure correct positioning and then clamped in position. After about one hour the adhesive spew was removed and the splice joint was left to cure for a further 23 hours. After curing the clamps were removed and the joint was inspected visually.

In this manner the two halves of each beam were reconnected to create three beams with two-plate splice joints of lengths, 210mm, 410mm and 610mm (including the 10mm gap between the ends of each half beam) at their centres.

Test Setup, Instrumentation and Test Procedure

The three beams with two-plate splice joints were tested in a symmetric simply supported four-point bending arrangement, as shown in Fig. 1, so that the splice joint was subjected to pure bending.

Fig. 1 also shows part of the instrumentation used to record deformations during the four-point flexure tests. A dial gauge with 50mm travel and a displacement resolution of 0.01mm was used to record the mid-span deflection at E. Four electronic clinometers each with a rotation range of 60° and an angular resolution of 0.001° were fixed to the longitudinal centreline of the beam web. Two of the clinometers were located above the simple supports at A and B. The other two were located in line with the ends of the splice plates at C and D. Not shown on Fig. 1 are four uniaxial strain gauges (gauge length 10mm and gauge resistance 120 Ohms) which were bonded to the outer surfaces of the top and bottom splice plates at mid-span. They were inset 10mm from the longitudinal edges of the splice plates with their sensitive axes parallel to the length of the beam. Thus, mid-span deflections, splice plate surface strains, support rotations and splice end rotations could be recorded throughout each spliced beam test.

As the design of PFRP structures tends to be dominated by serviceability rather than ultimate limit state criteria, it was decided to carry out four-point flexure tests only up to the deflection limit state. This had the advantage that it was unlikely that the spliced beams would be damaged during testing, so that repeat tests would be possible and that other types of test (not reported here) could be carried out on the same beams. The deflection limit chosen for the four-point flexure tests was set at

$1/200^{th}$ of the span for the major-axis tests. This value is slightly larger than that given in [6]. The value was reduced by $1/3^{rd}$ for the minor-axis tests.

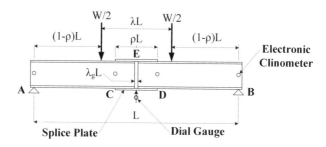

Fig. 1: Four-point bending test setup for a simply supported beam with a central two-plate splice joint.

The beams were loaded such that the deflection at mid-span increased in 2mm/1mm increments and at each deflection increment the load, rotations and strains were recorded. When the maximum deflection was reached (15mm for major and 10mm for minor-axis flexure respectively) the beam was unloaded in 3mm/2mm decrements. Each beam was subjected to three load – unload tests. The load - deflection responses were linear with good repeatability, as were the load - end rotation responses. However, the repeatability of the moment - splice end rotation responses showed some scatter. The strains recorded by individual gauges on the outer surfaces of the top and bottom splice plates showed good repeatability. Average top surface strains were about 50% of the bottom surface strains for the two longer splice joints and only 21% for the shortest splice joint. Strains recorded at opposite edges of the same splice plate sometimes differed significantly. Unsurprisingly, the correlation between experimental and theoretical strains was poor. Space limitations preclude a discussion of possible reasons for the poor strain correlations.

Test Results

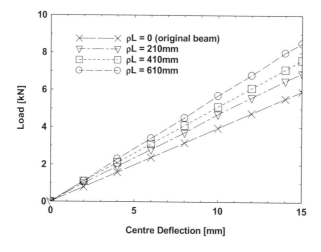

Fig. 2: Comparison of the load – deflection responses of beams with two-plate splice joint lengths of 210mm, 410mm and 610mm (bending about the major-axis).

The load – centre deflection responses of the three spliced beams are compared with the calculated response of the original beam in Fig. 2. It is evident that the response is linear for all of the beams up to the deflection limit and that they are stiffer than the original beam. Straight line fits to the load – deflection/support rotation/splice end rotation data obtained from the tests have enabled the transverse stiffnesses of the spliced beams and the rotational stiffnesses of the splice joints to be determined with respect to their major and minor-axes. These stiffnesses are given in Table 1.

Table 1

Transverse stiffnesses of spliced beams and rotational stiffnesses of two-plate bonded splice joints

Length of Splice Joint [mm]	Major-Axis Transverse Beam Stiffness [kN/mm]	Minor-Axis Transverse Beam Stiffness [kN/mm]	Major-Axis Splice Rotational Stiffness [kNm/mrad]	Minor-Axis Splice Rotational Stiffness [kNm/mrad]	Major-Axis Splice Rotational Stiffness/Unit Length [kNm/m/mrad]	Minor-Axis Splice Rotational Stiffness/Unit Length [kNm/m/mrad]
210	0.473	0.138	0.767	0.255	3.652	1.214
410	0.505	0.150	0.700	0.154	1.707	0.376
610	0.569	0.183	0.627	0.144	1.028	0.236
0	0.395	0.126	-	-	-	-

Spliced Beam Analysis

The deformation behaviour of a simply supported beam with a splice joint at mid-span, shown in Fig. 1, may be analysed using simple beam bending theory in conjunction with Mohr's Moment-Area theorems and the Method of Transformed Sections [7]. Hence, using the notation in Fig. 1, the expression for the mid-span deflection may be written as:-

$$\delta_E = \frac{WL^3}{96EI}(1-\lambda)\left\{\left(2+2\lambda-6\rho-\lambda^2+3\rho^2\right)+3\frac{\rho(1-\rho)}{(1+\phi_I)}\right\}. \tag{1}$$

Likewise, the expressions for the support and splice end rotations may be written respectively as:-

$$|\theta_A|=|\theta_B|=\frac{WL^2}{16EI}(1-\lambda)\left\{(1+\lambda-2\rho)+2\frac{\rho}{(1+\phi_I)}\right\}. \tag{2}$$

$$|\theta_C|=|\theta_D|=\frac{WL^2}{8EI}\frac{\rho(1-\lambda)}{(1+\phi_I)}. \tag{3}$$

In Eqs. (1) – (3) I and ϕ_I denote respectively the second moment of area of the original PFRP section and the additional second moment of area factor due to the two splice plates.

The geometries of the original and transformed cross-section for major-axis bending are shown in Figs. 3(a) and 3(b) respectively. λ_b, λ_f and λ_w denote the ratios of the flange width, flange thickness and web thickness respectively to the depth of the WF cross-section. Similarly, β_a and β_p denote the ratios of the adhesive and splice plate thickness respectively to the flange thickness. In addition, γ_a and γ_p denote the ratios of the adhesive elastic modulus and the longitudinal splice plate modulus respectively to the longitudinal elastic modulus of the beam. The values of these factors for the 152 x 152 x 6.4mm WF and 6.4mm plate sections and the 1mm thickness of the adhesive layers are:-

$$\lambda_b = 1; \; \lambda_f = \lambda_w = 0.42; \; \beta_a = 0.156; \; \beta_p = 1; \; \gamma_a = 0.174; \; \gamma_p = 0.721. \tag{4}$$

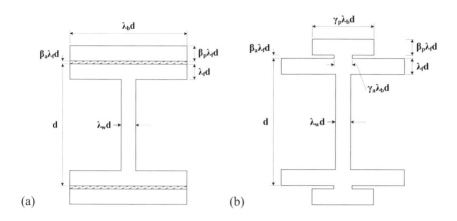

Fig. 3: (a) Section through the splice joint and (b) transformed section

Expressions have been derived for both I and ϕ_I in terms of the lambda, beta and gamma factors for both major and minor-axis flexure. Ignoring the effect of the adhesive layers of the splice joint, i.e. setting $\beta_a = 0$, the values of the additional second moment of area factors ϕ_I were evaluated as:-

$$\phi_I = 0.748 \, (\text{major-axis}); \; \phi_I = 0.720 \, (\text{minor-axis}). \tag{5}$$

Using these values together with values of $\lambda = 0.267$ and $\rho = 0.070, 0.137, 0.203$ for each of the three splice lengths, the mid-span deflection, end rotation and splice end rotation at E, B and D respectively were evaluated for the maximum load of each test. These values are compared with the experimental values in Table 2.

Table 2

Comparison of experimental and theoretical deflections and rotations for simply supported beams with a central two-plate splice joint

(a) Major-Axis Bending

Maximum Load (W) [kN]	Splice Plate Length (ρL) [mm]	Centre Deflection (δE) [mm]	Support Rotation ($|\theta_A|, |\theta_B|$) [mrad]	Splice End Rotation ($|\theta_C|, |\theta_D|$) [mrad]
8.5	610	17.41 (16.1%) 15.00	19.09 (14.2%) 16.71	4.065 (8.1%) 3.761
7.6	410	16.68 (11.2%) 15.00	17.96 (11.0%) 16.18	2.443 (-17.2%) 2.950
6.9	210	16.23 (8.2%) 15.00	17.11 (5.0%) 16.30	1.136 (-45.8%) 2.095

(b) Minor-Axis Bending

Maximum Load (W) [kN]	Splice Plate Length (ρL) [mm]	Centre Deflection (δ_E) [mm]	Support Rotation (|θ_A|,|θ_B|) [mrad]	Splice End Rotation (θ_C|,|θ_D|) [mrad]
1.9	610	12.27 (18.5%) 10.00	13.43 (-5.3%) 14.14	2.896 (-23.5%) 3.578
1.5	410	10.36 (3.6%) 10.00	11.14 (-10.0%) 12.25	1.537 (-61.8%) 2.487
1.4	210	10.35 (3.5%) 10.00	10.90 (-8.6%) 11.84	0.735 (-86.5%) 1.371

Note: Experimental values are the average values for the third test and are the lower values in each row. (..%) = [(theory – experiment)/experiment] x 100.

Concluding Remarks

The test results show that the transverse stiffnesses of all the spliced beams are greater than the original beam for both major and minor-axis flexure and that transverse stiffness increases with splice length. In contrast, splice rotational stiffness decreases with increasing splice length.

Finally, the closed-form equations, developed for four-point flexure of simply supported beams with central two-plate bonded splice joints have been shown to provide good to reasonable estimates of mid-span deflection and support rotation, but poor estimates of splice joint rotation.

Acknowledgements

The author wishes to acknowledge the contribution of Mr. A. Echegut, who carried out the test work under the author's direction during an eight-week Summer Internship in the Engineering Department. The assistance of the Engineering Department's Technician Staff with the experimental work is also acknowledged.

References

[1] G.J. Turvey: Prog. Struct. Eng. Mat. Vol. 2 (2000), p. 146

[2] G.J. Turvey and C. Cooper: Proc. I.C.E. (Struct. & Build.) Vol. 157 (2004), p. 211

[3] T. Keller and T. Vallée: Comp. Pt. B (Eng.) Vol. 36 (2005), p. 331

[4] T. Keller and T. Vallée: Comp. Pt. B (Eng.) Vol. 36 (2005), p. 341

[5] Anon: *EXTREN Fiberglass Structural Shapes Design Manual*, Strongwell, USA (1989).

[6] J.L. Clarke: *Structural Design of Polymer Composites: EUROCOMP Design Code and Handbook*, E & F N Spon, London (1996).

[7] J. Case and A.H. Chilver: *Strength of Materials*, Edward Arnold, London (1959).

Applied Mechanics and Materials Vols. 24-25 (2010) pp 407-412
© *(2010) Trans Tech Publications, Switzerland*
doi:10.4028/www.scientific.net/AMM.24-25.407

Assessment of Quasi-Static and Fatigue Loaded Notched GRP Laminates Using Digital Image Correlation

W. R. Broughton[1, a], M. R. L. Gower[1, b], M. J. Lodeiro[1, c], G. D. Pilkington[1, d] and R. M. Shaw[1, e]

[1] Materials Division, National Physical Laboratory, Teddington, Middlesex, TW11 0LW, UK

[a]bill.broughton@npl.co.uk, [b]michael.gower@npl.co.uk, [c]maria.lodeiro@npl.co.uk, [d]gordon.pilkington.@npl.co.uk, [e]richard.shaw@npl.co.uk

Keywords: Digital image correlation (DIC), fatigue, glass fibre-reinforced plastics (GRP), mechanical properties, strain measurement

Abstract. This paper presents the results of an experimental study carried out to assess the effects of cumulative damage on the strain response of open-hole (notched) quasi-isotropic $[45/0/-45/90]_{4s}$ E-glass/913 epoxy laminate specimens under quasi-static tension, and constant amplitude tension-tension, compression-compression and tension-compression cyclic loading conditions. In-plane surface strain was measured as a function of applied load and loading cycles for quasi-static and cyclic loading respectively using digital image correlation (DIC) and compared with strain gauge, fibre Bragg grating (FBG) optical sensor and contact extensometer strain measurements. DIC proved successful in monitoring local (near the hole) and global strains, providing critical information on the changes in strain distribution resulting from damage formation and growth incurred.

Introduction

The measurement of displacements and displacement gradients (i.e. strains) is now recognised as an important field of experimental stress analysis. Optical techniques, such as moiré interferometry, holographic interferometry and electronic speckle pattern interferometry (ESPI) are well established and used extensively for research and field inspection. However, these techniques have inherent sensitivity and resolution limitations along with issues relating to system stability and processing of fringe patterns, although computational algorithms for processing the data have been developed. An alternative approach is to use digital image correlation (DIC), which enables the operator to directly compare high quality images of the component to measure deformation rather than having to analyze complex fringe patterns. The technique has been successfully applied to both isotropic and anisotropic materials to study strain deformation, residual stress (e.g. hole drilling) and damage formation (e.g. crack growth) [1-3]. The work presented in this study uses DIC for measuring surface deformation under quasi-static and constant amplitude cyclic loading conditions. The results are compared with strain gauge, fibre Bragg grating (FBG) optical sensor and contact extensometer strain measurements.

Material and Experimental Procedure

Quasi-static and constant amplitude (sinusoidal waveform) cyclic fatigue tests were carried out on open-hole tension (OHT) and open-hole compression (OHC) coupon specimens machined from quasi-isotropic $[45/0/-45/90]_{4s}$. E-glass/913 epoxy laminates. The laminates (600 mm x 300 mm) were autoclave manufactured at the National Physical Laboratory using pre-impregnated tapes (0.125 mm thick) supplied by Hexcel Composites Ltd. The cure cycle was in accordance with the supplier's specifications. The fibre volume fraction of the laminates, as measured using the burn-off technique specified in ISO 1172 "Textile-glass-reinforced plastics - Prepregs, moulding compounds and laminates - Determination of the textile-glass and mineral-filler content - Calcination methods", was 54.0 ± 0.2%. OHT specimens were 250 mm in length, 36 mm wide and nominally 4 mm (32 ply) thick with a centrally located 6 mm diameter hole (Fig. 1). OHC specimens are a shorter version of the OHT specimen (i.e. 125 mm long with a 25 mm gauge-section) – see Fig. 2. The specimens were gripped via end tabs (50 mm long) manufactured from a plain-woven glass fabric/epoxy laminate (1.6 mm thick), with the fibre axes of the fabric set at ±45° to the specimen axis, and adhesively bonded to the specimen. The tab angle is 90° (i.e. not tapered). Specimen geometry and loading arrangement was in accordance with ASTM D 5766 D "Standard test method for open-hole tensile strength of polymer matrix composite laminates" and ISO/WD 12817 "Carbon-fibre-reinforced composites - Determination of open-hole compressive strength".

OHT tensile tests (Fig. 3) and OHC cyclic fatigue tests were conducted using an Instron 5500 servo-hydraulic test machine with the load introduced via fatigue-rated hydraulic wedge-action grips. A test frequency of 5 Hz was used for the fatigue tests. A lateral grip pressure of 200 bar was applied. Instron MAX software was used to control the test machine, and for data capture. The OHC strength was obtained by loading the specimens via end-loading blocks between two parallel, hardened stainless steel platens within a four-post die set to minimise buckling (Fig. 4). Testing was carried out under standard laboratory conditions (23 ± 2 °C, 50 ± 10% relative humidity).

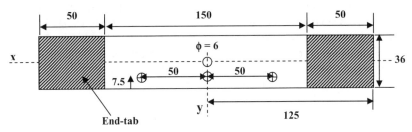

Figure 1. Open-hole (notched) tension specimen (units: mm)

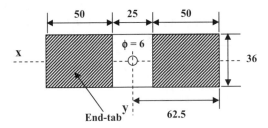

Figure 2. Open-hole (notched) compression specimen (units: mm)

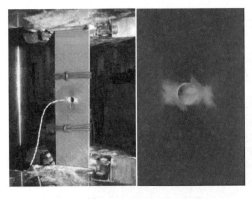
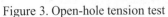

Figure 3. Open-hole tension test   Figure 4. Open-hole compression test

In-plane strain was measured using DIC, strain gauges adhesively bonded to the specimen surface, fibre Bragg grating (FBG) optical sensors embedded in the laminate and a contact (clip-gauge) extensometer with a gauge-length of 50 mm located at the mid-section of the specimen (see Fig. 3). DIC was the only strain measurement technique used for monitoring strain distributions in the OHC tests. A LaVision® DIC system with a single megapixel (1280 x 1024 pixel) video camera was used to obtain two dimensional (2D) maps of the strain distributions (ε_{xx}, ε_{yy} and ε_{xy}) along the length and across the width of notched specimens that had been subjected to either quasi-static or fatigue loading. One face of the coupon specimens was sprayed with black, grey and white paint to produce a specular random pattern for the image correlation. Images were recorded at a frequency of 1 Hz throughout the tests and all strain results were calculated relative to the first image recorded at zero load. LaVision® Strainmaster software was used for data capture and analysis (conversion of displacement to strain). For quasi-static testing, the load was increased in incremental steps at a displacement rate of 1 mm/min and the resultant displacement field was measured. This was repeated until an applied load was reached beyond which the damage in the vicinity of the hole was too extensive to obtain sensible strain measurements. In the case of fatigue testing, the test was stopped at set intervals of 10,000 cycles, and the resultant strain field was measured at a constant static load equivalent to the maximum load level of the fatigue cycle. The normal and shear strain distributions were determined along the central axes in the x- (loading) and y- (tranverse) directions of the notched specimens as a function of applied load or loading cycles.

Quasi-Static Tensile Tests

The un-notched and notched laminate tensile strength, measured at an equivalent displacement rate as the fatigue tests (i.e. 500 mm/min), was 484 ± 18 MPa and 346 ± 5 MPa, respectively. The corresponding tensile modulus, measured using the contact extensometer, was 21.9 ± 0.4 GPa and 20.6 ± 0.3 GPa. Each set of results represents five tests. The contact extensometer was only used to measure strain at low levels (0.05 – 0.25%) as the violent nature of the failure that occurs could have damaged the device. Strain in the OHT specimen was also measured using strain gauges located 7.5 mm in from the specimen edge at the centre-line, and at locations 50 mm above and 50 mm below the centre-line (see Fig. 1). The corresponding tensile modulus and failure strains obtained were 20.2 GPa, 19.9 GPa and 21.9 GPa, and 3.12%, 2.14 % and 2.01 %, respectively. The strain gauges located 50 mm from the centre-line are in low strain regions. The higher failure strain value obtained at the centre-line was due to a loss of stiffness resulting from localized damage near the hole (see Fig. 3). The strains measured using the different techniques were in good agreement as shown in Fig. 5. The strains tend to diverge at higher loads due to damage growth around the hole.

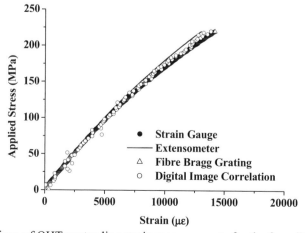

Figure 5. Comparison of OHT centre-line strain measurements for the four different techniques

Failure involves a number of fracture mechanisms commencing with matrix cracking of the 90° plies occurring near the hole boundary at an applied stress of ~60 MPa (coinciding with non-linearity of the axial stress/strain response) followed by matrix cracking of the ±45° plies (surface cracking occurs at ~243 MPa); splitting in the 0° plies and local delaminations at the interfaces between the 45° and 90° plies; and finally by fibre fracture and pull-out. Fig. 6 shows the ε_{xx} strain distribution across the specimen width as a function of increasing applied load measured using DIC. It can be seen that the distribution is not symmetrical about the hole due to the non-uniformity of damage formation (see Fig. 3). Fig. 7 compares the average value of strain concentrations $\varepsilon_{xx}/\varepsilon^{\infty}_{xx}$ (where ε_{xx} and $\varepsilon^{\infty}_{xx}$ denote local strain and applied strain) measured from the edge of the hole to the edge of the specimen at 50%, 60% and 67% UTS (ultimate tensile strength) with the analytical solution predictions for an orthotropic plate with a circular hole [4]. Again, the results are in reasonable agreement. The effects of damage become more apparent as the load increases. Beyond a stress level of approximately 233 MPa (67% UTS) damage became too extensive to be able to reliably measure displacement gradients near the hole using DIC.

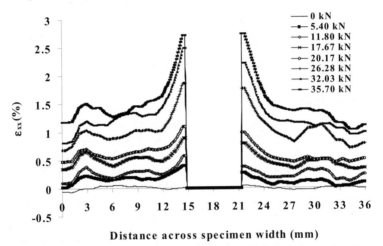

Figure 6. ε_{xx} strain distribution around the hole as a function of applied load

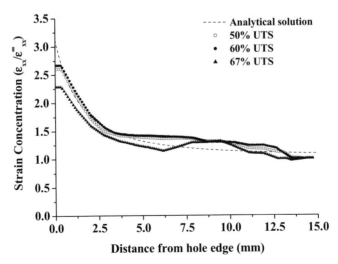

Figure 7. Strain concentration $\varepsilon_{xx}/\varepsilon^{\infty}_{xx}$ across the centre-line of the specimen

Cyclic Fatigue Tests

The ability to monitor damage growth using DIC becomes apparent when examining the fatigue behaviour of OHT and OHC specimens. OHT specimens were subjected to constant amplitude tension-tension (T-T) and the OHC specimens to either compression-compression (C-C) or tension-compression (T-C) cyclic loading conditions. The stress ratio R ($\sigma_{min}/\sigma_{max}$) values, and maximum applied stresses for the three loading conditions were 0.1, 10 and -1, and 156 MPa, -215 MPa and 123 MPa. The tests were stopped at set intervals of 10,000 cycles, unloaded and then statically re-loaded to a constant load to enable DIC imaging of the specimen surface. The static load applied for imaging was 20 kN (133 MPa) for T-T, -25 kN (-164 MPa) for C-C and 15kN (100 MPa) for T-C. Strain maps of ε_{xx} as a function of loading cycles for the three fatigue cases are shown in Fig. 8.

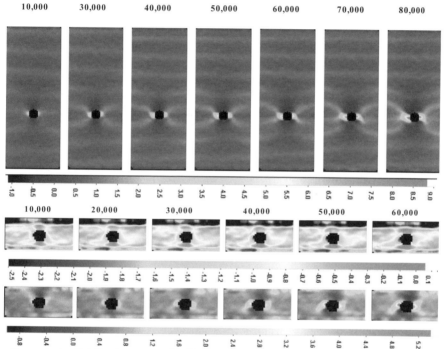

Figure 8. ε_{xx} strain distribution versus loading cycles: T-T (top), C-C (middle) and T-C (bottom)

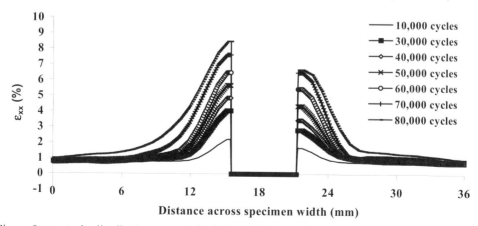

Figure 9. ε_{xx} strain distribution around the hole of OHT specimen as a function of loading cycles

The DIC images shown in Fig. 8 indicate damage growth increases around the hole with increasing number of loading cycles (N). Damage growth is difficult to detect both visually and using non-destructive evaluation methods, such as ultrasonic C-scan and X-radiography. Pulse thermography can be used to monitor damage growth, but the images, although informative, are not as easy to interpret compared with those images produced using DIC. Strain maps of ε_{xx}, ε_{yy} and ε_{xy} when used in combination provide a full picture of the effect of the different damage mechanisms on the in-plane surface strains. The maximum value of ε_{xx} strain was taken as the average of the two peak strains either side of the hole (see Fig. 9). This was carried out to account for non-symmetric damage formation around the hole. These averaged peak strains increase linearly with loading cycles (see Fig. 10). Global stiffness is less sensitive to damage growth with dramatic reduction in stiffness occurring near the onset of failure (i.e. last 1-2% of fatigue life).

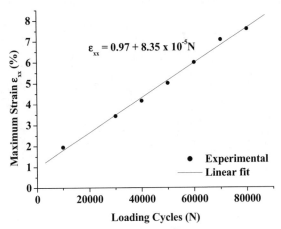

Figure 10. Average peak ε_{xx} strain at the hole-edge as a function of loading cycles for OHT

Conclusions

DIC proved successful in monitoring local and global strains, providing critical information on the changes in strain distribution around the hole of the notched laminates resulting from damage formation and growth incurred through either increasing load or number of loading cycles.

Acknowledgements

The authors acknowledge the financial support provided by United Kingdom Department for Business, Innovation and Skills (National Measurement Office), as part of the Materials 2007 Programme. The authors would also like to thank Hexcel Composites Limited, and Dr F Surre and Dr T Venugopalan at City University London for their technical support and advice.

References

[1] B. Wattrisse, A. Chrysochoos, J-M. Muracciole and M. Némoz-Gaillard: Exp. Mech. Vol. 41(1) (2001), p. 29.

[2] J.D. Lord, D. Penn and P. Whitehead: App. Mech. Mater. Vols. 13-14 (2008), p. 65.

[3] F. Lagattu, J. Brillaud and M-C. Lafarie-Frenot: Mater. Charact. Vol. 53(1) (2004), p. 17.

[4] J.M. Whitney and R.J. Nuismer, in: *Fracture Mechanics of Composites*, ASTM STP 593 (1975), p. 117.

Applied Mechanics and Materials Vols. 24-25 (2010) pp 413-418
© (2010) Trans Tech Publications, Switzerland
doi:10.4028/www.scientific.net/AMM.24-25.413

Comparison of the Drop Weight Impact Performance of Sandwich Panels with Aluminium Honeycomb and Titanium Alloy Micro Lattice Cores

R. Hasan [a], R. Mines [b], E. Shen, S. Tsopanos, W. Cantwell, W. Brooks and C. Sutcliffe

Department of Engineering, University of Liverpool, The Quadrangle, Liverpool L69 3GH, UK

[a]rafidah@liv.ac.uk, [b]r.mines@liv.ac.uk

Keywords: impact, sandwich, honeycomb, micro-lattice

Abstract. This paper is a study of the drop weight impact behaviour of small sandwich panels of carbon epoxy skins with aluminium honeycomb and titanium alloy micro-lattice cores. A series of experimental tests have shown that the specific impactor penetration behaviours are similar for both cores. The reasons for this are a result of the detailed deformation and rupture behaviour of the two types of core. The deformation and rupture mechanisms of honeycomb and micro-lattice structures will be discussed in general terms, and these observations will be used to inform discussion of actual deformation and rupture in the panel tests. In this way, micro energy absorbing mechanisms will be related to panel performance, and conclusions on the way forward for improved penetration performance using other core materials and geometries will be identified.

Introduction

Foreign object impact, such as dropped tools and bird strikes, is one of the important subjects to be focused on in the studies of sandwich structures and many discussions have been done on the drop weight impact performance of sandwich panels using various core materials [1-4]. Sandwich core behaviours as an energy absorbing materials have extensively been studied, in order to obtain better crash performance of sandwich structures. The most studied core is honeycomb, specifically aluminium honeycomb, which offers a high stiffness and strength-to-weight ratio, especially in the out-of-plane direction [5]. It is important to understand the mechanics of honeycombs as the basics to other complex core materials of sandwich panels such as foams and lattice structures.

The studies of micro-lattice structures as core materials in sandwich construction have been initiated from the approximation approach of microstructural models for foams. Mines has broadly discussed the behaviour of various foams and micro-lattice structures [6]. Advanced manufacturing techniques such as the rapid prototyping selective laser melting (SLM) technique [7] has strongly contributed to the initial study on the metallic micro-lattice structure at the University of Liverpool. Work has been carried out in studying the mechanical properties, crush behaviours, as well as impact properties of the SLM SS316L stainless steel and Ti64 titanium alloy micro-lattice [6-10]. However, it should be noted that the mechanical properties of the SLM micro-lattice structures are very much influenced by two processing parameters that are laser power (in Watt [W]) and laser exposure time (in micro seconds [µs]) [11].

Interestingly, it is found that the SLM micro-lattice structure manufactured from the Ti64 is competitive with aluminium honeycomb in specific strength [12]. It is also shown that strength and impact failure load of Ti64 micro-lattice is better than that of SS316L micro-lattice. There was a reduction in dent depth for a given impact energy for the Ti64 [13]. However, a major issue with micro lattice technology is the synthesis of optimal configurations. Therefore, theoretical and numerical parametric models are required to systematically vary such parameters as micro strut architecture, strut aspect ratio, cell size, parent material and rate of loading [14]. Also, a closer comparison study with the main competitor as sandwich construction core, which is the aluminium honeycomb, is also required in understanding the energy absorption capabilities and damage mechanisms of the micro-lattice structure.

In this paper, the results from drop weight impact test of small sandwich panels with aluminium honeycomb and Ti64 micro-lattice cores using carbon epoxy skins are going to be discussed. The experimental results have shown that the specific impactor penetration behaviours are similar for both cores [15]. Therefore, comparison and understanding on their failure deformations and micro mechanisms is important in future improvement on performance of different core materials and geometries.

Impact Test on Sandwich Structures

Impact Test. The drop weight impact test has been carried out on 100mm x 100mm x 20mm core size sandwich panels with two different core materials as detailed in Table 1. In this study, the full definition of manufacturing parameters [11] for the Ti64 micro-lattice is: architecture = BCC, parent material = Ti64, laser power = 200W, exposure time = 1000μs, strut diameter = 400μm, cell size = 2.5mm and unit cell = cubic (BCC/Ti64/200/1000/400/2.5/Cubic).

Table 1: Details of sandwich panels

Skin details					
Prepreg material	Skin lay-up	Areal density [g/m^2]	Tensile modulus (8 ply laminate) [GPa]	Tensile strength (8 ply laminate) [MPa]	Supplier
Plain weave carbon fibre / epoxy matrix (CFRP)	4 ply - Nominal thickness 1.1mm	410 ± 15	58	850	Gurit (EP121-C15-53)
Core details					
Name	Parent material	Density [g/cm^3]	Cell type / edge length [mm]	Manufacturing parameters	Supplier
Aluminium honeycomb	Aluminium alloy 5056	2.63	Hexagon open / 8	-	Hexcel (CRIII-1/4-5056-.001N-2.3)
Titanium alloy micro-lattice	Titanium alloy Ti 6Al 4V	4.43	BCC open / 2.5	SLM 200W x 1000μs	TLS Technik (Grade 5 ASTM)

The sandwich structures were formed by stacking four plies of CFRP prepreg as the upper and lower skin of each core material, and compression moulded using a hot press machine. The pressure was maintained below the yield stress of the core materials at a controlled temperature of 125°C for two hours. There was no adhesive used at the skin-core interface, since the process had allowed the prepreg resin to flow over the core material, and bonded them together.

The impact test has been conducted on the instrumented drop-weight impact tower. An instrumented carriage with the hemispherical tip impactor of 10mm diameter was released from a certain height, in order to give a certain amount of impact energy on the sandwich panel. The velocity of the impactor during the impact event was measured using a high speed camera, and the velocity versus time traces from the velocimeter was recorded by a computer to produce displacement versus time traces. The impacted sandwich panels were supported at their four corners.

Impact Test Results. Table 2 summarizes the impact test results of aluminium honeycomb and Ti64 micro-lattice cores which led to the calculation of specific impact energy. On the other hand, Figure 1 compares graphically, the difference in specific impact energy versus dent depth for aluminium honeycomb and Ti64 micro-lattice cores, together with the other two types of cores that had been compared elsewhere [15], the stainless steel micro-lattice (BCC/SS316L/140/500/200/2.5/Cubic) and aluminium foam (Alporas) cores. By looking at the dent depth, it is shown that at lower specific impact energy, the aluminium honeycomb is more than two times better in impact resistance, as compared to the Ti64 micro-lattice. However, at higher specific impact energy, the gap difference between the dent depths of both cores is becoming smaller.

Table 2: Impact test results of sandwich panels with aluminium honeycomb (honeycomb) and titanium alloy (Ti64) micro-lattice cores

Core	Drop height [m]	Dent depth [mm]	Absorbed energy [J]	Panel mass [g]	Panel thickness [mm]	Drop mass [kg]	Volumetric density [kg/m^3]	Specific impact energy [Jm^3/kg]
Honeycomb	0.50	19.08	11.90	42.28	21.10	2.07	200.4	0.051
	0.25	2.25	5.13	41.82	21.55	2.03	194.1	0.026
	0.37	10.77	7.85	41.87	21.50	2.03	194.7	0.038
	0.45	15.69	9.47	41.70	21.60	2.03	193.1	0.046
Ti64	1.00	18.80	21.29	96.10	21.90	2.08	438.8	0.046
	0.25	3.27	5.58	95.90	21.30	2.07	450.2	0.011
	0.50	4.78	10.68	95.90	21.30	2.07	450.2	0.023
	0.75	12.03	16.03	94.70	21.10	2.07	448.8	0.034

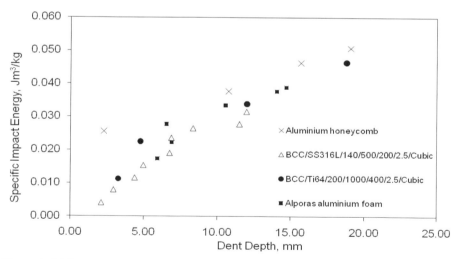

Figure 1: Difference in specific impact energy versus dent depth of Ti64 micro-lattice and aluminium honeycomb cores, together with SS316L micro-lattice and Alporas foam

Discussions on Deformation of Impacted Sandwich Panels

Aluminium Honeycomb Panel. Figure 2 shows CT scan images of side, top and 3-D view for impacted aluminium honeycomb sandwich panels. From the 3-D image, it can be observed that the affected diameter of the impacted region would be approximately twice that of the impactor size. This is supported by the side view observation that shows that there seems to be slight skin bending and honeycomb panel buckling around the impacted area. The impacted skin was fractured and cells in the region experienced similar crush behaviour with honeycomb cells under compression. There was also no visible skin-core separation observed in the honeycomb panel. Although it is reported that delamination occurred in the upper skin when the incident impact energy was more than 5J [3], further micrographic analysis needs to be done in this study for observing the delamination behaviour of the skin. For the core, around three to four hexagonal cells at the surrounding region were affected. This is due to the buckling and bending phenomena especially at the top region of the honeycomb panels, under the upper skin. From the deformation that can be observed in the CT scan images, it is suggested that the impact energy was absorbed in the aluminium honeycomb panel by upper skin bending, skin fracture, core crushing, core tearing, core buckling, core bending and core folding in the impacted and surrounding region. Figure 3 illustrates the deformation mechanisms of a unit cell for aluminium honeycomb, to further assist the understanding.

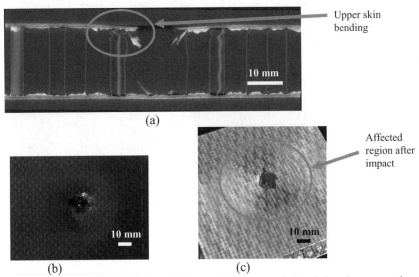

Figure 2: CT scan images of (a) side, (b) top and (c) 3-D view for impacted aluminium honeycomb sandwich panel

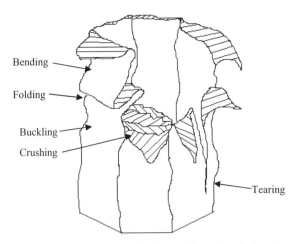

Figure 3: Illustration on the deformation mechanisms of a unit cell for aluminium honeycomb

Titanium Alloy Micro-Lattice Structure Panel. Figure 4 shows CT scan images of top and side view for impacted Ti64 micro-lattice core sandwich panels. It can be observed that localised skin damage and indentation occurred in this panel, rather than globalised damage as occurred in the aluminium honeycomb panels. The damage area is about the same size of the impactor which penetrated and fractured the skin. The area beyond that seems to be not affected by the impact incidence. No evidence of delamination in the skin was observed but there was debonding of the skin from the core at the fractured region. It can be also seen that the struts under the impacted skin have been fractured and fragmented in brittle manner. From the deformation of this impacted Ti64 micro-lattice core panel, it can be suggested that the impact energy was absorbed by the skin fracture, core-skin debonding and micro struts fractured in brittle manner. There were also a few micro struts that experienced slight bending, which most likely occurred prior to strut failure, since there was no strong evidence indicating ductile failure occurred at the same time as brittle failure. Figure 5 illustrates the brittle fracture of impacted micro struts for a unit cell Ti64 micro-lattice structure, to further assist the understanding.

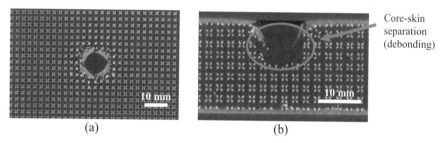

Figure 4: CT scan images of (a) top and (b) side view for impacted Ti64 micro-lattice core sandwich panel (impact energy = 22.98J)

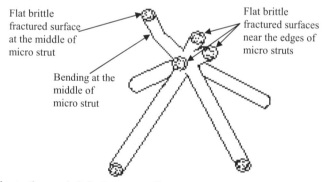

Figure 5: Illustration on brittle fracture of impacted micro struts for a unit cell Ti64 micro-lattice structure

Concluding Remarks

In comparison, both aluminium honeycomb and Ti64 micro-lattice cores show comparable impact resistance, from the similar result obtained for dent depth at high specific impact energy. However, damage area seems to be more localised for the Ti64 micro-lattice as compared to that of aluminium honeycomb. It is due to the different mechanisms during impact energy absorption in both cores. This can be an advantage for the Ti64 micro-lattice, and it needs to be further studied and analysed for the potential as one of the aluminium honeycomb competitors that can be used in aerospace applications. There is a requirement outlined by the aircraft manufacturer that the impact damage is preferably to have similar diameter with the impactor [16]. The localised impact damage means that there would be less replacement area needed for a particular sandwich structure.

Besides showing a good impact resistance which is comparable with aluminium honeycomb, the Ti64 micro-lattice shows unfavourable brittle failure, and this issue needs to be addressed and improvement should be done in order to have a more stable impact energy absorption behaviour. It should be noted that there is potential for SLM Ti64 micro-lattice structure to be optimized and have graded topology [6, 14] for better impact resistance.

Acknowledgements

Financial support from grants sponsored by EPSRC/EP/C009525/1, EPSRC/EP/009398/1 and EU FP6 CELPACT is gratefully acknowledged. The authors would like to acknowledge Dr A Johnson from DLR Stuttgart, Germany for CT scans and Prof. X. Wu from University of Birmingham for post manufacture HIP process of Ti64. R. Hasan would like to thank Malaysian Government and Universiti Teknikal Malaysia Melaka for her PhD study sponsorship.

References

[1] R.A.W. Mines, C.M. Worrall and A.G. Gibson: Composites Vol. 25 (1994), p. 95-110

[2] C.G. Kim and E.J. Jun: J. Compos. Mater. Vol. 26 (1992), p. 2247-2261

[3] M. Akay and R. Hanna: Composites Vol. 21(4) (1990), p. 325-331

[4] S. McKown and R.A.W. Mines: High Performance Structures and Materials II Vol. 7 (2004), p. 37-46

[5] L.J. Gibson and M.F. Ashby: *Cellular solids: structure and properties* – 2nd ed. (Cambridge University Press, Cambridge United Kingdom 1999).

[6] R.A.W. Mines: Strain. Vol. 44(1) (2008), p. 71-83

[7] M. Santorinaios, W. Brooks, C.J. Sutcliffe, R.A.W. Mines: WIT Trans. Built Environ. Vol. 85 (2006), p. 481-490

[8] S. McKown, W.J. Cantwell, W.K. Brooks, R.A.W. Mines, S. Tsopanos, C.J. Sutcliffe: Proceedings of the 28th International European *SAMPE*, Europe (2007), p. 396-401

[9] R.A.W. Mines, S. McKown, S. Tsopanos, Y. Shen, W.J. Cantwell, W.J. Brooks, C. Sutcliffe: Applied Mechanics and Materials Vol. 13-14, (2008), p. 85-90

[10] S. McKown, Y. Shen, W. Brooks, W.J. Cantwell, C. Sutcliffe, G. Langdon, G.N. Nurick: Int J. Imp. Engrg. 35, (2008), p. 795-810

[11] S. Tsopanos, R. Mines, S. McKown, Y. Shen, W. Cantwell, W. Brooks, C.J. Sutcliffe: Submitted to Journal of Manufacturing Science and Engineering ASME (2009)

[12] R. Mines, Y. Girard, V. Fascio: Proceedings of the International European SAMPE, Europe (2009), p. 248-256

[13] R. Mines, S. Tsopanos, E. Shen, S. McKown, W.J. Cantwell: Proceedings of the ICCM17 Conference, Edinburgh United Kingdom (2009), Paper No. B6.1

[14] K. Ushijima, W.J. Cantwell, R.A.W. Mines, S. Tsopanos, M. Smith: To be published Journal of Sandwich Structures and Materials (2010)

[15] Y. Shen: PhD Thesis, University of Liverpool, United Kingdom (2009)

[16] E. Morteau and C. Fualdes: FAA Workshop for Composite Damage Tolerance and Maintenance, National Institute for Aviation Research, Chicago USA (2006)

Applied Mechanics and Materials Vols. 24-25 (2010) pp 419-423
© (2010) Trans Tech Publications, Switzerland
doi:10.4028/www.scientific.net/AMM.24-25.419

Viscoelastic Characterization of Short Fibres Reinforced Thermoplastic in Tension and Shearing

A. Andriyana[1, a], L. Silva[1,b] and N. Billon[2,c]

[1]Dep. Of Mech Eng., Univ. Malaya, 56603 Kuala Lumpur, Malaysia

[2]MINES-ParisTech CEMEF, BP 207 06904 Sophia Antipolis, France

[a]andri.andriyana@um.edu.my, [b]Luisa.Silva@mines-paristech.fr, [c]noelle.billon@mines-paristech.fr

Keywords: Short fibres composites; Non-linear behaviour; Video extensometer; Tension; Sheraing.

Abstract. The present work can be regarded as a first step toward an integrated modelling of mould filling during injection moulding process of polymer matrix composites and the resulting material behaviour under service loading conditions. More precisely, the emphasis of the present research is laid on the development of a mechanical model which takes into account the processing-induced microstructure and is capable to predict the mechanical response of the material. In the Part I, a set of experiments which captures the mechanical behaviour of an injection moulded short fibre reinforced under different strain histories is described. Three mechanical testing are conducted: Dynamic Mechanical Analysis (DMA), uniaxial tension and simple shear. Tests show that the material exhibits complex responses mainly due to non-linearity, anisotropy, time/rate-dependence, hysteresis and permanent strain. Moreover, the relaxed state of the material is characterized by the existence of a so-called anisotropic equilibrium hysteresis independently of the prescribed strain rate.

Introduction

360 x 100 x 3mm^3 plates of a thermoplastic polyamide reinforced by 20 and 30 wt% of short glass fibres are used in this study. At the macroscopic level, such fibre reinforced thermoplastics are known to exhibit strong directional dependencies, inelastic responses such as time/rate-dependence, temperature dependence, hysteresis and permanent strain (see for example the work of [1-3]).

To gain additional insight into the time-dependent anisotropic behaviour of the composite of interest in this study, experiments probing the mechanical behaviour are conducted using machined samples.

The first step consists in assessing for the mechanical heterogeneities of the plates due to injection moulding to draw some guidelines for samples machining. This is achieved by means of the Dynamic Mechanical Analysis (DMA).

As plates appeared to be homogenous at the macroscopic scale, i.e. no significant variations of the *averaged mechanical properties over plate thickness* for different location in the injected plates *larger scale* mechanical testing (uniaxial extension and simple shear) are conducted. For larger strain tension and shearing are performed.

Low strain viscoelasticity and plate homogeneity

Homogeneity of the injection moulded plate can be illustrated through a set of DMA testing conducted on 45 x 8 mm^2 rectangle-shape specimens of 3 mm in thickness machined from the plate at different locations as depicted in Fig. 1. DMA is performed in 3 points flexural loading mode for three different frequencies: 0.1, 1 an 10 Hz. Two orientations are retained: perpendicular and parallel to the flow direction during injection process. During the DMA test, the storage modulus, the loss modulus and the loss factor of the specimen are measured as a function of temperature. The temperature is varied from 20 to 110 °C using a ramp set to 2 °C/min.

From these analyses (Fig. 2) it can be seen that loss modulus is not zero indicating that inelastic effects are not neglect able in this material. Moreover, the two modules depend on frequency (i.e. strain rate) over the entire temperature range. In fact, α transition of the composite takes place at 55 °C so that the composite a visco-elastic body close to room temperatureS. This will make strain rate a important parameter to analyse.

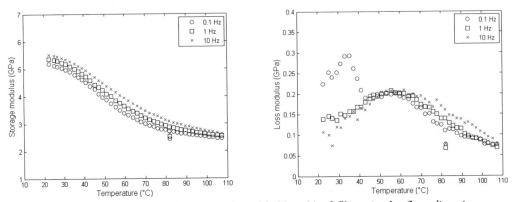

Fig. 2: Evolution of the modulus of a composite with 20 wt% of fibres *in the flow direction* as a function of temperature for different frequencies. Left) Storage modulus; Right) Loss modulus.

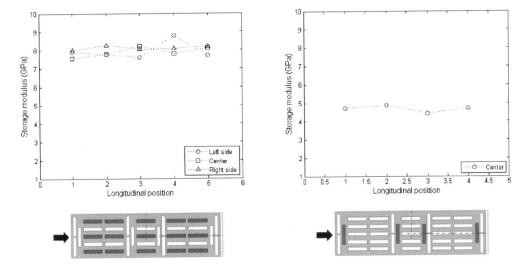

Fig. 1: Evolutions of the storage modulus at temperature of 20 °C as a function of specimen *longitudinal positions* in the plate. The arrow denotes the flow direction during injection process. Left) Storage modulus *in the flow direction*, plotted for three lateral sides called Left, Centre and Right. Right: Storage modulus *perpendicular to the flow direction*, plotted only for Centre side. The prescribed frequency is 1 Hz. The tested specimens contain 30 wt% of short fibre and are illustrated by dark coloured specimens on the plates.

Plates can be reasonable assumed to be homogeneous (Fig. 2). Conversely modules parallel and perpendicular to the flow direction are different (see modules on the right-Fig.1 and on the left-Fig.1). This will be confirmed with shearing tests.

Tension and shearing

Specimens were obtained by machining the plates to get a final geometry as illustrated in Fig. 3. All tests are carried out at room temperature and the specimens are subjected to various strain histories. Stresses measured in this work are presented in normalized values unless otherwise specified.

In case of tension a video extensometer is used to measure true strain in the central zone. Some 3D measurements allow assessing for compressibility of the material under loading. This is done measuring the 3 strain components.

In case of shearing a full 2D strain field analysis is performed using image correlation and random pattern. This latter is done using paints (Fig. 4). To be complete shear test is performed following Iosipescu (1967) and Walrath and Adams (1983) fixtures [4, 5].

Different orientations of test specimen are used and referred to with their angle with respect to the injection direction (Fig. 5). Using strain field measurements it is possible to check whether strain field are correct or not (i.e., true shearing) in any case. This only ensures that overall strain is close to pure shearing in the ligament. It is than assumed that loading is pure shearing.

However, it is demonstrated that the shear strain fields of specimens having different orientations are not identical suggesting that fibre orientation modifies the strain field in the material under loading. To further investigate the local strain field, the evolutions of the local quantities during monotonic simple shear for specimens having different orientations are plotted: lateral displacement (along $v2$) and shear strain. To this end, the lateral displacement across the middle longitudinal section of the specimen is plotted for different specimen orientation as depicted in Fig. 6. The displacement is plotted at time 75s and 135s. At time 75s, the lateral displacement varies linearly independently of the specimen orientation. However, the role of the specimen orientation, i.e. fibre orientation, becomes more significant at higher prescribed displacement. In fact this displacement field gives access to the shear strain field since the latter is obtained from the slope of the former.

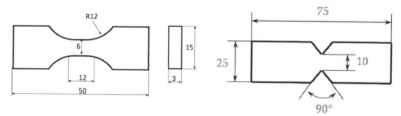

Fig. 3: Specimen for tension tests (left) and shearing tests (right). All dimensions are in mm.

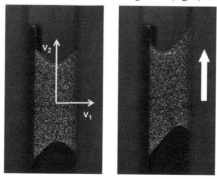

Fig. 4: Undeformed (left) and deformed (right) states of the specimen in shear test.

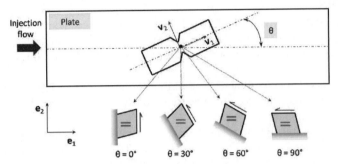

Fig. 5: Different orientations of simple shear test specimen machined from an injection moulded plate.

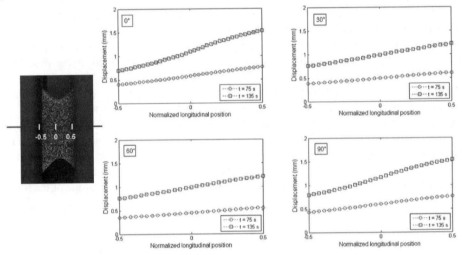

Fig. 6: Evolution of displacement across the middle longitudinal (horizontal) section at time stages of 15 and 27 (75 s and 135 s respectively) for specimens having different orientations.

Tests consist of loading – unloading tests at different strain rate and up to different totals strain (Fig. 7). This enables to emphasise the visco elastic characteristic of the material.

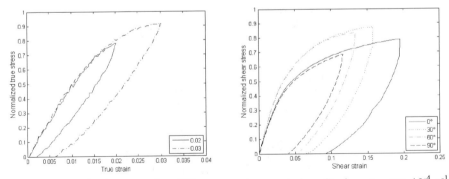

Fig. 7: Upload-unload curves for different maximum strains. Strain rates are 10^{-4} s^{-1} during upload and 10^{-4} s^{-1} during unload. Left in tension tests; right in shearing test.

Evidence for a limiting hysteresis loop, i.e. permanent strain, is deduced from tests combining loading and relaxation steps followed be unloading and relaxations step (Fig. 8).

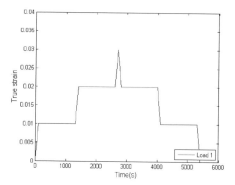 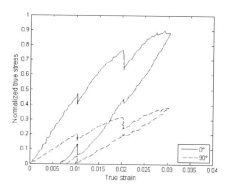

Fig. 8: Uploading and unloading test interrupted by several relaxations illustrating the anisotropic nature of the equilibrium hysteresis. Left: Prescribed strain. Right: Stress response.

Conclusions

To summarise, the experimental investigation has shown that:

1. No significant heterogeneity in the plate is observed.

2. The material is slightly compressible. This was obviously not decduced from shearing tets but from tensile tests using 3D extensometer (not described here).

3. The material responses are strongly anisotropic and non-linear.

4. The time-dependent behaviour is clearly marked. At relatively small strain, the dependence during the unloading is higher than during the uploading. An opposite trend is observed for higher strain.

5. The existence of a unique equilibrium state cannot be verified using our experimental time scale. Instead, the relaxed state is characterized by an equilibrium hysteresis. Furthermore, the size of the equilibrium hysteresis depends on the fibre content and orientation.

6. Since the equilibrium hysteresis is anisotropic, the total response (stress) cannot be modelled by a simple summation of an isotropic viscoplastic behaviour of the ground matrix and the anisotropic elastic behaviour of the fibres.

References

[1] Schapery, R. A: *Stress analysis of viscoelastic composite materials* (Technomic Publishing Co., London, UK 1968).

[2] C. A. Weeks and C. T. Sun: Compos. Sci. Technol. Vol 58 (1998), p. 603

[3] S. V. Thiruppukuzhi and C. T. Su,: Compos. Sci. Technol. Vol. 61 (2001), p. 1

[4] N. Iosipescu J.Mat. Vol. 2(3) (1967), p.537

[5] D. E. Walrath and D. F. Adams Exp. Mech. Vol. 23(1) (1983), p. 105

Session 10: Education

Applied Mechanics and Materials Vols. 24-25 (2010) pp 427-432
© *(2010) Trans Tech Publications, Switzerland*
doi:10.4028/www.scientific.net/AMM.24-25.427

The role of Experimental Stress Analysis at graduation and post graduation courses - A Brazilian case.

R.J.P.C. Miranda[1,a], P. Domingues [2,b], L.M. Zamboni[3,c], J. C. Salamani[4,d]

[1]Head Professor on Mechanics of Materials, Solid Mechanics, Finite Element Method and Experimental Stress Analysis in Centro Universitario da FEI(FEI), São Bernardo do Campo, and CEUN-IMT, São Paulo, Brazil.

[2]Mechanical of Materials Student and Monitor in Centro Universitario do Instituto Mauá de Tecnologia(IMT), São Caetano do Sul, São Paulo, Brazil.

[3]M.Sc. in Mechanical Engineering in São Paulo University(USP) B.Sc. in Mechanical Engineering(FEI).

[4]M Sc. in Naval Engineering (USP) - Professor on Automotive Engineering in FEI in undergraduate and graduate courses.

[a]renatocmiranda@yahoo.com.br; [b]pedro.modomingues@hotmail.com; [c]lmzamboni@gmail.com; [d]jcsala@uol.com.br;

Reviewer: D.R. Maiuri- M.Sc. in Energy (USP) ; Specialization on Fluid and Thermal Studies for Industry(University of Bristol, U.K.); B Sc. in Mechanical Engineering (University of Illinois, USA); Head Professor on Applied Thermodynamics and Coordinator of specialization courses in FEI.

Keywords: Experimental Stress Analysis, Solid Mechanical, Mechanical of Materials, Mechanical Engineer, Structures, Strain-gauges , Residual Stress Measurement.

Abstract. This paper presents an alternative to the theoretical classic teaching methods generally used in Strength of Materials or Solids Mechanics. It introduces a dynamic and interactive process, with emphasis in the use of ESA-Experimental Stress Analysis, especially strain gauges and photo-elasticity. In addition to the didactic and practical experiments, the students used all their energy and creativity to develop the interesting results presented here, such as, experimental device for measuring displacement, Buckling, Spaghetti and Popsicle sticks Bridges, Stress analysis in a cylinder block and Mini Baja project.

Introduction

This work is a continuation and an updating on the paper mentioned in Reference [1], presented on 2003 in the Annual Conference and Exposition on Experimental and Applied Mechanics, sponsored by Society for Experimental Mechanics in Charlotte, North Caroline, USA.

Most Brazilian Mechanical Engineering Schools still use an instructor-based method for teaching Strength of Materials. This leads to small participation and interaction, and, consequently, to high failure indexes.

It must be recognized the importance of encouraging practical activities, especially in laboratory, as a mean to help future engineers to absorb new theoretical concepts, to increase their sensibility, to get acquainted with measuring equipment, to develop a critical vision, to develop team work, to stimulate their creativity, to acquire abilities for preparing reports, to compare experimental results with those obtained from classical, analytical and numerical methods, and, finally, to arrive to the correct conclusions for the solutions of modern engineering problems [1,20].

Looking from this viewpoint, it is necessary that engineering schools introduce experiments, research and services in many knowledge areas.

We understand that all subjects should be related to a lab center and maintain systematic practical activities throughout the course.

In the case of FEI – Centro Universitário da FEI, the Solid Mechanical Laboratory, also called Experimental Stress Analysis Laboratory, was inaugurated in 1984, after some years of research, performed by professors and students in Brazil and abroad, specially through an agreement between FEI and ENSAM – Ecóle Nationale Superieure d'Arts et Metiers, Paris, France. This Laboratory is related to the subject Solid Mechanics, with two obligatory hours a week, in which all junior mechanical engineering students have a global vision of experimental stress analysis techniques, mainly electric extensiometry and Photo-elasticity.

The Solid Mechanical Lab is in constant updating, and, together with the Materials Lab and other labs, constitute the CLM – Mechanical Laboratories Center, which provides conditions for scientific initiation projects, undergraduate projects, researches and services for industry.

Instructors and students from the undergraduate and graduate courses present their statements and didactic equipment to illustrate the importance of experimental stress analysis in the mechanical engineering teaching [2,3,5,7].

Some works presented in this paper aim to prove the importance of experimental stress analysis:

Experimental devices for measuring displacement and rotation in flexion
Device for analysis of buckling in bars
Spaghetti and Popsicle sticks Bridge
Experimental stress analysis in a cylinder block
Mini Baja project

The importance of teaching material

As mentioned in previous works, the success of a good course depends on the discussion and preparation of an appropriate teaching plan for instructors, students and staff. Everybody must be prepared for the use of laboratories and acquainted with the corresponding literature.

In the case of Experimental Stress Analysis, which is a part of our course in Solid Mechanics, we use as a basic reference the book *Mechanics of Materials* [6], which is now in its fifth edition.

Our library maintains about 300 copies of this reference. It has excellent supplementary learning materials for students and teachers, specially the interactive tutorial and a great site "On line learning center". It also presents theoretical applications on 50 single and multiple gauges.

Complementary bibliography, such as Hibbeler's [16] and other works, are also used.

Considering only Experimental Stress Analysis, we used Dally and Riley [18], and the Encyclopedie D'Analyse des Constraints [19].

We extensively use as references catalogs and web sites of Brazilian and international manufacturers, as MM-Vishay [22], Kyowa [24], HBM [21] and Excel [23], as well as magazines and newspapers, such as BSSM Strain [17] and EMS-Experimental Mechanics [25].

Our work is supported by the intensive use of computational tools, such as Moodle and others e-learning tools.

Experimental devices for measuring displacement and rotation in flexion

When we apply a given bending moment on a beam, it undergoes a deformation, which causes displacement and rotation. To make that clear, during the teaching of displacement inflexion, it's important that the student sees what is happening. This will help him to understand the process.

To help the student to visualize the effects, can be developed devices to simulate the action of efforts in a bar, making visible the deformations and allowing their measurement.

Students created a device that allows the visualization of displacements on the bending of a simply supported beam, which receives the action of two forces acting on specific points. Thus, the student can visualize the displacements of the bar and also confirm the accuracy of the calculations. The layout of this beam is shown in Figures 1 and 2.

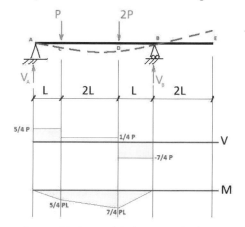

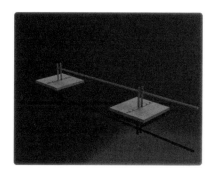

Layout of the beam and the forces applied (Figure 1)   Device build by a group of students (Figure 2)

This work allowed the students, using the theory of integration of elastic line, with or without the use of singular functions, to calculate the displacement and rotation in sections A, B, C, D and E. The theoretical results were then compared with the experimental results, obtained through the device built by the group. We observed that the deviation of results was acceptable [4,6,16].

Device for analysis of buckling in bars.
This interesting device, inspired by references [10,11,12,13], allows the student to determine experimentally the buckling load for the 4 major cases of linking, which are illustrated on Figures 3 to 7:

• Pin-connected at both ends (Fig. 3)   • Two fixed ends (Fig. 5)
• One end fixed and the other pin-connected (Fig. 4)   • One end fixed and the other free (Fig. 6)

The equipment allows us to view the deformed (elastic line) on each case, depending on the conditions of linking.
The experimental determination of the load can be done by calibrated weights or other process. When the results obtained experimentally are compared with those obtained theoretically, it is possible too demonstrate the accuracy of the theory.

(Figure 3)   (Figure 4)   (Figure 5)   (Figure 6)   Buckling Device (Figure 7)

Traditional Spaghetti and Popsicle sticks Bridges
The Spaghetti Bridge Contest is a traditional competition in many Brazilian and international universities. It aims to develop in the students the use of their creativity, theoretical

and practical knowledge to build a bridge that can withstand a load much greater than its own weight.

Some other universities have designed, built and developed bridges using popsicle sticks, instead of the traditional spaghetti.

The main elements that are developed are the lattice, since most bars only withstand normal forces, and buckling, because some bars suffer compression and tend to suffer instability of balance.

The students develop a report, where they analyze the forces of tension and compression that act on each bar and the maximum strength supported, taking into account the maximum allowable stresses. When the bar suffers buckling we should analyze the parameters involved, such as moment of inertia, area, radius of gyration, the equivalent length of buckling, effective slenderness ratio of the column, slenderness limit of material and the elastic(Euler) or plastic (other theories) buckling.

Consequently, the students have the opportunity to compare theoretical concepts to the real thing, and to see the problems and difficulties that occur during the construction process, as well as their influence in the final result.

As an example, it can be verified how a problem of alignment or symmetry, which make the forces to be differently distributed than calculated, can change the maximum load to be resisted by the bridge.

Through this competition the students develop a real ability to reason, leaving aside the mathematical estimative and learning to analyze and develop a structure in a different way. This helps the absorption of the involved concepts.

It must be added that the student's participation in the contest develops skills of team working, leadership and critical vision [4,20].

To illustrate this event we annex Figure 8 and 9.

The truss developed by the group (Figure 8)

Popsicle sticks Bridge of a student's group (Figure 9)

Experimental stress analysis applied to internal combustion engine design

Under the new scenery of green house effects, modern internal combustion engine   design must consider the effects of exhaust gas and emissions. To attain new requirements and legislation, low emission levels are very important.

The design of rings, piston, and cylinders are fundamental in this new development concept. Manufacturers are applying efforts to develop geometries to improve performance in reference to emissions. However, the cylinder liner plays an outstanding role. Circularity, linearity and the cylinder liner shape are responsible for the working stability of the piston and rings, also affecting their durability.

Oil consumption and combustion gas exhaustion are fundamental for emission control. The geometrical deformations are, among other reasons, due to residual stress. Such stresses come from manufacturing processes but also as a residual effect, created at foundry or during casting processes, and must be properly evaluated.

X – Ray surface analysis is one of the methods for such evaluation. However, many times it is not available and is always very expensive. Strain gauge evaluation of a cylinder liner is easy and simple. Just three strain gauges installed on the surface of the liner (rosette shape) will make possible to take a cross section cut of the liner. It is then possible to obtain the strain level before and after the cut and have a data bank of strain values with and without stress relieve, at several situations. It is then possible to organize a very simple and accurate procedure for stress control in the cylinder liner.

Mini Baja – FEI three times world champion and Formula FEI

In Reference [8] it is shown that the Mini Baja vehicles developed in FEI are instrumented with several strain gauges and systems of data acquisition, which lead to improving projects and reducing their weights. The result was that FEI's truggy was considered the world's best in Montreal, and Formula FEI was considered among the world's Top 10 in 2009, in Detroit. (See Figures 10 e 11)

Mini Baja (Figure 10)

Fórmula FEI (Figure 11)

Vehicles developed by IMT – Instituto Mauá de Tecnologia were also analyzed experimentally and obtained good results in national and international competitions.

Conclusion

The pedagogic technique, that use contest or competitions of projects, prototypes or similar, to stimulate the creativity and responsibility on the students, in activities like Mini Baja, Formula, Aero design, Spaghetti bridge, Young SA, among others, is being successful . This has involved many companies or entities as SAE, SEM, BSSM, Petrobrás, Alcoa and others.

We try to emphasize the modern concept of interaction between Solids Mechanics and other Mechanical Engineering subjects. Through Physics, Mechanics, Strength of Materials, Experimental Stress Analysis, Finite Element Method, Computer Aided Project and other disciplines and practical activities, the student starts to become a professional with a broad view of the Mechanical Engineering.

This paper enforces the use of laboratories. In the case of Strength of Materials, we recommend the construction of a didactical laboratory, with simple models, to introduce the student in the subject. Later, it is important to develop an Experimental Stress Analysis Laboratory, with well prepared equipments and experiences, where the future engineer can develop a critical view.

Acknowledgements

The authors are grateful for the collaboration of Professors R. Bortolussi, M. Lucato, A. Bernardini, S. Delijaicov, G. Donato, R. Montefusco, K. Amann and others, who have contributed, directly or indirectly, to the conclusion of this work. Our special acknowledgement to the CLM, the Mechanical Engineering Department, the Rectory and the Presidency of FEI and to the Board of Directors of EEM/CEUN-IMT.

References

[1]   R. J. P. C. Miranda, et al. *Experimental Stress Analysis and the Mechanical Engineers Courses* (2003 SEM Annual Conference and Exposition on Experimental and Applied Mechanics, North Carolina, USA, 2003)

[2]   R. J. P. C. Miranda, C. A. M. Santos, *Equipamento de Ensaio de Flexão – Uma contribuição para o ensino e aprendizado experimental de Resistência dos Materiais* (COBEM-CIDIM, Minas Gerais, Brazil, 1995)

[3]   R. J. P. C. Miranda, *O uso do Laboratório e Computador no processo de ensino e aprendizado de Resistência dos Materiais* (XXIII – COBENGE, Pernambuco, Brazil, 1995)

[4]   R. J. P. C. Miranda, et al. *Resistência dos Materiais* Edited by author, São Bernardo do Campo, SP, Brazil (2002)

[5]   G. H. B. Donato, R. F. V. Bôas, *Desenvolvimento de Experimentos e Modelos Físicos Quali e Quantitativos para Ilustração de Conceitos de Estática, Resistência dos Materiais e Mecânica dos Sólidos durante Aulas Teóricas e Práticas de Graduação do Centro Universitário da FEI,* São Bernardo do Campo, SP, Brazil, (2010)

[6]   F. P. Beer, E. R. Johnston, J. T. DeWolf, D. F. Mazurek, *Mechanics of Materials* 5. Ed. New York: McGraw-Hill, (2009).

[7]   R. J. P. C. Miranda, S. Delijaicov, *Laboratório de Mecânica dos Sólidos,* Edited by author, São Bernardo do Campo, SP, Brazil, (2006)

[8]   R. Bortolussi, A. C. Prado, *BAJAFEI, Um projeto vitorioso,* Revista Pesquisa & Tecnologia FEI, (2004), p. 2

[9]   *Modern Stress and Strain Analysis: A state of the art guide to measurement techniques,* BSSM, Published by Eureka magazine, (2009)

[10]   J. A. Santos, *Sobre a Concepção, o Projeto, a Execução e a Utilização de Modelos Físicos Qualitativos na Engenharia de Estruturas,* Escola Politécnica da USP, São Paulo, SP, Brazil (1983).

[11]   Information on http://www.usdidactic.com

[12]   Information on http://www.tecquipment.com

[13]   Information on http://www.heliodon.com.br

[14]   Information on http://www.maua.br

[15]   Information on http://www.fei.edu.br

[16]   R. C. Hibbeler, *Resistência dos Materiais*, 5 ed., São Paulo, SP: PEARSON PRENDICE HALL, (2004).

[17]   Information on http://www.bssm.org

[18]   Dally, J. W., Riley, W. F.; *Experimental Stress Analysis*, 3 ed., New York: McGraw-Hill, (1991).

[19]   J. M. Avril, *Encyclopédie D'Analyse des Constraintes*, Ed. Vishay Micromessures, (1984).

[20]   M. F. Giorgetti, et al.; *O concurso de Projetos e Protótipos como atividade educacional e seleção para o mercado de trabalho*, Edited by ABENGE, vol. 28, N. 01, (January – July 2009)

[21]   Information on http://www.hbm.com

[22]   Information on http://www.vishay.com

[23]   Information on http://www.excelsensor.com.br

[24]   Information on http://www.kyowa-ei.co.jp

[25]   Information on http:/www.sem.org

Keyword Index

3D Scanning 167, 287

A

Acetabular Cup 275
Acoustic Emission 45, 221
Adhesive Bonding 189
Alumino-Thermic Weld 305
Analytical Shape 359
ANSYS 393
Austenitic Stainless Steel 207
AZ91D Alloy 325

B

Bayesian Sensitivity Analysis 25
Beam Bending 401
Bend Test 385
Biaxial 115
Biaxial Ratcheting 207
Biaxial Stress State 213
Biomechanics 287
Bolted Langevin Transducer 65
Bonded Joint 401
Bridge Monitoring 173
Bridging Law 245
Buckling 393
Buckling Behaviour 103

C

Cantilever 371
Cement Paste 1
Cervical 287
CFRP Composite 233
Chromatic Monitoring 123
Civil Engineering 161
Coating Processes 83
Coating Stress 317
Cohesive Law 189
Composite 109, 195, 393
Composite Material 239
Compressive Failure 15
Compressive Load 287
Concrete 1
Crack Compliance Method 253, 261
Crack Initiation 253, 261
Crack Measurement 161
Cross-Correlation 135
Cross Ply Polymer Composite
    Laminate 91

Cross-Spectrum 135
Crossover 343
Cruciform 115
Curve Detection 359
Cyclic Loading 213

D

Damage 109
Damage Detection 51
Damping Coefficient 77
Deflectometry 109
Deformation 275
Delamination 393
Delamination Growth 245
Diametral Contraction 141
Digital Holographic Interferometry 147
Digital Image Correlation 103, 115, 129,
    161
Digital Image Correlation (DIC) 91, 141,
    161, 167, 201, 221, 267, 275, 331,
    385, 407
Digital Image Correlation (DIV) 15
Digital Photoelasticity 239
DP Steel 201
Dynamic Buckling 331
Dynamic Energy Absorption 349

E

Edge Crack 91
Edge Detection 141
Effect of Preload 65
Elevator Car System 77
Engine 71
Experimental 71
Experimental Stress Analysis 427

F

Fast Breeder Reactor 207
Fatigue 305, 407
Fatigue Fracture 221
FIB 267
Fibre 359
Fibre Bragg Grating 173
Filament 359
Finite Element Analysis (FEA) 129, 281, 331
Finite Element Method 275
Flexural Strength 385

Fourier Transform Misalignment
  Analysis (FTMA) 15
Fracture Energy 179
Fracture Process Zone FPZ 155
Frequency Response Function 337
Friction 343
Full-Field 109
Full Field Measurement 135, 141, 379
Full-Field Strain Pattern 365

G
G-Flute Corrugated Board 103
Gas Turbine 317
Gear Teeth 45
GFRP 91
Glass Fibre Reinforced Plastics
  (GFRP) 407
Graphite 385

H
High Strain Rate 325
High Strength Concrete 155, 179
High Strength Steel 299
High Temperature 1
Hole Drilling 267
Holography 147
Honeycomb 413
Hopkinson Bar 325, 349
Hybrid 393
Hydrostatic Compression 97

I
Impact 413
Impact Damage 233
Impact Excitation 337
In Situ 161, 299
Interface Element 189, 245

K
Kinkband Formation Models 15

L
Lamb Wave 51
Laminated Composite 245
Large Deflection 371
Laser Speckle 135
Laser Ultrasonics 281
Lightning 129
Lightweight Structures 129
Loading Rate 155, 179

Loop 359
Low-High Speed 343

M
Magnesium Alloy 325
Material Testing 371
Mechanical Behavior 207
Mechanical Engineer 427
Mechanical of Materials 427
Mechanical Property 371, 379, 407
Metal Forming 311
Micro-Lattice 413
Micromechanical Measurements 239
Micron-Scale 267
Microstructure 201, 305
Mixed-Mode Fracture 189
Modal Analysis 71
Model Updating 337, 365
Modified Iosipescu Test 15
Modified SEN Specimen 253, 261
Modulus 385

N
Non-Destructive Evaluation (NDE) 57
Non-Destructive Testing (NDT) 147
Non-Linear Behaviour 419
Nonlinear Model 25
Notched 91
Nuclear 385

O
Optical Measurement 167
Own-Weight Deformation 371

P
Permeability 1
PFRP Material 401
Photoelasticity 123
Photogrammetry 167
Piezoelectric Paint 83
Piezoelectric Sensor 83
Piezoelectric Transducer 51
Polymer 195, 349
Polymer Matrix Composite 115
Porcine Vertebrae 287
Powertrain 71
Printing Processes 83
Pulsed Phase Thermography (PPT) 57
Pultrusion 15

R
Rail 305
Rectangular Rosette 207
Relaxation 195
Residual Stress 253, 261, 267, 305
Residual Stress Measurement 427
Response Function 65

S
Sandwich 413
Scanning Electron Microscopy (SEM) 233
Shape Features 365
Shear Lag Theory 317
Shear Test 97
Sheraing 419
SHM 51
Short Fibres Composites 419
Single Fibre Fragmentation Test 239
Skin 281
Softening Effect 213
Solar Radiation Sensors 173
Solid Mechanical 427
Spot Weld 299
Stiffened Panel 331
Stiffness Coefficient 77
Strain 1, 109, 201
Strain Fields Measurement 103
Strain Gauge 427
Strain Gauge Flexible Rollers 83
Strain Measurement 123, 407
Strain Rate 349
Strain Rate Sensitivity 325
Stress Fields Measurement 103
Stress Intensity Factor 227
Stress-Strain Loop 349
Structural Health Monitoring 45, 51, 173
Structure 427
Superconducting Coils 379

Syntactic Foam 97

T
Temperature Compensation 173
Tensile Property 325
Tension 419
Thermal Barrier Coating (TBC) 317
Thermal Effect 281
Thermoelastic Stress Analysis (TSA) 57,
    91, 227
Thermoelasticity 227
Thin Material 371
TSA 91

U
UD CFRP 15
Ultrasonic Forming Tests 311
Ultrasonics 311
Uncertainty Analysis 25

V
Vibration 71
Vibration Monitoring 83
Vibro-Thermography (VT) 57
Video Extensometer 419
Virtual Fields Method 379
Virtual Image Correlation 359
Viscoplasticity 195

W
Wedge Test 299

Y
Yield Surface 213
Young's Modulus 371

Z
Zernike Moment 365

Author Index

A

Abdul Aziz, S. 311
Ahmad, I.R. 325
Alexeenko, I. 147
Andriyana, A. 419
Anyfantis, K.N. 189, 245
Arai, M. 317
Asquith, D. 141
Auradou, H. 359

B

Balzano, E. 343
Barrington, P.E. 393
Barton, E.N. 173
Battams, G.P. 91
Battipede, M. 25
Becker, W.E. 25
Bellet, M. 135
Beltrán-Fernández, J.A. 253, 261, 287
Bergheau, J.M. 299
Bigsby, R. 275
Billon, N. 419
Blobel, S. 195, 239
Boyd, S.W. 91
Brooks, W. 413
Broughton, W.R. 407
Burguete, R.L. 331

C

Cámara, M. 155
Cantwell, W. 413
Cardoni, A. 65
Cédric, T. 147
Celotto, S. 201
Chastel, Y. 135
Chellapandi, P. 207
Choqueuse, D. 97
Clarke, A. 45
Cognard, J.Y. 97
Cole, M. 129
Cooper, R. 337

D

David-West, O.S. 337
Davies, P. 97
Deakin, A.G. 123
del Viso, J. 179
Delprete, C. 71

Desloges, I. 103
Devivier, C. 109
Di Liberto, F. 343
Domingues, P. 427
Dulieu-Barton, J.M. 57, 91, 227
Dumont, P.J.J. 103

E

Eaton, M.J. 45, 129, 221, 331
Eberhardsteiner, J. 1
Esfahani, M.M.N. 393
Evans, S.L. 45, 129, 221, 275
Everitt, H. 275

F

Featherston, C.A. 129, 331
Flahaut, P. 305
François, M.L.M. 359
Fruehmann, R.K. 57

G

Galeazzi, A. 71
Galek, A. 1
Garza, C. 123
Georges, M. 147
Ghadbeigi, H. 201
Ghasemnejad, H. 393
González-Rebatú, A. 287
Gower, M.R.L. 115, 407

H

Hariri, S. 305
Harkness, P. 65
Hasan, R. 413
Hebb, R.I. 227
Hensman, J.J. 221
Hernández-Gómez, L.H. 253, 261, 287
Herrera, I. 77
Hillbrand, H.H. 83
Hodzic, A. 233
Holford, K.M. 45, 129, 221
Holt, C.A. 275
Hon, K.K.B. 123
Huang, Z.H. 281

J

Jezzini-Aouad, M. 305
Johns, R. 331

Jones, G.R. 123

K
Kaczmarczyk, S. 77
Kästner, M............................... 195, 239
Kermouche, G. 299
Khan, I.. 275
Kim, J.H. 379
Klein, M. 167
Klimaytys, G.M... 385
Klöcker, H.. 299
Kowalewski, Z.L.. 213
Kratmann, K.K.. 15

L
Lackner, R... 1
Lacroix, R... 299
Lakshmana Rao, C. 207
Lammen, B.. 83
Lens, A.. 299
Li, C.H. 281
Li, S.A... 281
Lingham, I.J. 385
Lodeiro, M.J.. 407
Lord, J.D. 161, 385
Lucas, M.. 65, 311

M
Malinowski, P.H. 51
Maniu, I... 83
Mauret, E... 103
McCormick, N.J.............................. 161, 385
McCory, J.P. 45
Mines, R.A.W. 413
Miranda, R.J.P.C.. 427
Monatte, J... 299
Mortimer, J.. 331
Moser, H.. 1
Mottershead, J.E.. 365

N
Nakai, K. 349
Nunio, F. 379

O
Obst, M. 195
Ohtsuki, A. 371
Orgéas, L... 103
Ostachowicz, W.M..................................... 51
Osten, W... 147

P
Patki, A. 365
Patterson, E.A. 365
Pedrini, G. 147
Perreux, D. 97
Peruggi, F. 343
Pierron, F. 109, 379
Pilkington, G.D... 407
Pinna, C.. 201
Porras, R. 179
Poveda, E. 179
Pradille, C. 135
Pregno, F.. 71
Pullin, R. 45, 129, 221

Q
Quinn, S.. 57

R
Ring, T. 1
Romero-Ángeles, B. 253, 261
Ruiz, G... 155, 179
Russell, J.. 65

S
Salamani, J.C. 427
Schmidt, R. 83
Semin, B. 359
Serpico, M.. 343
Shaw, R.M. 115, 407
Shen, E. 413
Shu, D.W. 325
Silva, L. 419
Sohier, L. 97
Spencer, J.W. 123
Stanley, A. 167
Staszewski, W.J. 233
Su, H.. 77
Sultan, M.T.H. 233
Surace, C... 25
Suresh Kumar, R... 207
Sutcliffe, C.J. 413
Szymczak, T. 213

T
Tai, Y.H. 141
Tarifa, M. 155
Tatum, P... 227
Thielsch, K... 195, 239
Thompson, D. 109
Thomsen, O.T. 15
Torres-Torres, C. 253, 261

Tsopanos, S. 413
Tsouvalis, N.G. 189, 245
Turvey, G.J...401

U
Ulbricht, V. 195, 239
Urquhart, J.M. 385
Urriolagoitia-Calderón, G. 253, 261, 287
Urriolagoitia-Sosa, G. 253, 261, 287

V
Vacher, P..103
Vandenrijt, J.F...147
Vedrine, P..379
Viguié, J. 103
Voicu, M.C..83
Vollheim, B..147

W
Wandowski, T. 51
Wang, J..337
Wang, W.Z..365

Winiar, L..305
Winiarski, B..267
Wisnom, M.R...109
Withers, P.J..267
Worden, K.........................25, 221, 227, 233

X
Xu, W.B...281

Y
Yates, J.R..141, 201
Yokoyama, T. 349
Yu, R.C. 155, 179

Z
Zakrzewski, D..305
Zamboni, L.M...427
Zanganeh, M. 141
Zeiml, M. 1
Zhang, B. 173
Zhang, X.X. 155, 179